THE ESTUARY AS A FILTER

Produced by

The Estuarine Research Federation

Sponsored by

Baltimore Gas & Electric

Carolina Power & Light Company

Marine Sciences Research Center,
State University of New York at Stony Brook

National Science Foundation

U.S. Department of Interior,
Minerals Management Service

U.S. Environmental Protection Agency,
Chesapeake Bay Program

U.S. Fish and Wildlife Service

U.S. Geological Survey

U.S. National Oceanic and Atmospheric Administration,
Ocean Assessments Division

Proceedings of the Seventh Biennial International
Estuarine Research Conference,
Virginia Beach, Virginia, October 23–26, 1983.

THE ESTUARY AS A FILTER

Edited by

VICTOR S. KENNEDY

Biology Department
St. Francis Xavier University
Antigonish, Nova Scotia, Canada
and
University of Maryland
Center for Environmental and Estuarine Studies
Horn Point Environmental Laboratories
Cambridge, Maryland

ACADEMIC PRESS 1984
(Harcourt Brace Jovanovich, Publishers)
Orlando San Diego New York London
Toronto Montreal Sydney Tokyo

Academic Press Rapid Manuscript Reproduction

ACADEMIC PRESS, INC.
Orlando, Florida 32887

United Kingdom Edition published by
ACADEMIC PRESS, INC. (LONDON) LTD.
24/28 Oval Road, London NW1 7DX

Library of Congress Cataloging in Publication Data

Estuarine Research Federation. Biennial Conference (7th : 1983 : Virginia Beach, Va.)
The estuary as a filter.

Papers presented at the seventh biennial conference of the Estuarine Research Federation, held in Virginia Beach, Va., Oct., 1983.
Includes index.
1. Estuarine ecology--Congresses. 2. Estuaries--Congresses. I. Kennedy, Victor S. II. Title.
QH541.5.E8E87 1984 574.5'26365 84-24625
ISBN 0-12-405070-0 (alk. paper)

PRINTED IN THE UNITED STATES OF AMERICA

84 85 86 87 9 8 7 6 5 4 3 2 1

CONTENTS

LIST OF CONTRIBUTORS

Numbers in parentheses indicate the pages on which the authors' contributions begin.

Robert B. Biggs (107), College of Marine Studies, University of Delaware, Newark, Delaware 19716

Donald F. Boesch (447), Louisiana Universities Marine Consortium, Chauvin, Louisiana 70344

K. F. Bowden (15), Sailways, 21 Undercliff Road, Wemyss Bay, Renfrewshire PA18 6AJ, Scotland

Walter R. Boynton (367), University of Maryland, Chesapeake Biological Laboratory, Solomons, Maryland 20688

D. E. Bretschneider (67), Pacific Marine Environmental Laboratory, National Oceanic and Atmospheric Administration, 7600 Sand Point Way N.E., Seattle, Washington 98115

G. A. Cannon (67), Pacific Marine Environmental Laboratory, National Oceanic and Atmospheric Administration, 7600 Sand Point Way N.E., Seattle, Washington 98115

H. H. Carter (81), Marine Sciences Research Center, State University of New York, Stony Brook, New York 11794

Peter Casapieri (489), Directorate of Scientific Services, Thames Water Authority, Reading RG1 8DB, England

Robert R. Christian (349), Biology Department, East Carolina University, Greenville, North Carolina 27834

Thomas M. Church (241), College of Marine Studies, University of Delaware, Lewes, Delaware 19958

Luis A. Cifuentes (241), College of Marine Studies, University of Delaware, Lewes, Delaware 19958

J. Kirk Cochran (179), Marine Sciences Research Center, State University of New York, Stony Brook, New York 11794

Deborah A. Daniel (349), Institute for Coastal and Marine Resources, East Carolina University, Greenville, North Carolina 27834

John W. Day, Jr. (447), Center for Wetland Resources, Louisiana State University, Baton Rouge, Louisiana 70803

Nicholas A. Funicelli (435), Florida Fishery Research Station, Homestead, Florida 33090

Stephen W. Hager (221), U.S. Geological Survey, Menlo Park, California 94025

Dana D. Harmon (221), U.S. Geological Survey, Menlo Park, California 94025

George R. Helz (131), Department of Chemistry, University of Maryland, College Park, Maryland 20742

Frederick A. Hoffman (313), University of Georgia Marine Institute, Sapelo Island, Georgia 31327

J. R. Holbrook (67), Pacific Marine Environmental Laboratory, National Oceanic and Atmospheric Administration, 7600 Sand Point Way N.E., Seattle, Washington 98115

Charles S. Hopkinson, Jr. (313), University of Georgia Marine Institute, Sapelo Island, Georgia 31327

Barbara A. Howell (107), College of Marine Studies, University of Delaware, Newark, Delaware 19716

W. Michael Kemp (367), University of Maryland, Horn Point Environmental Laboratories, Cambridge, Maryland 21613

Victor S. Kennedy (1), University of Maryland, Horn Point Environmental Laboratories, Cambridge, Maryland 21613

Kate Kranck (159), Atlantic Oceanographic Laboratory, Bedford Institute of Oceanography, Dartmouth, Nova Scotia B2Y 4A2, Canada

Daniel R. Lynch (131), Thayer School of Engineering, Dartmouth College, Hanover, New Hampshire 03755

Thomas C. Malone (291), University of Maryland, Horn Point Environmental Laboratories, Cambridge, Maryland 21613

Scott W. Nixon (261), Graduate School of Oceanography, University of Rhode Island, Narragansett, Rhode Island 02882

Charles B. Officer (131), Earth Sciences Department, Dartmouth College, Hanover, New Hampshire 03755

Christopher P. Onuf (415), Marine Science Institute, University of California at Santa Barbara, Santa Barbara, California 93106

Jonathan R. Pennock (241), College of Marine Studies, University of Delaware, Lewes, Delaware 19958

David H. Peterson (221), U.S. Geological Survey, Menlo Park, California 94025

Michael E. Q. Pilson (261), Graduate School of Oceanography, University of Rhode Island, Narragansett, Rhode Island 02882

D. W. Pritchard (27), Marine Sciences Research Center, State University of New York, Stony Brook, New York 11794

Laurence E. Schemel (221), U.S. Geological Survey, Menlo Park, California 94025

J. R. Schubel (1, 81), Marine Sciences Research Center, State University of New York, Stony Brook, New York 11794

George H. Setlock (131), Department of Chemistry, University of Maryland, College Park, Maryland 20742

Jonathan H. Sharp (241), College of Marine Studies, University of Delaware, Lewes, Delaware 19958

Catherine A. Short (395), Jackson Estuarine Laboratory, University of New Hampshire, Durham, New Hampshire 03824

Frederick T. Short (395), Jackson Estuarine Laboratory, University of New Hampshire, Durham, New Hampshire 03824

Charles A. Simenstad (331), Fisheries Research Institute, College of Ocean and Fishery Sciences, University of Washington, Seattle, Washington 98195

Donald W. Stanley (349), Institute for Coastal and Marine Resources, East Carolina University, Greenville, North Carolina 27834

J. Court Stevenson (367), University of Maryland, Horn Point Environmental Laboratories, Cambridge, Maryland 21613

Virginia K. Tippie (467), U.S. Environmental Protection Agency, Chesapeake Bay Program, 839 Bestgate Road, Annapolis, Maryland 21401

John M. Tramontano (241), College of Marine Studies, University of Delaware, Lewes, Delaware 19958

R. Eugene Turner (447), Center for Wetland Resources, Louisiana State University, Baton Rouge, Louisiana 70803

Robert R. Twilley (367), University of Maryland, Horn Point Environmental Laboratories, Cambridge, Maryland 21613

M. E. C. Vieira (27), Marine Sciences Research Center, State University of New York, Stony Brook, New York 11794

Larry G. Ward (367), University of Maryland, Horn Point Environmental Laboratories, Cambridge, Maryland 21613

Robert C. Wissmar (331), Fisheries Research Institute, College of Ocean and Fishery Sciences, University of Washington, Seattle, Washington 98195

Joy B. Zedler (415), Biology Department, San Diego State University, San Diego, California 92182

FOREWORD

It is ironic that estuaries, which have always been at the human doorstep and are one of the most heavily utilized and most productive zones in our planet, should be so long in evoking recognition as unique ecosystems. Perhaps it is because they represent a first-order interface. Their integrative processes, tying together terrestrial, freshwater, and marine biomes, weave a web of complexity far greater than that of their three contributor systems and far out of proportion to their occupation of less than 1% of the planet's surface. It has been easier to ignore the estuaries and to concentrate on more tractable studies within the other systems.

With "The Estuary as a Filter," the Estuarine Research Federation completes its first 10 years of a series of publications incorporating papers presented at its Biennial International Estuarine Research Conferences. These conferences and peer-reviewed publications constitute a major effort by the Federation to serve national and international science. What has characterized these efforts, and what portends for their future?

The first conference volume was "Estuarine Research"* (1975). It unquestionably established a multidisciplinary scope and style which embraced estuaries in all their aspects, from natural systems to used and abused human resources. Five subsequent symposia have followed that lead, establishing the national and international role of the Estuarine Research Federation as a society dedicated to estuarine science as a unique discipline.

In these five conferences and in the resultant volumes, the Estuarine Research Federation has succeeded in focusing attention on various aspects of estuarine complexity: "Estuarine Processes" (1977),† "Estuarine Interactions"† (1978), "Estuarine Perspectives"** (1980), "Estuarine Comparisons"** (1982) and now "The Estuary as a Filter" (1984). For those of us who have been involved in these symposia from the beginning, there has been a growth which is reflected

*Cronin, L. E., Editor. Academic Press, New York.
†Wiley, M. L., Editor. Academic Press, New York.
**Kennedy, V. S., Editor. Academic Press, New York.

in the printed volumes but which has been even more apparent through first-hand experience within the sessions: the growth of interdisciplinary sophistication in the participants. We have gradually developed from a time when "interdisciplinary symposium" meant "concomitant sessions in each discipline" to a time of multidisciplinary delivery and discussions within our sessions. We have accomplished this in an era when burgeoning information within each discipline has placed ever-increasing demands on our intellect, and in a field where the terms "basic" and "applied" are inherently indistinguishable, where solutions increasingly mix socioeconomics with physics and chemistry, and where scientists and politicians must deal with each other as part of the daily equation of reality. These developments have made us increasingly dependent on one another. The "we" of the team effort is replacing the "I" of the ivory tower.

Does this movement "buck the tide" of increasing specialization in disciplinary science? Does it "come full cycle" and return us to the plenary sessions which produced that great predecessor volume "Estuaries"* (1967)? The 1983 conference reflected these aspects in part, but it seemed to signify much more. The study of estuaries has itself become a specialty, but one in which the estuary's characteristics, not our own, have demanded that the science be developed on a multidisciplinary level. The conference at Virginia Beach is a tribute to its chairman, Jerry Schubel, and to all who organized, conducted, and participated in the sessions for helping further this development with the perspectives on "filtration" set forth in this volume. Estuarine science seems to have come of age, and it may now be taking a leading role in holistic, interdisciplinary accomplishment.

Barbara L. Welsh
President
Estuarine Research Federation
1981–1983

*Lauff, G. H., Editor. American Association for the Advancement of Science, Washington.

PREFACE

When the Estuarine Research Federation held its seventh biennial conference at Virginia Beach, Virginia, in late October, 1983, nearly 600 researchers, managers, and students from around the world gathered to consider the theme of the estuary as a filter. Whereas each of the earlier biennial meetings had had a unifying theme, this was the first conference in which it was so carefully focussed and integrated. Thus, in five invited sessions, scientists and managers considered the physical, geological, chemical–geochemical, and biological processes involved in the "filtering" role of estuaries and reflected on management implications of these matters. Most of their presentations and reflections are included in this book in order to demonstrate what is known and what needs to be explored further.

The papers in this volume are grouped as they were presented at the conference. Thus, physical oceanographers begin the work by considering turbulence, mixing, and circulation processes in estuaries. Geologists then examine estuarine sedimentation, including the roles of flocculation and bioturbation in accelerating this process. Chemists and geochemists describe the interactions among and effects of inputs of nutrients, metals, and organic matter into estuaries, and the fate of radionuclides in these systems. Biological and biochemical processes involving surface foam, microbes, sea grasses, and wetlands are considered, along with carefully derived nutrient budgets of selected estuarine regions. Finally, some of the problems facing managers of estuarine ecosystems in three areas of the United States are described, along with the success story of the ongoing rehabilitation of the Thames Estuary in England.

The 27 invited papers that were submitted for publication were each considered by two or three referees to ensure careful peer review. This volume includes those papers that passed this important review process. As with previous proceedings in this series, publication has occurred within about a year of the meetings. The timely presentation of this material should be beneficial to scientists and managers concerned with estuaries, their structure and function, and the impact of human use of these ecosystems. Because of the interdisciplinary nature of these proceedings, coupled with the effort taken to focus each paper on the theme, the book should also be useful as a text for students who are studying estuarine science.

As before, I am indebted to the referees who assisted me. I appreciate the care taken by authors, and especially by their secretaries, in the difficult task of preparing camera-ready copy. The staff of Academic Press were always helpful in providing guidance. St. Francis Xavier University and its Biology Department provided space, and Mary Murphy gave secretarial assistance as I worked on this book during a sabbatical leave. Finally, Deb, Jen, and Chris were as supportive as ever during this endeavor: My thanks to all.

Victor S. Kennedy

THE ESTUARY AS A FILTER: AN INTRODUCTION

J. R. Schubel

Marine Sciences Research Center
State University of New York
Stony Brook, New York

Victor S. Kennedy

Horn Point Environmental Laboratories
University of Maryland
Cambridge, Maryland

"Estuaries and coastal areas trap significant quantities of material and thus act as filters between land and the oceans. It is important to quantify the capacity of these sedimentation basins as filters for different materials and elements. Biological processes may also play an important role in trapping and mobilization of material carried by rivers and land runoff."
Ocean Science for the Year 2000

INTRODUCTION

It was during an international meeting in Rome in 1979 on "River Inputs to Ocean Systems" that one of us (JRS) realized that "The Estuary as a Filter" would make an interesting theme for an Estuarine Research Federation (ERF) biennial conference. As Program Chairman for the 1983 ERF conference, Schubel was able to implement this idea. An attempt was made to ensure a well-balanced set of presentations that would cover the full range of physical, geological, chemical, and biological processes that contribute to the "filtering" actions of estuaries and which would be directed squarely at this theme. Not all

ISBN 0-12-405070-0

of those who presented papers orally submitted manuscripts for publication, and not all of the manuscripts were accepted. Still, the papers presented in this volume represent the first attempt to deal with the important topic of the estuary as a filter in any systematic way. We present here a brief introduction to the topic and to the papers in this volume.

Webster's Third International Dictionary of the English Language (1976) defines a filter as a device through which fluids are passed to separate out matter in suspension. Using an analogy from electronics, a filter may also be defined as a device or process which operates on an input signal to produce an output signal which is altered in a manner prescribed by the filter. These processes occur efficiently and effectively in estuaries, at least during some stages in the development of estuarine systems.

For example, estuaries modify the suspended matter signals they receive from their rivers and from the ocean (Schubel and Carter)[1] as well as the chemical signals they receive from a variety of natural and anthropogenic sources (Sharp et al.). Estuaries may also filter the physical input signals they receive from rivers, wind, and tide (e.g. Pritchard and Vieira describe the way in which Chesapeake Bay filters the effects of wind stress as an input signal on the velocity field as an output signal).

The ability of estuaries to remove and to retain materials in suspension and in solution has important practical as well as scientific implications. It leads, or at least contributes in a significant way, to many of the most serious estuarine pollution and management problems with which we are confronted: the accumulation of contaminants in sediments; dredging and dredged material disposal, particularly of contaminated sediments; nutrient enrichment; dissolved oxygen depletion; degradation and loss of benthic habitat; loss of submerged aquatic vegetation; and many others. Fortunately, it also allows an amelioration of estuarine water quality despite high loading rates and promotes recycling of nutrients and organic matter and the maintenance of high rates of primary and secondary production.

Estuarine Circulation

Pritchard (1967) defined an estuary as "a semi-enclosed coastal body of water which has a free connection with the open sea and within which sea water is measurably diluted with

---

[1]References to authors without accompanying dates indicate that the paper cited is found in this volume.

fresh water derived from land drainage" (but see also Hopkinson and Hoffman who argue that the estuarine influence may extend to nearshore coastal waters where seawater is diluted by land drainage). Pritchard's classic definition describes a class of coastal water bodies in which mixing of freshwater and seawater produces density gradients which drive distinctive estuarine (gravitational) circulation patterns. These in turn lead to the eventual discharge of freshwater to the sea. The mixing may be caused primarily by the action of the river, the tide, or the wind. There has been described a sequence of estuarine circulation types displaying different degrees of mixing. The position that an estuary occupies in this sequence depends primarily upon the relative magnitudes of river flow and tidal flow and upon the geometry of the estuarine basin (Pritchard 1955). Changes in any of these factors may alter the mixing and the estuarine circulation. At one end of this range is the river-dominated, poorly mixed (highly stratified) salt wedge estuary (Type A estuary). At the other end is the thoroughly mixed, sectionally homogeneous estuary (Type D). Two intermediate types which have been described are the partially mixed estuary (Type B) and the vertically homogeneous estuary (Type C).

Mixing processes determine the strength and character of an estuary's gravitational circulation pattern. This, in turn, along with the size of the estuary, controls the strength and character of that estuary's ability to remove and retain particulate matter. In other words, the gravitational circulation and basin geometry set the limits on an estuary's filtering efficiency for fine-grained sediments (Schubel and Carter). In the sequence of estuaries proposed by Pritchard (1955), partially mixed (Type B) estuaries have the strongest gravitational circulation patterns and are the most effective filters for fine-grained suspended matter, particularly when the limit of sea salt penetration is sufficiently far inside the estuary so that it cannot be displaced from the basin even during periods of high river discharge (Schubel and Carter).

Fjords also can be extremely effective filters for particulate matter. Cannon et al. describe the variations in transport in Puget Sound and how entrainment is able to trap suspended sediment and associated contaminants in that fjord estuary. They point out that mixing of surface water flowing seaward with deeper new water entering the estuary over the entrance still increases the filtering efficiency of Puget Sound for suspended and dissolved constituents. As they put it, "The downward mixing near the sill of seaward-flowing water is probably unique to the fjord-like Puget Sound estuary compared to other estuaries in the lower 48 states. Most other U.S. estuaries mix incoming water upward into the seaward-flowing surface water continuously along the estuary."

Using quite different approaches, Schubel and Carter (1977; this volume) and Officer et al. have demonstrated that the partially mixed Chesapeake Bay traps essentially all of the fluvial sediment it receives and that it actually imports sediment from the ocean. These investigators show that filtering is related to gravitational circulation and is concentrated at two locations. The most effective filtration occurs in the upper Bay near the upstream limit of sea salt penetration; the other zone of geological filtration is in the lower Bay near the mouth of the estuary.

As an estuary evolves geologically and its basin is filled with sediments, its depth, volume and surface area decrease. Consequently, its intertidal volume decreases and the strength of the river flow relative to the tidal flow increases, causing the estuarine circulation to shift in the direction of a river-dominated (Type A) estuary. The length of the estuary also decreases as it evolves. Depending upon where the estuary is in the sequence of estuarine circulation types, shifting its circulation in the direction of a river-dominated estuary either could increase or decrease its filtering efficiency. Ultimately, however, the estuary reaches a size beyond which further ageing results in a decrease in this efficiency. Eventually, the estuary is filled to the point where it no longer traps suspended and dissolved materials. The size of the basin is so small and the role of the river is so strong that seawater no longer penetrates the semi-enclosed coastal basin and the estuary ceases to exist (Schubel and Carter). Throughout the world, we can find estuaries in various stages of geological evolution with widely varying filtering efficiencies (Schubel and Carter; Biggs and Howell).

Geological and Geochemical Processes

Biggs and Howell and Schubel and Carter review a number of ways in which the filtering efficiency of estuaries for fine-grained suspended matter can be estimated. Biggs and Howell argue for the importance of the contribution that filter-feeding organisms (planktonic and benthic) make to estuarine sedimentation (referred to also by Kranck) and to the high filtering efficiency of estuaries. To estimate this efficiency, they also apply to estuaries the simple capacity-inflow ratio which was developed to predict the useful lifetime of a reservior before it filled with sediment.

The strength and character of the estuarine circulation not only control the removal of suspended particles by physical processes but also play major roles in the removal of dissolved, colloidal, and suspended matter through regulation of a variety of geochemical processes including precipitation, flocculation, and sorption. The estuarine circulation pattern controls the location and, in part, the strength of the turbidity maximum (Festa and Hansen 1976; Bowden; Schubel and Carter). Turbidity maxima, well-known features typical of partially-mixed estuaries, are zones within which concentrations of suspended particulate matter are greater than those either farther upstream in the source river or farther downstream in the estuary. Usually they are located near the null zone, i.e. the zone in which the upstream flow of the lower layer dissipates until finally the net flow is downstream at all depths. The turbidity maximum is a zone of convergence in the lower layer where upward vertical velocities favor the accumulation of suspended matter. The high concentration of particles within this zone provides major sites for physical, chemical, and biological reactions between dissolved and particulate species and for interactions among particulate species. As a result, the zone acts as a chemical/geochemical filter for removal of dissolved and colloidal species as well as for suspended species. Electrochemical precipitation (flocculation) of colloidal and dissolved materials plays important roles in the removal and retention of substances such as iron. Flocculation may also contribute to the removal and retention of fine suspended particles (Krank). There is disagreement among investigators as to whether flocculation is a consequence of the turbidity maximum (Schubel 1968), a cause of it, or both a cause and a consequence (Krank).

Geochemical removal of dissolved and colloidal substances is affected by physical processes, particularly by those which control the concentration of suspended particles. These include not only mixing and gravitational circulation, but also the alternate deposition and resuspension of bottom sediment by tidal scour and wind waves. Scavenging of substances from solution by suspended particles is most effective where concentrations of particulate matter are high and where there is a continual supply of "new" particles, i.e., particles whose carrying capacity for particular dissolved species has not been reached. Particle-reactive chemical species are concentrated in estuarine areas in which fine-grained sediments are accumulating and particle reworking is rapid and deep (Cochran). In some cases the removal from solution may be temporary, and materials may be released from the particles back into solution. This may occur while the particles are in the water column or after they have been deposited on the estuary floor (Cochran; Sharp et al.). Thus, the estuarine filter is "leaky."

Sharp et al. demonstrated that geochemical filtration in Delaware Bay is most rapid in its upper reaches where suspended sediment concentrations are highest--in the well-developed turbidity maximum. They report that the effectiveness of the filter is revealed by "an abrupt drop in the concentrations of some dissolved materials in the low salinity region . . .".

The estuarine filter is selective, particularly for dissolved species (Cochran; Sharp et al.). As Sharp et al. point out, "Much of the dissolved material carried into the estuary is simply diluted upon mixing and delivered to the sea. In contrast, some dissolved chemicals react within the estuary and are removed from the dissolved phase, thus making the estuary a 'filter'." Species whose concentrations are affected only by dilution are described as having conservative behavior. Other species exhibit strongly non-conservative behavior which may be either positive or negative. Some materials which enter the estuary adsorbed to suspended particles actually are mobilized (solubilized) within the estuary and are thus conveyed to the sea in this manner.

Successful understanding of the estuary as a geochemical filter requires knowledge not only of the distributions of dissolved and particulate species but also of their rates of introduction, discharge, and accumulation. Cochran points out that the radionuclides of the uranium and thorium decay series form a unique class of tracers which can be used to document both chemical behavior and the rates and processes which affect them. For the most part, uranium exhibits conservative behavior but may be removed in some organic-rich estuaries. On the other hand, radium is strongly non-conservative and exhibits concentrations in estuarine waters higher than those in river or open-ocean waters because of release of radium adsorbed to river-borne suspended sediment and mobilization and release from sediments deposited in estuaries (Cochran). Estuaries are effective filters for ^{234}Th and ^{228}Th and for ^{210}Pb and ^{210}Po. In Long Island Sound, ^{234}Th has a mean residence time of only about one day before it becomes adsorbed to particles which are subsequently deposited in the Sound (Aller and Cochran 1976) and ^{210}Pb is scavenged rapidly from estuarine waters by suspended particles (Benninger 1978; Santschi, Li and Bell 1979).

Biological and Biochemical Processes

Biological and biochemical filtration occurs primarily as the result of two processes: (1) filter-feeding activities of planktonic and benthic organisms and (2) biological fixation of dissolved materials into particulate form followed by removal during food chain transfer or sedimentation. The removal and retention of fine suspended particles and the dissolved and colloidal materials that become adsorbed to these particles is facilitated by the large populations of filter-feeding animals that inhabit estuaries (Biggs and Howell). Bacteria and organic coatings of particles may also contribute to this agglomeration and accelerate settling and retention of particles in estuaries.

Sharp et al. report abrupt declines in ammonium, phosphate, and nitrate levels in the middle reaches of the Delaware estuary during spring phytoplankton blooms. Similarly, materials coming into marshes and being taken up by marsh macrophytes are incorporated into organic matter (Hopkinson and Hoffman) which may be moved to bays and the nearshore open coastal environment adjacent to estuaries by tidal action. In Georgia, Hopkinson and Hoffman describe how carbon fixed in marshes in excess of that respired or lost to the sediment appears to be exported to these nearshore waters. Here it would help support a metabolic level that could not be sustained if the only primary production available were that of the pelagic region. Thus the estuary may produce an output signal wherein carbon coming into the estuary is fixed and made available to the nearshore system as a supplement to pelagic production.

In the lower Hudson estuary, sewage waste discharge provides new nitrogen throughout the year (Malone). Phytoplankton play a major role in assimilating this new nitrogen and are thus a structural component of the lower estuary's nutrient filtration system. In spring, input to the Hudson estuary and plume includes this sewage nitrogen plus new nitrogen from offshore. The output signal is phytoplankton biomass which is exported into nearby coastal water. In summer when the estuary's capacity to assimilate and metabolize this new nitrogen is at its greatest, most phytoplankton production is metabolized internally and there is little net export. The limited new production that forms the output signal is probably plankton food web material (biomass, feces). From November to February, assimilation of sewage nitrogen is low, so most is exported.

More such information is required on the exchange of organic matter between estuaries and offshore waters and on inputs of material from land to estuaries. That is, data on material budgets of estuaries are needed to assess the efficiency with which estuaries remove and retain material such as organic carbon, nitrogen, phosphorus, and pollutants. Nixon and Pilson move in this direction by estimating the total system metabolism of Narragansett Bay using nutrient ratios. About 75% of organic matter fixed in the Bay is consumed or retained within the system, so net production exceeds consumption and retention. Only a small portion of the organic matter accumulates in the sediments or is removed by humans in the fishery. Thus, an amount equal to 22-24% of the phytoplankton production is exported from the Bay. Nixon and Pilson outline components of the budget that require further elucidation.

Other mechanisms contribute to the effectiveness of estuarine filtering systems. For example, surface foam in shallow regions of Puget Sound entrains and concentrates constituents such as dissolved and particulate carbon, nitrogen, and phosphorus to levels several times higher than those of subsurface or neritic water (Wissmar and Simenstad). Bacteria, benthic algae, seagrasses, and enriched sediment particles seem to be involved in this phenomenon. Important food resources are thus concentrated, perhaps to be used by higher order consumers.

The freshwater-seawater transition zone has been looked upon as a biological filter within an estuary, removing living biomass (in this case, freshwater halophobic phytoplankton) and resulting in different biogeochemical interactions and properties downstream from the transition zone (Morris et al. 1978; Christian et al.). This seems to be true of the Tamar Estuary in England (Morris et al. 1978) but was not demonstrated in the Neuse River Estuary in North Carolina (Christian et al.). The subject requires further investigation.

We have seen that physical mixing, sediment adsorption, chemical exchange, and biological transformation of nutrients are among the components of a dynamic filter which may affect nutrients so that the kinds and quantities exported from an estuary are different from those the estuary receives. In addition, seasonal rainfall and discharge factors may be involved. Schemel et al. describe how river inflow variability (in addition to biological removal) control nutrient levels during the summer in Suisun Bay in northern San Francisco Bay. River discharge affects the estuarine circulation (the mixing and the location of the null zone) which in turn influences phytoplankton dynamics and, consequently, the kinetics of nutrient removal.

For arid regions of southern California, Zedler and Onuf show how biological (consumption, nutrient absorption) and physical (sedimentation, salt accumulation) filters are affected by wet and dry seasons. For example, in estuarine channels, winter streamflow may carry sediment sufficient to smother benthic invertebrates; thus an increase in a physical filtering activity may decrease components of a biological filtration system. The systems are complex and dynamic and the exact roles of many of the interactors are not clear. Nor is enough known about the oceanic contribution to budgets of inorganic and organic materials within the estuary when discharge varies over time.

As components of biological filtration systems in estuaries, submerged macrophytes can be very important (Kemp et al; Short and Short). The physical structures they provide as they grow towards the water surface aid in trapping of suspended matter. Thus, in the vicinity of beds of vegetation, water is less turbid because of increased removal and decreased resuspension of fine-grained suspended particulates. When this occurs through the warmer summer months, it can enhance photosynthetic activity because of increased light levels. Large quantities of allochthonous organic material can also be trapped to serve as food for secondary producers. Within beds of vegetation, fish abundance may be significantly greater than in nearby unvegetated areas. Beds of vegetation also remove dissolved nutrients from the water column, although in Chesapeake Bay this is a less significant process than is their role in trapping sediments.

Management

Management of dynamic systems such as estuaries can be as complex as the systems themselves. Managers must keep abreast of scientific insights to appreciate the structure and function of estuarine systems. The filtration role of estuaries can be of great importance to the well-being of the ecosystem.

The freshwater signal to estuarine ecosystems has obvious importance (Funicelli; Zedler and Onuf). Funicelli discusses the effects of reduced flow into two Texas estuaries. With the continued removal of nutrients by sedimentation and chemical processes (in addition to export from the system), ultimately the nutrient pool would be depleted unless replenished by renewed freshwater inflow. Thus, managers of inland Texas water resources (e.g. reservoirs) must take estuarine needs into account. Water may need to be released during below-average periods of natural inflow, or at biologically important periods of the year to encourage successful recruitment to the biota.

In the Mississippi Deltaic Plain, coastal wetlands are deteriorating rapidly (Boesch et al.). The situation is complex but changes in sedimentation and in salinity regime seem to be among the causes. In the past, sediment was trapped and new coastal landforms were built up naturally. These landforms were the base for extensive wetlands which were nourished by trapped nutrients and which in turn nourished the animal biota. Presently, the system appears to be becoming a less efficient sediment and nutrient filter as humans modify it by building canals, channels, and flood protection structures.

The role of estuaries in concentrating pollutants is considered by Tippie (Chesapeake Bay) and Casapieri (Thames Estuary). In both aquatic environments, a variety of pollutants enter from different sources. In Chesapeake Bay, the great size of the system has modulated the effects of the trapping and recycling of nutrients and toxic compounds (Tippie). However, toxic materials are increasing in certain rivers and estuaries that are near industrial centers. The need to control point and non-point sources of nutrients, toxic materials and sediments, is obvious but will require a diversity of approaches. Efforts to reduce nutrient input signals to the Thames have been successful enough so that dissolved oxygen levels have improved and numerous species of fish, including salmon, have returned to the estuary.

Conclusions

The locations of an estuary's most effective physical, geological, and geochemical filtering actions usually overlap and all are located in the upper reaches of the estuary in the turbidity maximum. The locations of an estuary's biological and biochemical filters usually are less well defined, but most active filtration occurs where biological productivity and benthic populations are highest. Note also that particulate materials are trapped but dissolved substances (e.g. nutrients) can be recycled to maintain higher productivity than would occur if inputs were simply passed through the system.

It appears that our knowledge of estuarine filtration mechanisms may be greater for non-biological systems than for biological systems, but even for the former our understanding is at best semi-quantitative. Because the filtering activities of estuaries determine their life-spans and play significant roles in controlling their "quality of life," it is important that we improve our understanding of the processes that control these activities. Also, effective management of estuarine resources should include consideration of how the filtering capacities of estuaries could be used to further management objectives. We hope that these proceedings will serve to stimulate further interest in these important topics.

Acknowledgments

We acknowledge the critical comments of those of our colleagues who read this manuscript as it was being prepared.

REFERENCES CITED

Aller, R.C. and J.K. Cochran. 1976. $^{234}Th/^{238}U$ disequilibrium in nearshore sediment:particle reworking and diagenetic time scales. *Earth Planet. Sci. Lett. 29*:37-50.

Benninger, L.K. 1978. ^{210}Pb balance in Long Island Sound. *Geochim. Cosmochim. Acta 42*:1165-1174.

Festa, J.F. and D.V. Hansen. 1978. Turbidity maxima in partially mixed estuaries: a two-dimensional numerical model. *Est. Coastal Mar. Sci. 7*: 347-359.

Morris, A.W., R.F. Mantoura, A.J. Bale and R.J.M. Howland. 1978. Very low salinity regions of estuaries: Important sites for chemical and biochemical reactions. *Nature 274*: 678-680.

Ocean Science for the Year 2000. 1984. UNESCO, Rome.

Pritchard, D.W. 1955. Estuarine circulation patterns. *Proc. Amer. Soc. Civil Engineers 81*:717/1-717/11.

Pritchard, D.W. 1967. What is an estuary: Physical viewpoint, pp. 3-5. *In*: G.H. Lauff (ed.), *Estuaries*, Amer. Assoc. Adv. Sci. Pub. No. 83, Washington, DC.

Santschi, P.H., Y-H. Li and J. Bell. 1979. Natural radionuclides in the water of Narragansett Bay. *Earth Planet. Sci. Lett. 45*:201-213.

Schubel, J.R. 1968. Turbidity maximum of the northern Chesapeake Bay. *Science 161*: 1013-1015.

Schubel, J.R. and H. H. Carter. 1977. Suspended sediment budget for Chesapeake Bay, pp. 48-62. *In*: M. Wiley (ed.), *Estuarine Processes, Vol. II*; Academic Press, New York.

PHYSICAL PROCESSES

TURBULENCE AND MIXING IN ESTUARIES

K.F. Bowden[1]

Oceanography Department
University of Liverpool
Liverpool, England

Abstract: Mixing and dispersion in an estuary result from a combination of advective and diffusive processes and are closely related to patterns of circulation. The intensity and scale of turbulence depend largely on the stability of the density distribution and so affect directly the vertical fluxes of momentum and matter. Indirectly, the state of turbulence also influences the longitudinal and transverse mixing. In setting up models to predict changes in estuarine conditions arising from natural external causes or human intervention, most attention in the past has been given to longitudinal fluxes across planes perpendicular to the axis of an estuary. The apparent simplicity of one-dimensional treatments, often used for this purpose, conceals the importance of contributions from physical processes in the vertical and transverse directions. It is likely that two- and three-dimensional models will be more widely used in future but their successful employment calls for a better understanding of the physical processes which are to be simulated.

INTRODUCTION

The mixing with which we are concerned in an estuary is that of the river water with the intruding sea water, or of a discharged substance with the ambient water. Mixing is brought about, in general, by the combined action of currents

---

[1]*Present address: "Sailways," 21 Undercliff Road, Wemyss Bay, Renfrewshire, PA18 6AJ, Scotland.*

ISBN 0-12-405070-0

and turbulent diffusion. The physical processes of mixing are of interest to workers in other disciplines, since the waters being mixed usually differ in dissolved chemical constituents, suspended matter, or biological populations, and the mixing influences the reactions taking place. The mixing processes in various parts of an estuary, of which the region of the turbidity maximum is an example, thus play a significant role in its action as a "filter." The flow of water in estuaries is usually turbulent, with few exceptions, and the turbulence has a major influence on the mixing processes. It is this aspect of turbulence with which this paper is concerned and not the dynamics of the turbulent flow itself.

DISPERSION AND MIXING

Dispersion and mixing are two aspects of the same process, since the dispersion of one water mass within another brings about the mixing of the two. Dispersion is usually produced by the combined effect of advection and diffusion. A certain degree of dispersion can be produced by advection alone, however, where there is a shear or other form of deformation in the mean flow.

A simple example is that of transverse shear in a tidal current, as illustrated in Fig. 1. A strip of water across the estuary is considered, marked in some way, as at AA, and it is assumed that the velocity at the axis of the estuary is greater than that nearer the sides. Starting at the slack water before a flood tide, the strip AA will become distorted during the flood tide, as shown at BB, and in a sense dispersed. If the flood and ebb are symmetrical, as in Fig. 1(a), the movement on the ebb will be the reverse of that on the flood and, in its final position, the strip will be at CC, coinciding in position and shape with AA. Dispersion by advection alone is, therefore, reversible in this case.

In the case shown in Fig. 1(b), the flood and ebb currents are not symmetrical, the flood being stronger in the centre of the estuary and the ebb stronger at the sides. In this case the distortion on the flood is increased during the ebb and the marked water is further dispersed, as shown by its final position CC compared with AA.

Diffusive processes always act so as to increase the dispersion of a group of particles and to reduce gradients of concentration of a constituent. They are, therefore, essentially irreversible. Even when the flood and ebb currents are symmetrical, any diffusion present would continue to act during the ebb, as well as on the flood, and the group of particles would end up occupying a larger volume than they did initially, as shown by the dotted lines in Fig. 1(a).

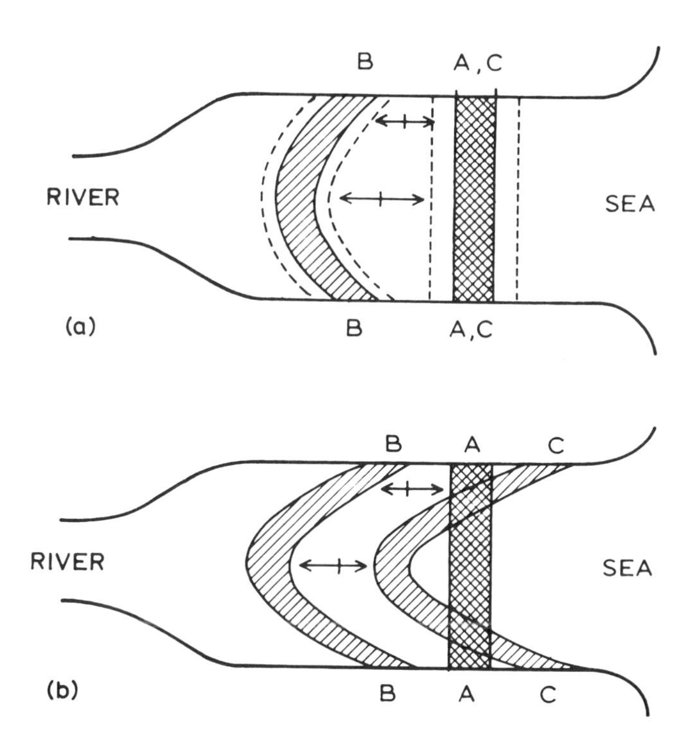

FIGURE 1. Dispersion arising from transverse shear in a tidal current: (a) flood and ebb currents symmetrical; (b) flood and ebb not symmetrical. The dotted lines in (a) indicate the effect of eddy diffusion.

Molecular diffusion is usually negligible and the effective diffusive process is nearly always turbulent diffusion, also known as eddy diffusion. The intensity of eddy diffusion in the vertical is dependent on stability, which is usually related to shear in the horizontal flow.

OVERALL TREATMENT OF MIXING

There are some overall methods of treating mixing in an estuary which do not require a knowledge of the processes involved. They use fresh water (or sea water) as a tracer and depend on the assumption that other conservative constituents will be mixed in the same way.

The most general parameter representing the mixing properties of an estuary is the flushing time, or residence time, calculated by using fresh water as a tracer. Assuming steady state conditions, the flushing time is simply the total fresh water content of the estuary divided by the rate of river discharge. The difficulty here is that the assumption of a steady state is seldom justified. Nevertheless, the concept of flushing time, applied to the estuary as a whole or to several segments into which it may be divided, is often a useful first approximation to its overall mixing properties.

A second overall treatment is to use the salinity distribution along an estuary to deduce directly the distribution of any conservative constituent, introduced at a given point. The method can readily be modified to allow for non-conservative processes, as described by Dyer (1973) or Officer (1976). There has been a revival of interest in this method in recent years, particularly in relation to chemical constituents, as described in a review paper by Liss (1976). Since then the physical basis of the method has been examined and clarified in papers by Officer (1979), Officer and Lynch (1981), and Rattray and Officer (1981). In a recent paper, Rattray and Uncles (1983) used a development of the method to predict the distribution of caesium 137 along the Severn estuary in Great Britain, allowing for its radioactive decay.

RELATION OF MIXING BEHAVIOUR TO CIRCULATION PATTERNS

Because of the advective element in mixing and the dependence of turbulence characteristics on the mean flow, there is a close relationship between the pattern of circulation and the mixing behaviour in an estuary. The three basic factors in the circulation pattern are: (1) river flow; (2) density current flow, seaward in the upper layer and upstream below it, and (3) tidal currents which, apart from moving the whole pattern to and fro, generate turbulent kinetic energy and so tend to increase vertical mixing.

An illustration of the interaction between the various factors, in the case of a partially mixed estuary, is the occurrence of the turbidity maximum, a region where there is a high concentration of suspended matter. In Fig. 2, the river flow alone is shown in Fig. 2(a), decreasing in velocity seaward as the cross-sectional area increases and remaining nearly uniform with depth. The density-driven circulation, shown in Fig. 2(b), is seaward in the upper layer and upstream in the lower layer, reaching its maximum development about mid-way along the estuary. The vertical water movements shown are necessary to maintain continuity of the flow.

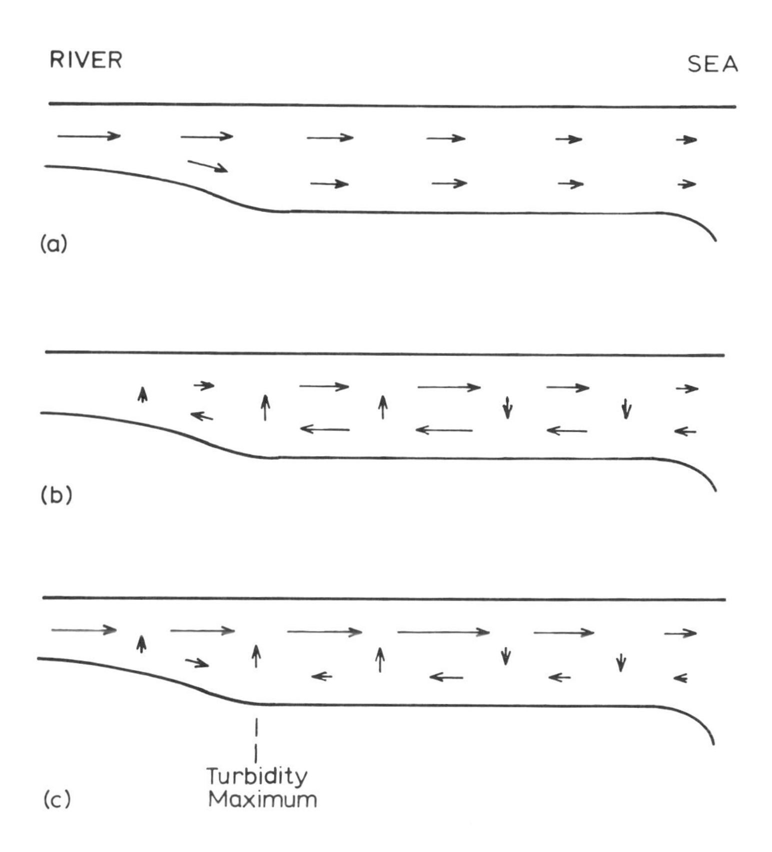

FIGURE 2. Superposition of river flow and density-driven circulation, giving rise to a turbidity maximum: (a) river flow alone; (b) density-driven current alone; (c) superposition of the two flows.

The superimposed river flow and density flow, shown in Fig. 2(c), indicate that in the lower layer there is a point where the resultant flow changes from being downstream to upstream. It is a region of convergence in the lower layer, with an upward vertical velocity, giving conditions favouring the accumulation of suspended matter there: i.e., the turbidity maximum. With the tidal flood and ebb, its position will oscillate along the estuary. An increase in river flow will tend to move its mean position in a seaward direction.

These simple physical considerations indicate where a turbidity maximum is likely to occur. Further information on the velocity convergence and the intensity of turbulent mixing in the region is needed to make quantitative estimates of the effect on particle concentration and on deposition or re-suspension. The high concentration of particles provides major sites for chemical and biological reactions. The turbidity maximum thus plays an important part as a filter, since it is a region where particulate matter may be removed from the estuary water by deposition and where dissolved constituents may undergo considerable chemical change. The phenomena which occur at the turbidity maximum are still only poorly understood and call for multidisciplinary programs of research.

LONGITUDINAL DISTRIBUTION OF PROPERTIES

In studies of mixing in estuaries, more attention has been given to the longitudinal distribution of properties than to any other aspect. The treatment of the longitudinal distribution is based on cross-sectional averages of the quantities concerned and usually deals with average properties over a tidal period.

The instantaneous flux of a constituent across a small area of the cross-section is simply the product of the longitudinal component of velocity and the concentration. However, both these quantities vary with depth and across the estuary. Moreover, they both have, in general, a steady part and an oscillatory part. When the flux is integrated over the whole cross-section and with respect to time over a tidal period, a large number of terms arise. Various workers have broken down the total flux into individual terms in different ways (e.g., Dyer 1973; Fischer 1976). A number of investigations have led to estimates of the relative magnitudes of these terms and it has been found that, in some particular estuaries, two or three of the terms have a dominant influence, but in other estuaries the situation is less clear.

Several theoretical workers have shown that, under certain conditions, the various transport terms are all proportional, to a first approximation, to the concentration gradient along the estuary. This has been taken to justify the use of a coefficient of horizontal dispersion D_x to represent the average dispersive flux across a given section due to the resultant effect of all the processes. This reasoning leads to the following one-dimensional longitudinal dispersion equation for the mean concentration $\bar{c}(x, t)$, averaged over the cross-section and over a tidal period.

$$\frac{\partial}{\partial t}(A\bar{c}) + \frac{\partial}{\partial x}(A\bar{u}\bar{c}) = \frac{\partial}{\partial x}\left(AD_x \frac{\partial \bar{c}}{\partial x}\right) + F(\bar{c},x,t)$$

The variation with time is the slowly varying part, after tidal averages have been taken. x is distance measured along the estuary, A(x) is the cross-sectional area, and $\bar{u}(x,t)$ is the mean velocity perpendicular to the section. D(x,t) is the longitudinal dispersion coefficient which, as indicated above, includes (1) the tidal variations of cross-sectional means of u and c, (2) tidally averaged shear effects in the vertical and transverse directions, (3) oscillatory shear effects, again both vertical and transverse, and (4) turbulent diffusion. The function $F(\bar{c},x,t)$ is a source or sink function, allowing for the rate of generation or decay of a non-conservative constituent.

A number of workers have attempted to predict the value of the dispersion coefficient D_x in terms of the geometry and other parameters of the estuary. This becomes a two- or three-dimensional problem, since the various terms contributing to the longitudinal flux involve the vertical and transverse variations in velocity and concentration (e.g., Fischer et al. 1979; Smith 1979, 1980).

In a recent paper, Smith (1982) has proposed the use of a "delay-diffusion" equation to take into account the fact that cross-sectional mixing does not take place simultaneously. If the time-scale of cross-sectional mixing is comparable with, or longer than, the tidal period, then after a flow reversal a cloud of dye, for example, may contract for a time before proceeding to disperse again. The delay-diffusion equation can cope with this effect, which has been observed in dye tracer studies and which would require the dispersion coefficient, as usually defined, to become negative at certain stages of the tidal cycle.

The use of a dispersion coefficient is avoided altogether in the random walk technique, which has been applied to estuary flows in a recent paper by Allen (1982). In the particular problem treated in her paper, the dispersion of a vertical line of particles in a tidal flow, with a specified shear, was considered. As each particle moved horizontally with the tidal velocity, it was acted upon by a random component of vertical velocity. The subsequent movement of a number of particles, of the order of 10,000, could be simulated by a computer program. The variation of the concentration of particles with distance from the release point reproduced a number of features of observed distributions, including a markedly non-uniform increase in dispersion, with a tendency for the cloud of particles to remain unchanged for a while after a reversal of flow. This effect was similar to that shown by Smith's delay-diffusion equation.

TWO- AND THREE-DIMENSIONAL TREATMENTS

The apparent simplicity of the one-dimensional model is deceptive, in that the effective dispersion coefficient D_x disguises a number of processes involving vertical and transverse variations. One may think it preferable, therefore, to consider a two- or three-dimensional model from the outset. The three-dimensional equation shown below represents the distribution of the concentration c, averaged over a short time interval such as a few minutes:

$$\frac{\partial c}{\partial t} + u\frac{\partial c}{\partial x} + v\frac{\partial c}{\partial y} + w\frac{\partial c}{\partial z} = \frac{\partial}{\partial x}\left(K_x\frac{\partial c}{\partial x}\right) + \frac{\partial}{\partial y}\left(K_y\frac{\partial c}{\partial y}\right) + \frac{\partial}{\partial z}\left(K_z\frac{\partial c}{\partial z}\right) + F\,(c,x,y,z,t)$$

In this equation it is assumed that c, u, v and w are averaged over a volume of dimensions which are small compared with those of the estuary itself, as well as over a time interval which is small compared with the tidal period. Fluctuations of the concentration and velocity components on smaller space and time scales, and the corresponding fluxes of the constituent, are regarded as "turbulent" or "sub-grid-scale" and it is assumed that they can be represented by the diffusion terms involving the eddy coefficients K_x, K_y and K_z.

Intermediate cases between one-dimensional and fully three-dimensional models include two-layer models and laterally integrated or vertically integrated two-dimensional models. These models will not be discussed here, except to say that each type has its advantages and may well be the most suitable for tackling a particular problem.

The main difficulty in formulating a two-or three-dimensional model is in specifying the eddy diffusion coefficients K_x, K_y and K_z. Some workers have used a turbulence closure model for this purpose, relating the eddy coefficients to the kinetic energy of the turbulence, rather than attempting to specify them directly (e.g., Leendertse and Liu 1978; Smith and Takhar 1981). The level of turbulent energy may lag behind that of the velocity shear terms by which it is generated, so that this approach can deal with the kind of effect mentioned above in relation to the delay-diffusion equation of Smith (1982).

SMALL-SCALE MIXING PROCESSES

The physical oceanographer is sometimes criticized by chemists and biologists for devoting too much attention to large-scale processes and long-term average conditions when they are more interested in what is happening in a limited region over shorter time scales. An example is the turbidity maximum, already mentioned, in which the concentration of particles varies, at any instant, with height in the water column, and changes with time during the tidal period. The geologist and chemist would like the physical oceanographer to provide details of upwelling velocities, turbulent intensities, and coefficients of eddy viscosity and diffusion, which take into account their vertical profiles and their variation with time during the tidal cycle. This is a challenge which physical oceanographers should heed.

Another region which is of multidisciplinary interest and which requires attention to small-scale features is the bottom boundary layer. It is through this layer that fluxes of chemical constituents from the water column to the sediment and vice versa take place, and it is within this layer that deposition and re-suspension of sediment occur. These processes are dependent on the physical properties of the flow, including turbulent intensities, and they react on the flow through the effects of high concentrations of suspended matter near the boundary.

Increased attention has been given recently to observations of relatively small-scale features in estuaries, such as plumes and fronts, associated with the influx of river water on the one hand and the intrusion of higher salinity water on the other. A particular example is the plume formed by the Connecticut River flowing into Long Island Sound, studied by Garvine and Monk (1974). The occurrence of a front of a different kind, in the River Seiont in North Wales, has been described by Simpson and Nunes (1981). There have been a number of studies of fronts in coastal waters (e.g., Bowman and Esaias 1978), both observational and theoretical, and some of the results could be applied to fronts in estuaries. A front, which is essentially a sharp transition in density, is often associated with changes in the concentration of chemical constituents and of phytoplankton. While it may appear from the accumulation of material at a front that its effect is to inhibit dispersion, the high velocity shears near a front may, in fact, tend to increase vertical mixing in the vicinity.

In a number of estuaries, mixing processes are dominated by localized, time-dependent events, which cannot be represented adequately by coefficients of eddy diffusion. It is necessary to take into account the details of the physical processes involved. Examples of such cases were described by Gardner, Nowell and Smith (1980), who pointed out the valuable assistance which can be obtained by applying ideas and results from investigations in fluid dynamics to estuarine problems of this type.

CONCLUSION

It is hoped that this review of the physical processes of mixing has indicated some ways in which the present state of the art can be of assistance to workers in other disciplines, including the geology, chemistry and biology of estuaries. At the same time, there are some aspects of the physics of mixing processes which should be highlighted as requiring further investigation. In this, observational programs and theoretical methods, including numerical modelling, should go hand in hand.

It would be of great value if a number of long time-series of measurements of currents, conductivity and temperature could be made in selected estuaries, in order to study more effectively the changes arising from varying river discharges and wind effects. The value of such a program would be much enhanced if it were made a multidisciplinary one, including measurements of suspended matter, chemical constituents and possibly some biological variables.

The direct measurement of the vertical flux of salt, given by $\overline{w'S'}$, where w' and S' are the fluctuating parts of the vertical component of velocity and the salinity respectively, would be extremely useful in experiments on vertical mixing. There are instruments, such as electromagnetic and acoustic flow meters, capable of measuring velocity fluctuations, but there does not yet seem to be a sensor commercially available for measuring rapid fluctuations of conductivity.

On the theoretical side, the representation of momentum and material fluxes, whether by eddy coefficients or in some other way, and the dependence of these fluxes on stability are topics which call for more detailed investigation. The value of turbulence closure models and how they might be improved is one aspect of this problem.

Finally, in applying general results to a particular estuary it is important to identify small-scale features, such as fronts and regions where critical hydraulic events, internal waves or trapping mechanisms may occur.

REFERENCES CITED

Allen, C.M. 1982. Numerical simulation of contaminant dispersion in estuary flows. *Proc. Roy. Soc. Lond. A 381*:179-194.

Bowman, M.J. and W.E. Esaias (eds.). 1978. *Ocean Fronts in Coastal Processes*. Springer-Verlag, Berlin. 114pp.

Dyer, K.R. 1973. *Estuaries: A Physical Introduction*. John Wiley & Sons, London. 140pp.

Fischer, H.B. 1976. Mixing and dispersion in estuaries. *Ann. Rev. Fluid Mech. 8*:107-133.

Fischer, H.B., E.J. List, R.C.Y. Koh, J. Imberger and N.H. Brooks. 1979. *Mixing in Inland and Coastal Waters*. Academic Press, New York. 483pp.

Gardner, G.B., A.R.M. Nowell and J.D. Smith. 1980. Turbulent processes in estuaries, pp.1-34. *In*: P. Hamilton and K.B. Macdonald (eds.), *Estuarine and Wetland Processes: with Emphasis on Modeling*. Plenum Press, New York.

Garvine, R.W. and J.D. Monk. 1974. Frontal structure of a river plume. *J. Geophys. Res. 79*:2251-2259.

Leendertse, J.J. and S.K. Liu. 1978. A three-dimensional turbulent energy model for nonhomogeneous estuaries and coastal sea systems, pp.387-405. *In*: J.C.J. Nihoul (ed.), *Hydrodynamics of Estuaries and Fjords*. Elsevier, Amsterdam.

Liss, P.S. 1976. Conservative and non-conservative behaviour of dissolved constituents during estuarine mixing, pp.93-130. *In*: J.D. Burton and P.S. Liss (eds.), *Estuarine Chemistry*. Academic Press, London.

Officer, C.B. 1976. *Physical Oceanography of Estuaries*. John Wiley & Sons, New York. 465pp.

Officer, C.B. 1979. Discussion of the behaviour of nonconservative dissolved constituents in estuaries. *Estuar. Coastal Mar. Sci. 9*:91-94.

Officer, C.B. and D.R. Lynch. 1981. Dynamics of mixing in estuaries. *Estuar. Coastal Mar. Sci. 12*:525-533.

Rattray, M. and C.B. Officer. 1981. Discussion of trace metals in the waters of a partially mixed estuary. *Estuar. Coastal Mar. Sci.* *12*:251-266.

Rattray, M. and R.J. Uncles. 1983. On the predictability of the ^{137}Cs distribution in the Severn estuary. *Estuar. Coastal Shelf Sci.* *16*:475-488.

Simpson, J.H. and R.A. Nunes. 1981. The tidal intrusion front: an estuarine convergence zone. *Estuar. Coastal Mar. Sci.* *13*:257-266.

Smith, R. 1979. Calculation of shear-dispersion coefficients, pp.343-362. *In*: C.J. Harris (ed.), *Mathematical Modelling of Turbulent Diffusion in the Environment.* Academic Press, New York.

Smith, R. 1980. Buoyancy effects upon longitudinal dispersion in wide well-mixed estuaries. *Phil. Trans. Roy. Soc. Lond. A 296*:467-496.

Smith, R. 1982. Contaminant dispersion in oscillatory flows. *J. Fluid Mech.* *114*:379-398.

Smith, T.J. and H.S. Takhar. 1981. A mathematical model for partially mixed estuaries using the turbulence energy equation. *Estuar. Coastal Mar. Sci.* *13*:27-45.

VERTICAL VARIATIONS IN RESIDUAL CURRENT RESPONSE TO METEOROLOGICAL FORCING IN THE MID-CHESAPEAKE BAY

D. W. Pritchard
M. E. C. Vieira

Marine Sciences Research Center
State University of New York
Stony Brook, New York

Abstract: Records of the nontidal current velocity obtained from current meters deployed for 20 days in two cross sections in Chesapeake Bay were used with an interpolation scheme to produce a vertical profile of the laterally-averaged longitudinal velocity component at one meter depth intervals at each section, at 3 h intervals over the 15 day truncated record. Various statistical procedures were used to relate variations in the residual current at each depth to wind variations. Surface layers down to about 8 m responded directly to wind with little time lag. A slope of the water surface was also set up by the wind, with a consequent barotropic pressure force directed opposite to the wind. The current near the bottom responded first to this pressure force, flowing opposite to the wind with a phase lag of about 8 h. This counter response proceeded up the water column, such that in the intermediate layers just below the pycnocline the negative response to the wind had a phase lag of about 20 h. A diagnostic analytical model, based upon a Fourier transform of the linearized equations of motion, was exercised for a four-layered simulation of the vertical response of the currents to inputs of the observed fluctuations in wind and surface slope. The major features of the time variations in the residual currents in the four layers were reproduced by the model.

ISBN 0-12-405070-0

INTRODUCTION

The most obvious way in which the estuary functions as a "filter" is in its retention of particulate matter and of substances bound to particulate matter. The estuary may also serve as a partial filter for certain dissolved constituents introduced in river inflow due to such biochemical processes as incorporation of calcium into shell material. In this paper, we have taken a different meaning for the term *filter*. From a mathematical standpoint, a filter is a process which operates on an input signal, producing an output signal which is altered in a manner prescribed by the filter. It is in this sense that we here consider the estuary as a filter.

The field of motion in an estuary responds to various input processes in a manner which is prescribed by the characteristics of the estuary. The input signals are rate of inflow of fresh water, rate of change of sea level at the mouth of the estuary (the astronomical and meteorological tide), and wind stress acting on the surface of the estuary. Geometric features of the estuary modify the manner in which internal and boundary stresses act to modify the effects of these input processes on the output velocity field.

In this paper we consider the way the estuary filters the effects of wind stress as an input signal on the velocity field as an output signal. Our concern is with the response of low frequency, subtidal velocity fluctuations to subtidal fluctuations in wind stress. We are not concerned here with random turbulent fluctuations in the velocity field due to high frequency fluctuations in wind stress, nor with the dominant motions at tidal frequencies produced by the astronomical tide.

DESCRIPTION OF DATA SET

The data set we have used for this study was previously described by Pritchard and Rives (1979). It consists of an extensive set of current meter, tide gauge, and wind records collected in the fall of 1977, in the middle reaches of Chesapeake Bay. The area involved is shown in Fig. 1.

In all, 30 current meters were deployed during the period 12 October to 2 November 1977 (Fig. 1). Thirteen of these meters were deployed on five vertical moorings on Section G, which extended from Kenwood Beach on the western shore of Chesapeake Bay to the northern tip of Taylors Island on the eastern shore. Nine meters were deployed on three similar moorings along Section H, which extended easterly across the

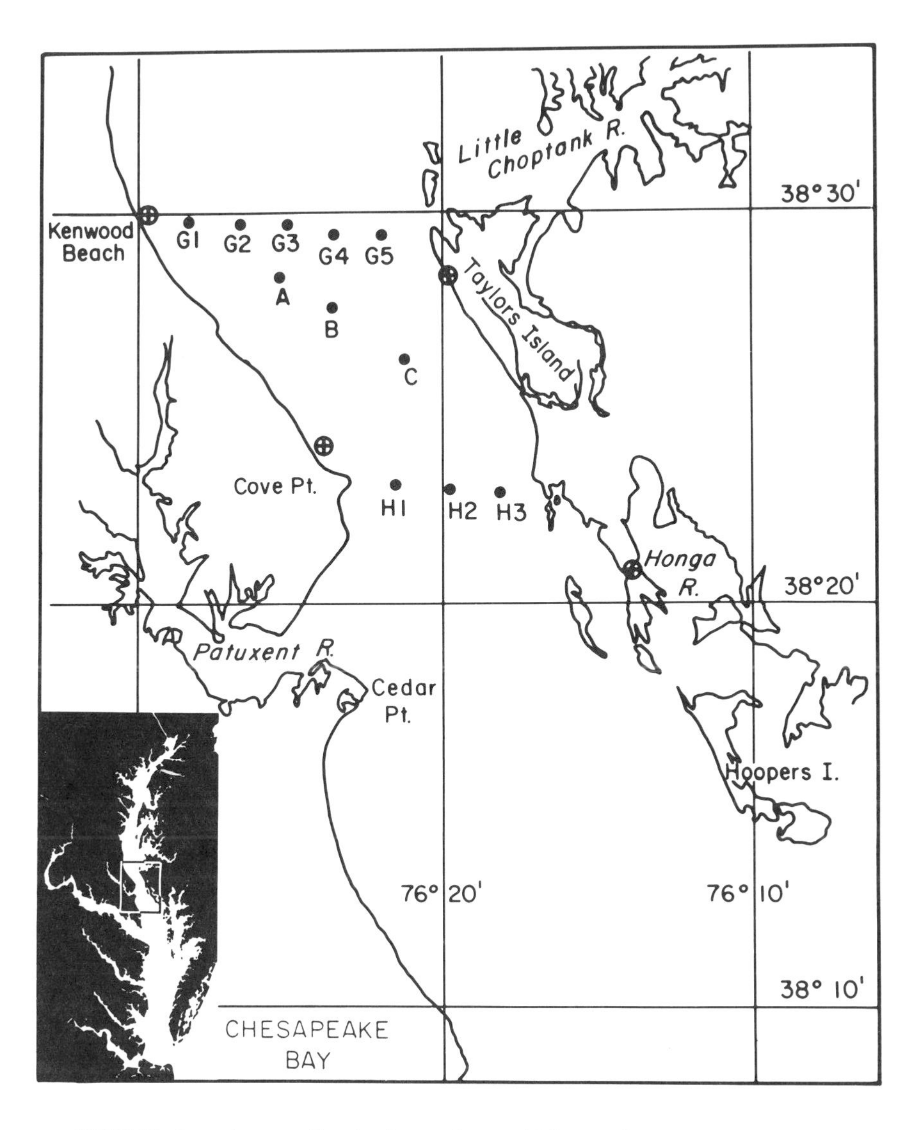

FIGURE 1. Area of study and position of current meter moorings ● and tide gauges ⊕.

Bay from Cove Point. Three intermediate stations, labeled A, B, and C, were located on a diagonal between the two cross sections, and contained the remaining eight meters.

Four temporary tide gauges were installed for the study at approximately the four corners of the study area; i.e., at Kenwood Beach, Taylors Island, Honga River, and Cove Point

(Fig. 1). In addition, use was made of tide gauge records from Havre de Grace, Baltimore, Annapolis, Solomons Island (all in Maryland) and Kiptopeake (Virginia) provided by the National Oceanic and Atmospheric Administration.

Wind data, in the form of hourly values of speed and direction at the Patuxent Naval Air Station, were obtained from that facility for the three month period September-November, 1977. Daily mean discharge data for the Susquehanna River at Conowingo were obtained from the U.S. Geological Survey for the same three month period for which the wind data were obtained. Twenty-six of the thirty meters deployed gave useful records. Two of the meters that failed were located on Section G and the other two on Section H. Figure 2 shows the two cross sections and the location of the eleven meters on Section G and of the seven meters on Section H which gave useable records.

DATA REDUCTION

Much of the material contained in this and subsequent sections of this paper represents a condensation of a detailed treatment given by Vieira (1983). The procedure utilized to reduce the current meter records was the same as that described by Pritchard and Rives (1979). Briefly, the predominant ebb and flood directions were found for each meter by obtaining an independent least square fit of the data for the ebbing and for the flooding stage of the current, respectively. The data were resolved about these directions to produce the longitudinal and transverse component of the current. For this study, only the longitudinal component was used.

The record of the longitudinal component of the current was then subjected to a Low Pass (LP) Lanczos filter with a half-power point of 2 h to suppress high frequency noise. This somewhat smoothed record was then subjected to a Low Low Pass (LLP) filter with a half-power point of 34 h, zero amplitude at 24 h and 95% amplitude at 48 h. The LLP filtered record was subtracted from the LP filtered record to produce the band pass (BP) record. We assumed that the astronomical tidal signal is contained in the BP record, while the LLP filtered record represents the subtidal signal.

We assumed that the stilling well of a tide gauge installation essentially provided a LP filtered record of the water surface elevation. Therefore the tide gauge records were passed through only the LLP filter described above. In this case, the BP tidal record was obtained by subtracting the LLP record from the raw tide gauge record.

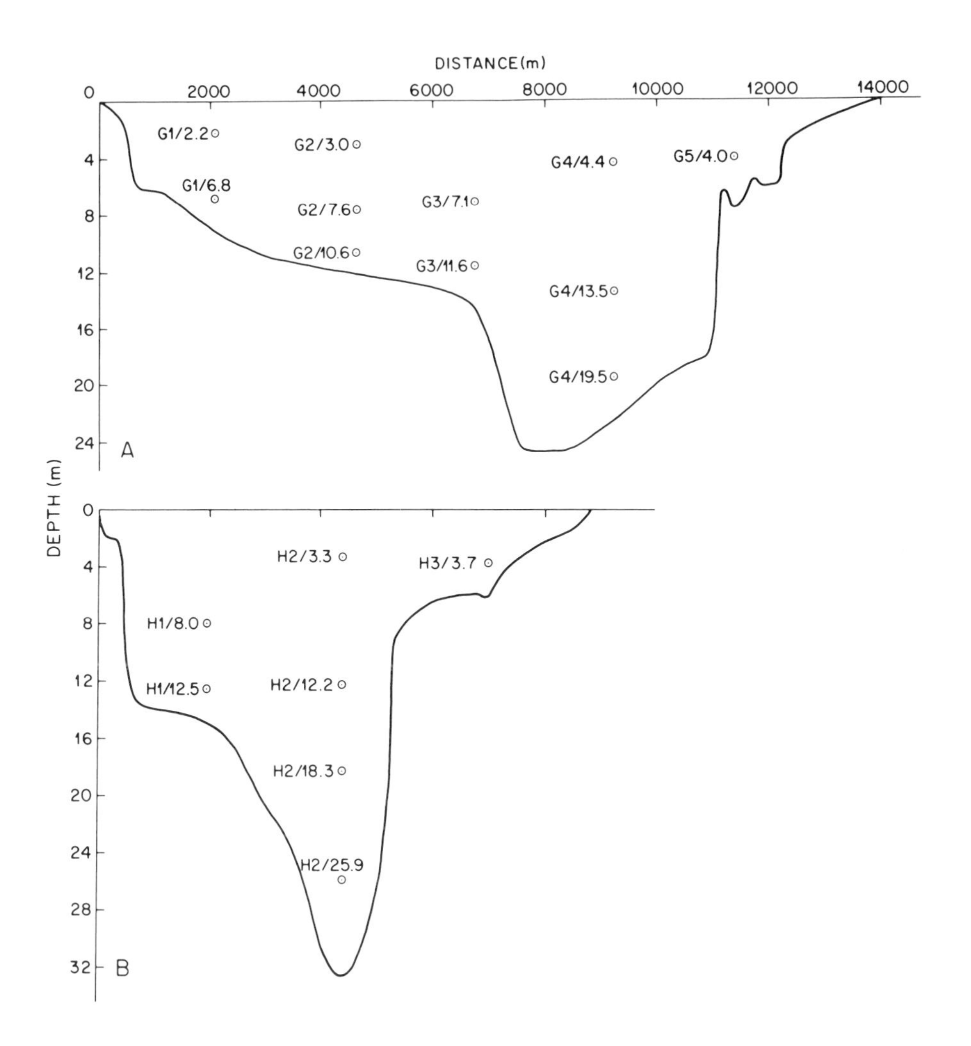

FIGURE 2. Schematic of current meter spatial distribution in cross-section G (A) and cross-section H (B).

The wind data from the Patuxent Naval Air Station have been shown to be representative of the mid-Chesapeake Bay (Elliott, Wang and Pritchard 1978). The direction and speed data were converted to wind stress using a quadratic law (Halpern 1976), with a drag coefficient of 0.025. The stress vector was then resolved along a set of orthogonal coordinates such that the longitudinal component bears on 320°/140° True and the transverse component on 230°/50° True. By our convention, wind stress is considered to be positive if directed seaward (towards 140°) and transversely towards 50°. The stress vector was subjected to the same LLP filter as the tide gauge and current meter records, but no BP wind record was obtained.

Due in part to vandalism and instrument failure, considerable gaps existed in records from temporary tide gauges installed at the four corners of the study area. Only records from Kenwood Beach and Honga River were utilized, and even these had gaps in them for part of the study period. However, a very strong coherence existed between LLP filtered records and BP filtered records from these gauges and corresponding records from permanent NOAA gauges at Solomons Island and Baltimore. The strength of these relationships allowed the missing records from the two temporary gauges to be filled in with a high degree of confidence, using the statistically determined transfer functions between these gauges and the NOAA gauges.

Figure 3 shows BP and LLP records for the Honga River for 23 September through 28 November, obtained as described above. Clearly, the BP record consists primarily of fluctuations of tidal period. The LLP record appears to be made up of fluctuations having a number of frequencies, and the amplitude of the variations in this LLP record is of the same order of magnitude as that seen in the BP record for the tidal signal.

The LLP longitudinal and transverse components of the Patuxent wind stress for 23 September through 28 November are shown in Fig. 4. The fluctuations in this LLP wind stress record are at least superficially similar to those seen in the LLP record of water surface elevation.

Figure 5(A) shows the BP and LLP longitudinal component of the current as observed at 3.3 m depth at Station H2. Again, as with the water surface elevation record, the variance of the LLP record is not markedly less than that of the tidal signal. The long term average of the LLP record is positive; that is, the mean flow as shown by this record was directed down the Bay. However, there were times when the tidal average flow at this depth was directed up the Bay (e.g., 18-19 September).

Figure 5(B) shows the BP and LLP longitudinal component of the current as observed at 18.3 m at Station H2. The amplitude of the fluctuations is somewhat less than for the near surface meter, but is still a significant fraction of the amplitude of the BP record. At this depth the record mean current is directed up the Bay (i.e., is negative). However, there were several periods during which the tidal average flow was directed down the estuary at this depth.

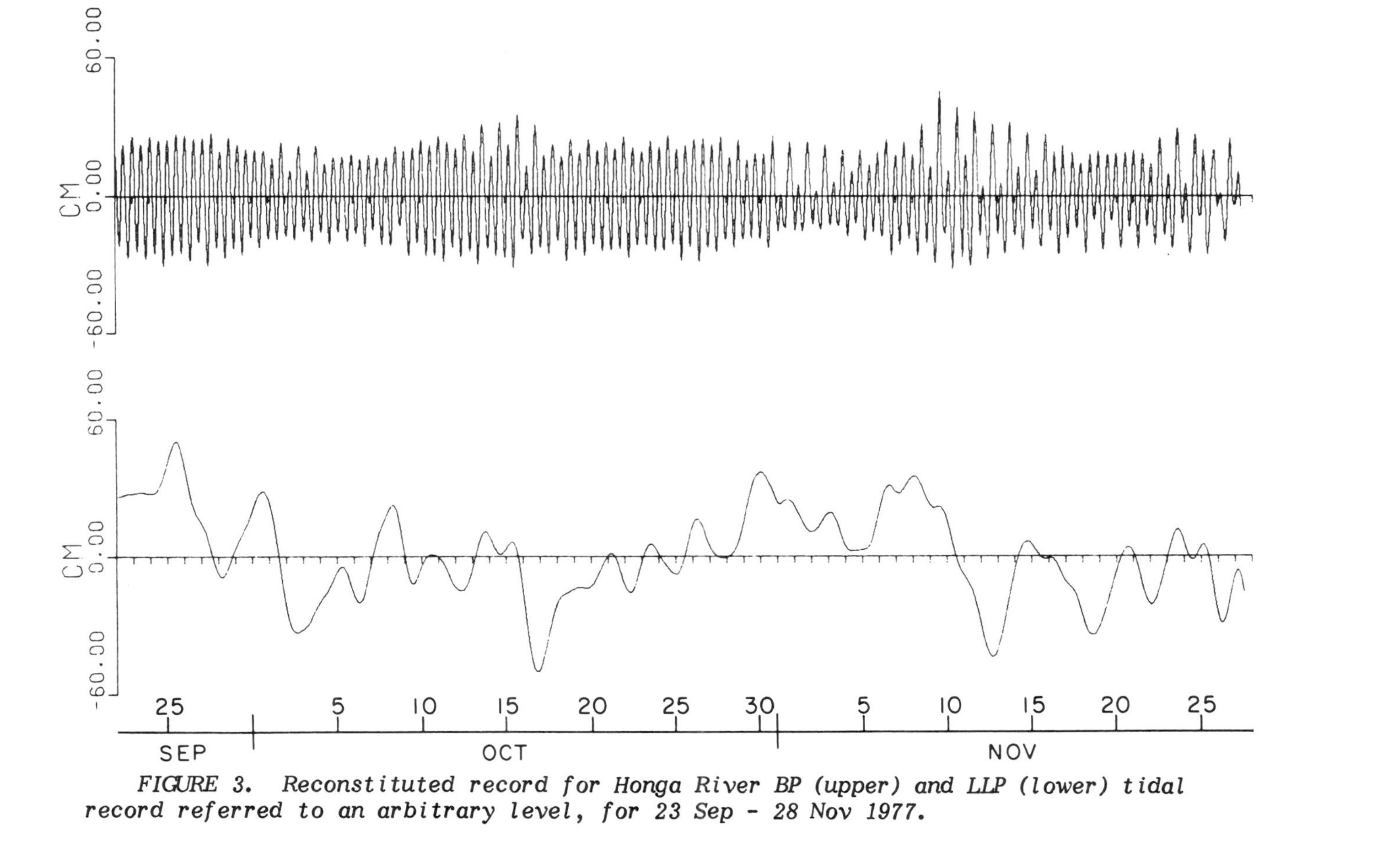

FIGURE 3. Reconstituted record for Honga River BP (upper) and LLP (lower) tidal record referred to an arbitrary level, for 23 Sep - 28 Nov 1977.

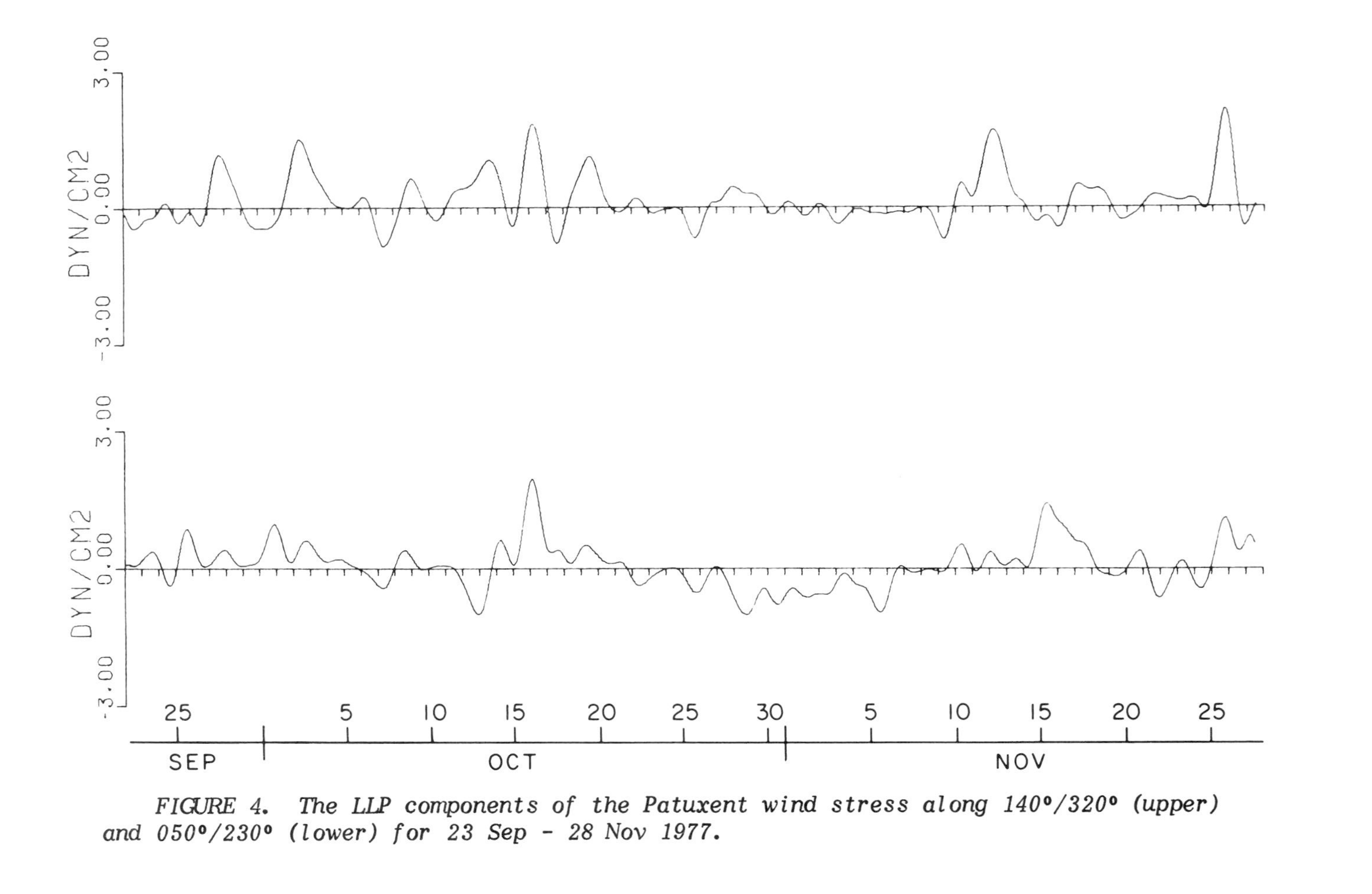

FIGURE 4. The LLP components of the Patuxent wind stress along 140°/320° (upper) and 050°/230° (lower) for 23 Sep - 28 Nov 1977.

In the process of filtering, the record is shortened. The sharper the filter, the larger the loss of data at each end of the record. The LLP filter used here resulted in a shortening of each current meter record by 120 h, divided equally between beginning and end. The filtered current meter records were trimmed to the longest length common to all, which is from 0900 16 October to 1800 30 October. Note, however, that some of the information content of the lost portions of the record is contained in the data points near each end of the filtered record, because of the finite size of the filtering window.

VERTICAL DISTRIBUTION OF THE RECORD LENGTH MEAN LLP VELOCITY

Even with the considerable effort required for the deployment, recovery and processing of the 30 current meters used in this field study, and despite the 87% data recovery rate, there are large gaps in the spatial coverage of the area of each cross section (Fig. 2). One purpose of the study described here was to develop and utilize a set of objective procedures for interpolating between the points of observation, in order to obtain sufficient velocity values in space to allow spatial integration and averaging.

Based on a study of the characteristics of the spatial distribution of velocity in other cross sections of the Bay, coupled with theoretical considerations, the procedure described below was utilized to obtain the vertical distribution of the laterally averaged LLP longitudinal component of the velocity in each cross section. The details of this interpolation scheme are described by Vieira (1983). Briefly, a vertical profile at each mooring was obtained first. For those stations having a sufficient number of functioning current meters in the vertical, a cubic spline was used to interpolate between meters. Extrapolation to the bottom was done using the no slip condition (zero velocity at the bottom) and the assumption that the first derivative of the velocity with depth varied linearly from the deepest point of observation to the bottom. Extrapolation from the topmost meter to the surface was based on the assumption that the first derivative of the velocity varied linearly from its value at the topmost meter, obtained from the cubic fit, to a value at the surface consistent with the assumption that the shear stress at the water surface was equal to wind stress. When there was only one current meter in the vertical, a fit was made assuming that the bottom velocity was zero and that the vertical gradient in the velocity varied linearly with depth from a surface value consistent with wind stress, with the constraint that the curve must pass through the observed point.

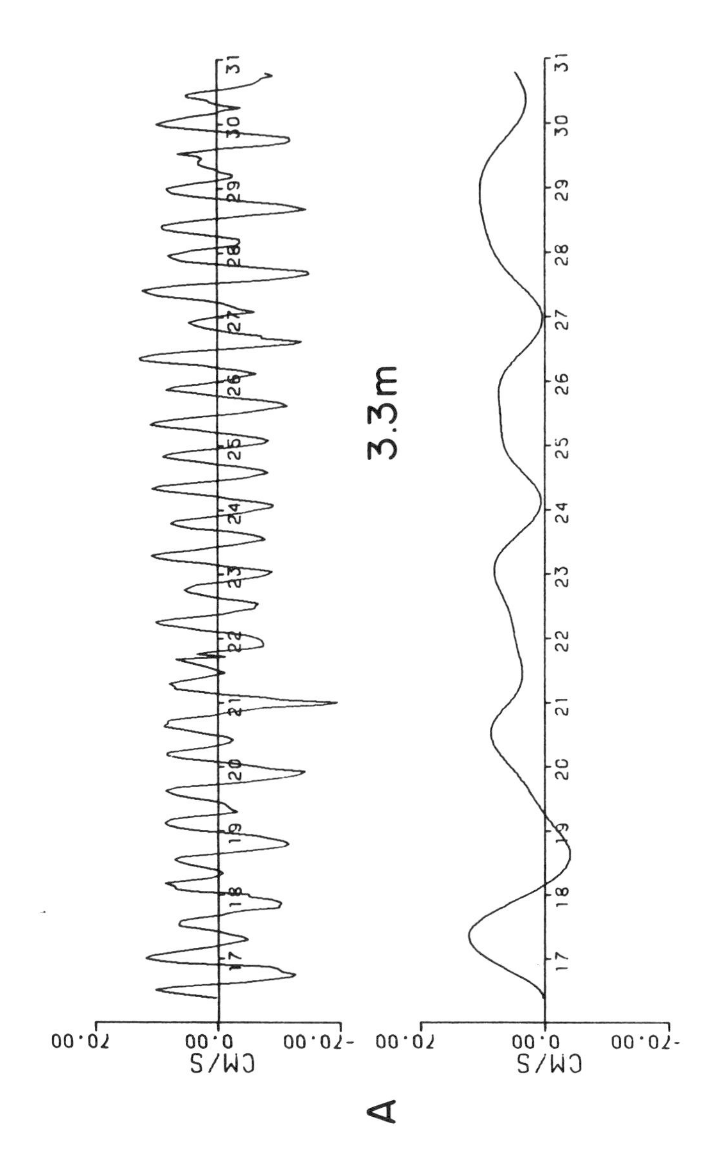
A
3.3m
CM/S
70.00
0.00
-70.00
CM/S
70.00
0.00
-70.00
17
18
19
20
21
22
23
24
25
26
27
28
29
30
31

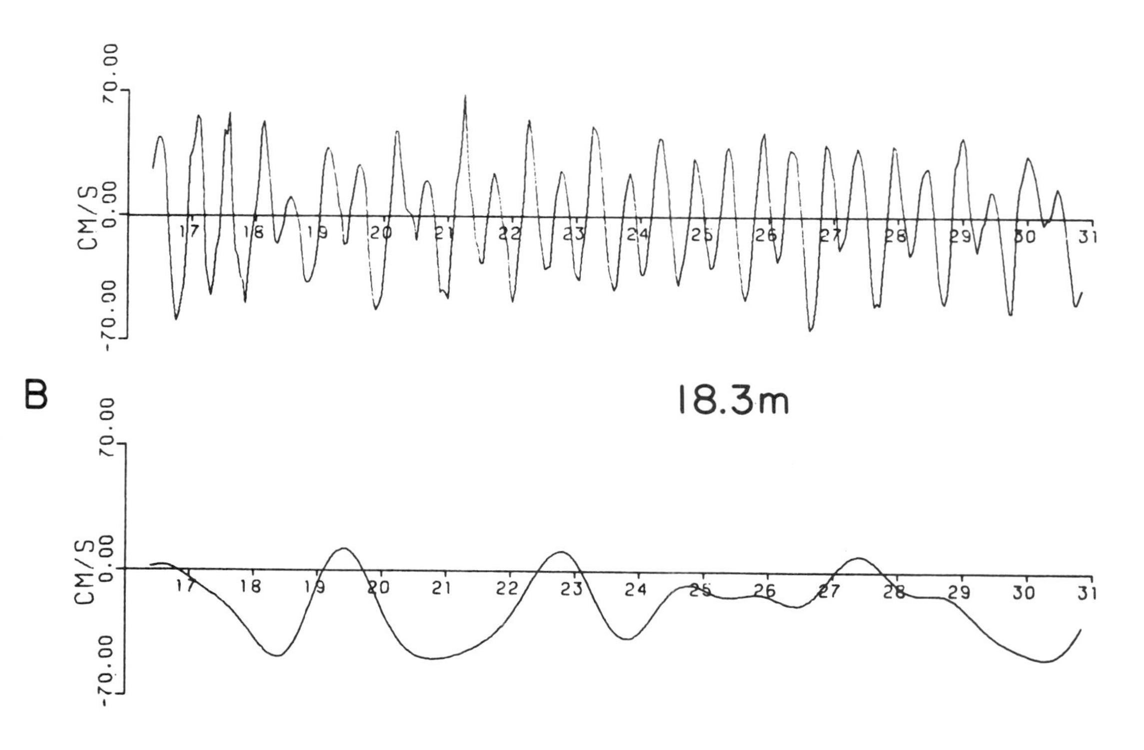

FIGURE 5. The longitudinal component of the current at H2/3.3 (A) and H2/18.3 (B) between 16-30 October 1977: BP record (upper) and LLP record (lower).

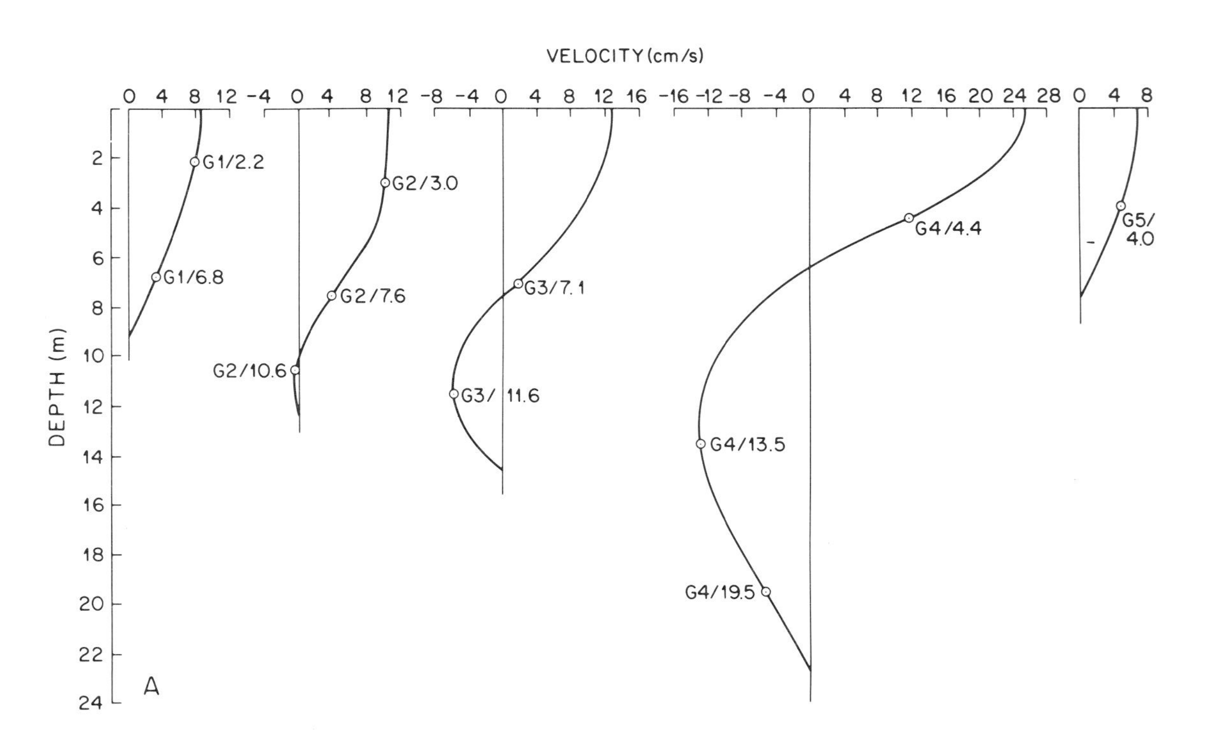

VELOCITY (cm/s)
DEPTH (m)
G1/2.2
G1/6.8
G2/3.0
G2/7.6
G2/10.6
G3/7.1
G3/ 11.6
G4/4.4
G4/13.5
G4/19.5
G5/ 4.0
A

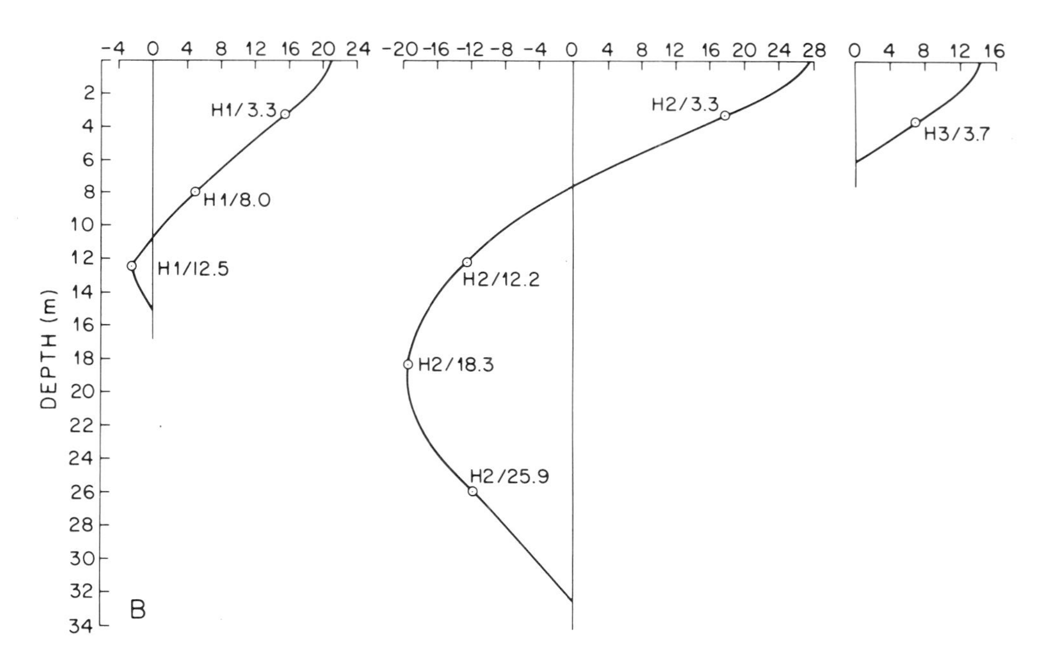

FIGURE 6. Vertical profiles for cross-section G (A) and cross-section H (B) obtained with the longitudinal component of the LLP currents averaged between 16-30 October.

Vertical profiles of the record mean LLP longitudinal velocity component for each of the moorings in Sections G and H are shown in Fig. 6. Velocity values for each meter of depth were computed using the numerical relationships for these vertical profiles for each station. For each meter of depth, the interpolated velocities were fitted to a cubic spline to obtain the lateral distribution between stations. For depths from 15 m to the bottom, the cubic spline fit extended to the solid side boundaries, assuming a velocity of zero at the boundaries. From zero to 6 meter depth, extrapolation to the side boundaries from the nearest station was accomplished assuming a linear variation in velocity to a zero value at the boundary. From 7 to 14 m it was found better to assume a linear variation in the first derivative of velocity with respect to lateral distance, again constraining the velocity to be zero at the boundary. The same procedure was used for the data sets for each 3 h interval over the length of the record.

For each meter of depth, the lateral distribution of velocity as determined using the above procedure was numerically integrated across the width of the section to obtain the volume flux per meter of depth. The resulting vertical profiles derived from the record length average of the longitudinal component of the LLP current velocity for Sections G and H are shown in Fig. 7. The particular shape of these profiles results from the fact that the vertical profile of the longitudinal component of the LLP velocity characteristic of strong positive estuaries, such as the vertical profile for Station H2 shown in Fig. 6, is, in effect, multiplied by the width of the cross section, which is a decreasing function of depth.

SIMPLE BOX MODEL OF VOLUME CONTINUITY FOR THIS BAY SEGMENT

Here we examine how well computations of volume flux through each section, based on measured current velocities, agree with each other, and how well an accounting of volume continuity in the Bay above these sections, based upon direct measurements of volume flux, agrees with an independent accounting based upon a direct measure of the time rate of change of volume of the upper Bay using tide gauge records.

The following definitions were used:

V = Volume of the Bay and tributaries in the segment between the head of tide and either of the two cross sections at time t.

Q_R = Volume flux of fresh water into the Bay above either of the two cross sections.

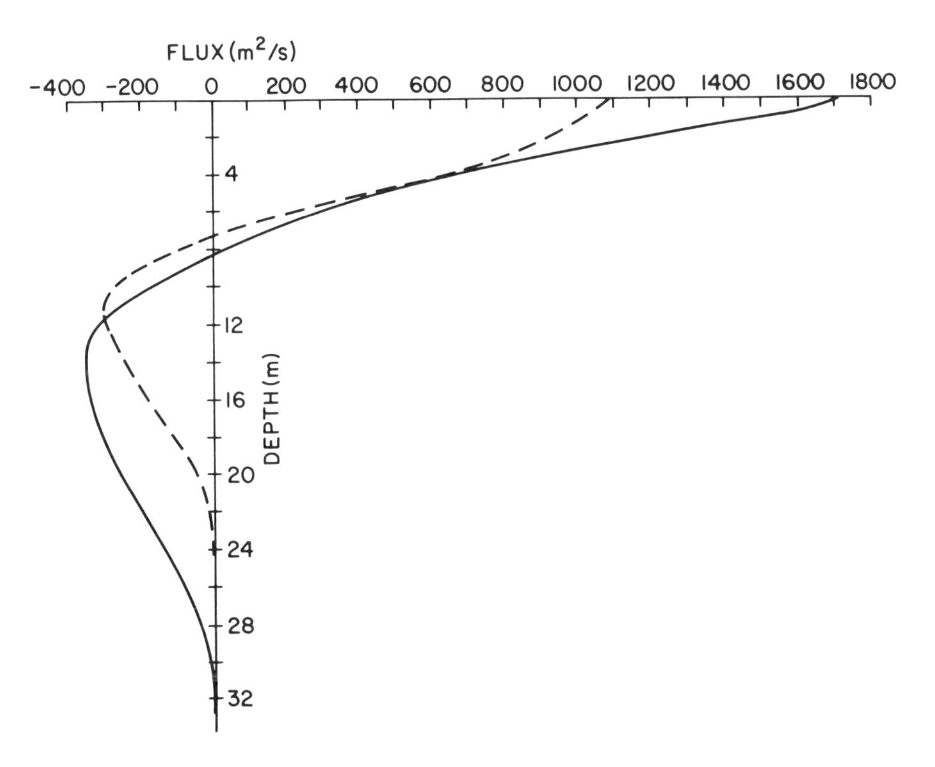

FIGURE 7. Vertical profile at cross-section G (dashed line) and cross-section H (solid line) of the laterally integrated flux per meter of depth derived from the longitudinal components of the LLP currents averaged between 16-30 October.

Q_x = Volume flux through either cross section below the plane of mean sea level (MSL).

Q_η = Volume flux through the tidally varying portion of either cross section, i.e. flux through that portion of the cross section bounded by the plane at $z = 0$ (mean sea level) and by the actual water surface (at $z = \eta$) at time t.

The time rate of change of the volume V is then given by:

$$\partial V/\partial t = Q_R - Q_x - Q_\eta \qquad (1)$$

Following the arguments given by Pritchard and Rives (1979) and more recently by Vieira (1983), the tidal mean of this equation is:

$$\partial V_\tau/\partial t = Q_{R,\tau} - Q_{x,\tau} - b_o \langle U_{b,\eta}\eta' \rangle_\tau \qquad (2)$$

where b_o is the estuary width at $z = 0$, $U_{b,\eta}$ is the zero centered, laterally averaged, oscillatory tidal velocity at $z = \eta$, i.e., at the time-varying water surface elevation, while η' represents that portion of the water surface elevation η which is zero centered and varies with tidal frequency. The subscript τ, such as in the term V_τ, indicates a time average over the tidal period.

The expression $\langle f \rangle_\tau$, where f is any time-varying parameter, also represents the tidal mean value of f. The last term in Equation (2) is then the tidal mean of the cross product of the tidal variations in water surface elevation and the tidal oscillations of the current velocity. This term arises from the covariance of $U_{b,\eta}$ and η'. Even though the tidal averages of these two variables are individually equal to zero, the tidal average of their cross product is not generally zero. This term is called the Stokes transport, and will be designated in the following by Q_{ST}. For further discussion of the tidally induced Stokes transport, see Pritchard (1980).

The tidal velocity $U_{b,\eta}$ was computed by taking at every point in time the lateral mean of the BP filtered records from the topmost meter in each cross section, and assuming this to be a reasonable estimate of the value at the surface. The time series thus obtained was then multiplied by the tidal (BP) elevation record obtained from the tide gauge nearest to the respective cross section. This record of the cross product of $U_{b,\eta}$ and η' was then subjected to the 34 h Lanczos LLP filter described earlier to give a time series record of $\langle U_{b,\eta}\eta' \rangle_\tau$. Daily values of the flow of the Susquehanna River at Conowingo, multiplied by a factor to account for the ungauged drainage area between Conowingo and each of the two cross sections, served as an estimate of the tidal mean record of the fresh water inflow, $Q_{R,\tau}$. The quantity $Q_{x,\tau}$ was obtained by summing the values of the tidal mean volume flux per meter of depth, obtained by the procedures described in the previous section, over the depth of the cross section.

An independent direct measure of the time rate of change of the volume of the segment of the estuary landward of either cross section can be obtained from tide gauge records. Note that V_τ can be expressed as the sum of a time-independent part, say, the volume below the long time mean sea level, designated by V_o, and a time-varying portion, designated by V_η. This time-varying volume is bounded in the vertical by the fixed surface of mean sea level (at $z = 0$) and by the time-varying water surface at $z = \eta_\tau$. The time-varying volume can be determined by numerically integrating the water surface elevation over the surface area, S, of the subject segment of the estuary. In practice this integration is accomplished by subdividing the segment into subsegments, the boundaries of which are determined by the location of tide gauges. Designating the tidal averaged volume of each such subsegment by $V_{\tau,i}$, and the tidal averaged water surface elevation of this segment by $\eta_{\tau,i}$, then, the time rate of change of the volume of the subsegment is given by:

$$\partial V_{\tau,i}/\partial t = S_i(\partial \eta_{\tau,i}/\partial t) \quad (3)$$

Evaluating this expression for each subsegment, using LLP records of water surface elevation obtained by processing records from tide gauges located at the boundaries of each subsegment, and then summing the individual results over the total number of subsegments, gives the independent estimate of $\partial V_\tau/\partial t$ for the full segment up estuary from either cross section.

Consider now the record length average of these expressions. Designating the record length average by an overbar, Equations (2) and (3) become

$$\partial \bar{V}_\tau/\partial t = \bar{Q}_{R,\tau} - \bar{Q}_{x,\tau} - \bar{Q}_{ST} \quad (4)$$

$$\partial \bar{V}_{\tau,i}/\partial t = S_i(\partial \bar{\eta}_{\tau,i}/\partial t) \quad (5)$$

The record length average tidal mean volume flux through the tidal mean cross section can be divided into two parts: (1) record length average flux through the tidal mean upper layer, defined as that part of the tidal mean cross section in which the record length average LLP current is directed down the estuary; and (2) record length average flux through the lower layer, defined as that part of the cross section in which the record length average LLP current is directed up the estuary. Thus

$$\bar{Q}_{x,\tau} = \bar{Q}_u + \bar{Q}_l \quad (6)$$

Table 1 shows the results of the evaluation of Equations (4) and (5) for each cross section. Values for each of the

TABLE 1. Evaluation of the terms on the right hand side of Equation (4) for the period 16-30 October, 1977, and the independent evaluation of $\partial \bar{V}_\tau/\partial t$ for the same period, using Equation (5), for Section G and Section H. Units are m^3/s.

Cross Section	$\bar{Q}_{R,\tau}$	$\bar{Q}_u$	$\bar{Q}_l$	Q_{ST}	$\partial \bar{V}_\tau/\partial t$ *EQ(4)*	$\partial \bar{V}_\tau/\partial t$ *EQ(5)*
G	3331	5393	-2524	-162	624	575
H	3339	7382	-4397	-259	613	580

terms on the RHS of Equation (4), including the terms arising from the substitution of Equation (6) into Equation (4), are included. The Stokes transport term appears relatively small compared to other terms on the RHS of Equation (4). However, the fresh water inflow for the period of this study was some seven times the long time average flow for this calendar period, and so for more normal conditions for this time of the year the Stokes transport would be close to the same magnitude as the fresh water inflow.

The value of $\partial \overline{V}_{\tau}/\partial t$ as computed using Equation 4 for Section G is remarkably close to that computed for Section H. The difference of 11 m^3/s is too large, and in fact of the wrong sign, to be accounted for by any actual time rate of change in the tidal mean volume of the segment between the two sections, over the length of the study period. Even so, considering the possible uncertainties in the determination of the various terms on the RHS of Equation (4), the agreement in the computations for the two sections is quite good.

The last column of Table 1 gives the values of $\partial \overline{V}_{\tau}/\partial t$ obtained using Equation 5. Considering the precision of the measurements of water surface elevation obtained from digitally recording tide gauges, these estimates of the net volume flux into the subject segment of the Bay are probably accurate to within ±10 m^3/s.

The net volume flux through Section H of 2726 m^3/s is just 19 m^3/s larger than the net flux through Section G. Runoff of fresh water from the land bordering the segment between the two cross sections would account for some increase in net seaward flow between the two sections. We estimate that this runoff would have amounted to about 8 m^3/s.

This two-layered box model of the flow into and out of the segment of the Bay bounded by Section G and Section H is shown schematically in Fig. 8. The upper diagram shows volume flux values in m^3/s. The vertical flux from the the lower layer to the upper layer is given in this diagram, as is the estimated fresh water runoff from the land boundaries of the segment. The lower diagram gives the mean velocities associated with the flux values. Note that the vertical velocity at the layer interface of slightly over 10^{-5} m/s is of the same order as determined in previous studies of other estuaries (Pritchard 1954).

TIME VARIATIONS IN NET VOLUME FLUX INTO AND OUT OF THE SEGMENT

Vertical profiles of the flux per meter of depth for each cross section were determined from the LLP longitudinal velocity records every 3 h over the length of the study. LLP-filtered tide gauge records, decimated to 3 h intervals, were also available. Fresh water inflow records were available for each day of the study period, and we assumed that the time variations in this term were sufficiently regular so that a smooth record of values at 3 h intervals could be constructed from the daily flow values of the Susquehanna River at Conowingo, corrected for the contribution of runoff from ungauged areas. The product of the BP velocity and tide gauge records at each cross section gave estimates of Stokes transport every 3 h. These data were used to evaluate Equation (2) each 3 h from 16 to 30 October, 1977. Tide gauge records throughout the upper Bay were also available allowing the evaluation of Equation (3) every 3 h over this same period.

Figure 9 shows the results of these computations for Section G and Section H. This seems to be the first time a comparison of computations of net volume flux into and out of a segment of an estuary has been made using two independent methods for a time series of data evaluated at such closely spaced intervals of time. The agreement between the results of the two methods was very good. Note that the square root of the variance of these records was on the order of ±10,000 m^3/s with individual events showing fluctuations over a 2 to 3 d interval of ±20,000 m^3/s. Fluctuations in nontidal flux through each of the cross sections was the major contributing

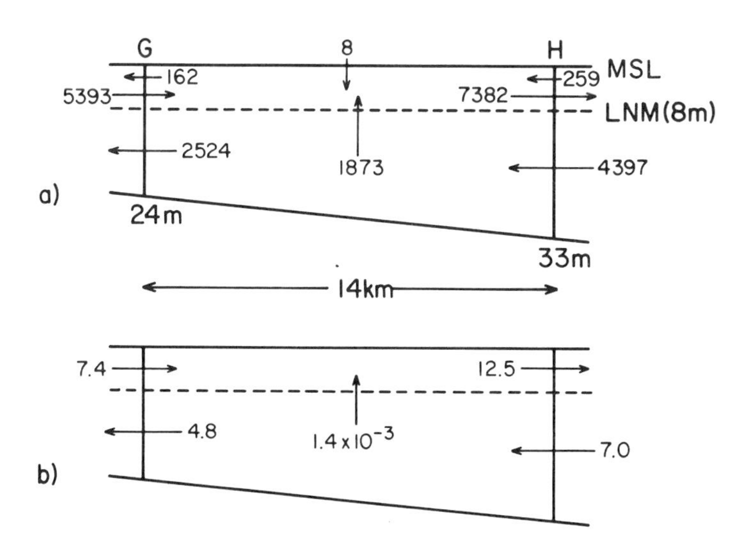

FIGURE 8. Two-layered box model of Eulerian mean non-tidal circulation between cross-sections G and H for the period 16-30 October 1977. a) Transports in m^3/s. b) Flows in cm/s. (LNM = level of no motion; MSL = mean sea level).

factor to this variation of net flux into and out of the segment upestuary from the sections. By comparison, the root mean square variation in fresh water inflow about its mean value during this period was about ±2,100 m^3/s.

TIME VARIATIONS IN VERTICAL PROFILES OF LLP VELOCITY

As noted earlier, the record length mean of the LLP laterally averaged velocity had a vertical distribution characteristic of strong, partially mixed estuaries, with a seaward flowing upper layer and a lower layer in which the flow is directed up the estuary. On a shorter time scale the vertical structure of nontidal flow was highly variable, a fact previously described by others (e.g., Elliott 1978; Elliott and Wang 1978; Elliott et al. 1978; Wang 1979a,b; Wang and Elliott 1978). The fact that these time variations in the vertical distribution of the nontidal flow result from the response of the estuarine system to meteorological forcing has been well established, but the relationship between the time variations of currents to the near field and far field wind is only partially understood. Here we intend to examine the variations in the vertical response of the laterally averaged residual flow to variations in the local wind forcing in a different way than has been tried in earlier studies.

Figure 10 shows the vertical profile of the LLP laterally averaged velocity for various times during the study period. In order to study the meteorological forced fluctuations in the current field, the effects of variations in the density driven flow field should be separated from the record. We consider that the variations in the density driven flow are of lower frequency than the meteorologically forced motions, and can be substantially reduced by removing the mean and linear trend from the velocity record. The profiles shown in Fig. 10(A), however, are based on data which include the mean and linear trend, in order to demonstrate that the meteorological forced motions can sometimes completely overcome the classical density driven flows.

Figure 10(B,C) shows examples of the vertical profiles of the nontidal flow with the mean and linear trend removed from the LLP velocity records. The residual flow with the mean and linear trend removed may be two-layered in the same sense as the classical estuarine circulation (0900, 17 Oct), or in the opposite sense (0300, 27 Oct), or may be directed seaward at all depths (1800, 22 Oct) or up the estuary at all depths (0600, 30 Oct). These profiles of the vertical distribution of the residual velocity at a given time represent a complex dynamic response of the estuarine system in which the filtering of the effects of the wind stress varies with depth.

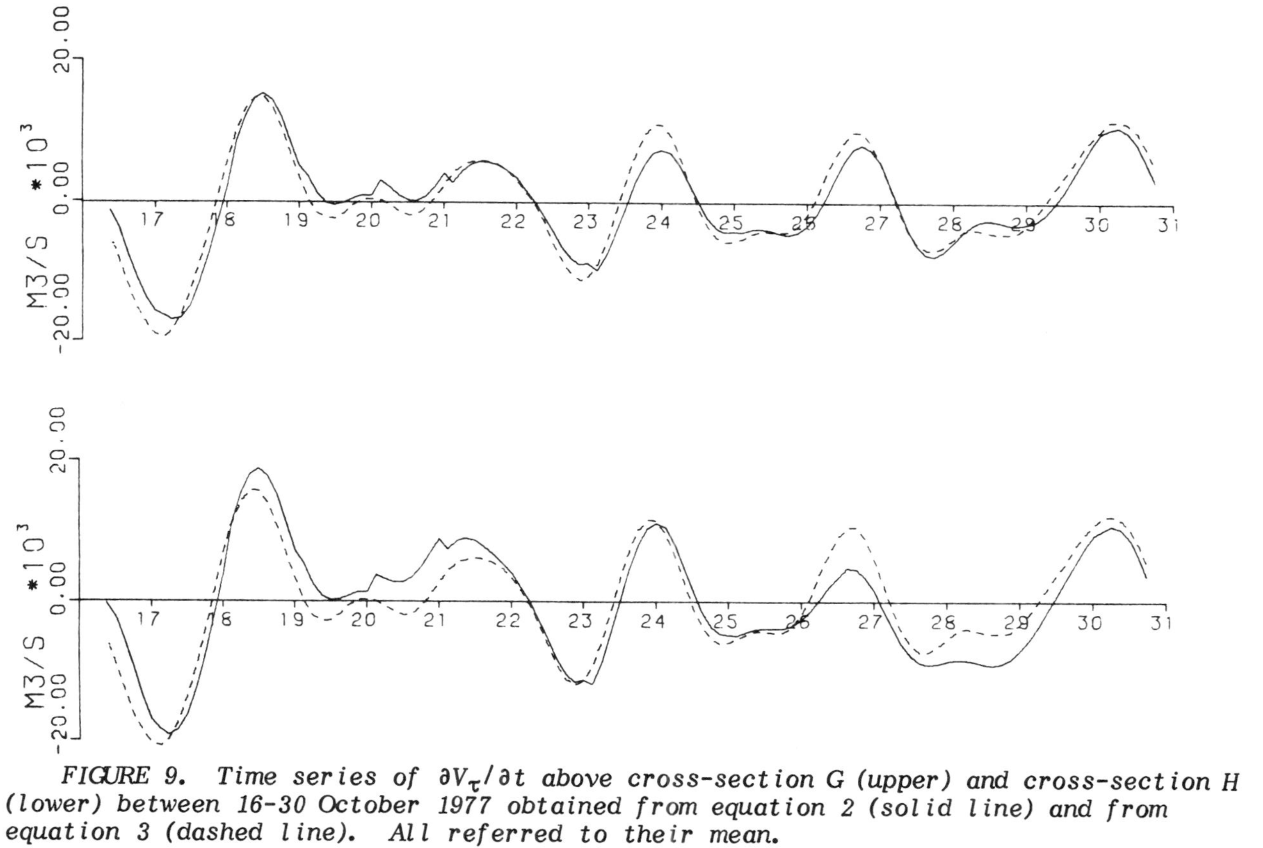

FIGURE 9. Time series of $\partial V_{\tau}/\partial t$ above cross-section G (upper) and cross-section H (lower) between 16-30 October 1977 obtained from equation 2 (solid line) and from equation 3 (dashed line). All referred to their mean.

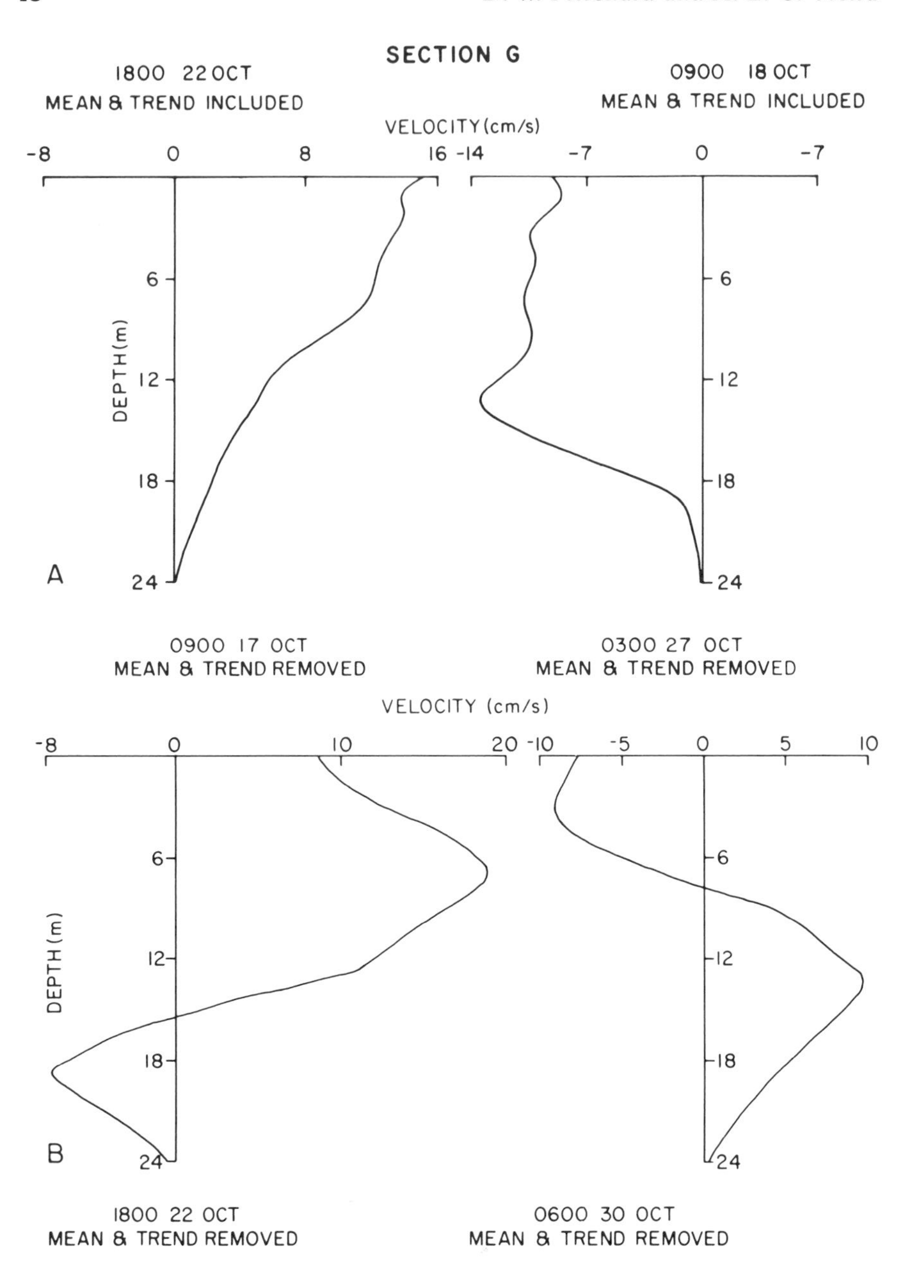

FIGURE 10. Vertical profile at cross-section G of the laterally averaged LLP longitudinal current, at the designated times; (A) with the mean and the linear trend included; (B,C) with the mean and linear trend removed.

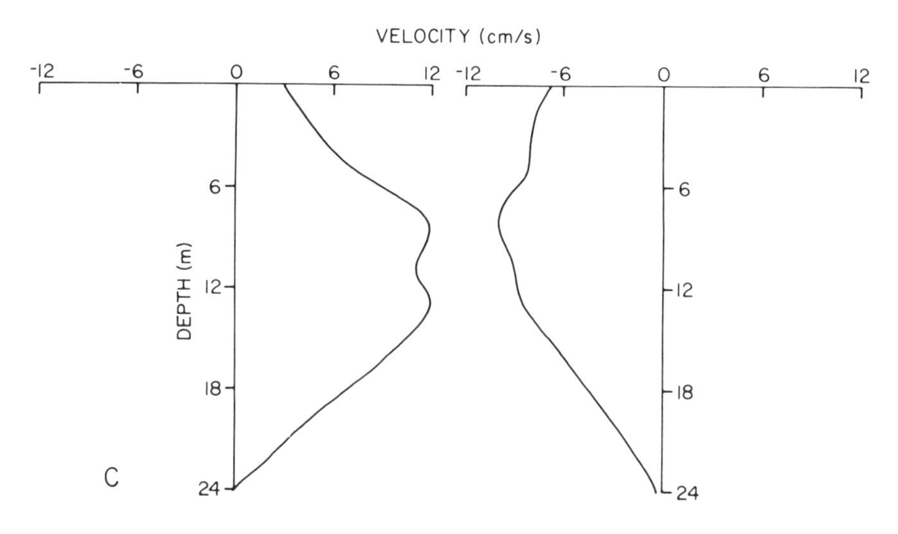

FIGURE 10 (C).

In the interest of brevity the following shortened terminology will be used for the remainder of this paper: (1) residual velocity in place of laterally averaged LLP velocity, with the mean and linear trend removed from the record; (2) axial wind stress in place of LLP axial component of the Patuxent wind stress; (3) slope of the water surface in place of LLP slope of the water surface as determined using tide gauge records from Annapolis and Solomons.

Inspection of a plot of time variations in residual velocity with that of axial wind stress suggests that the surface layers respond directly to the wind with no significant time lag, while in the near bottom layers the residual flow is in the opposite direction to the wind but with some time lag. There are various ways to quantify this apparent correlation of current and wind. Perhaps the simplest approach is to determine the lagged linear correlation function. Table 2 shows the results of such a determination for each of the 26 usable current meter records. The lag shown is that at which the maximum correlation occurred between the residual current and the wind stress records. A positive or negative sign in the fourth column indicates that the current response is in the direction of or opposite to the wind, respectively. Data in Table 2 have been grouped according to the depth at which the current meters were deployed. Lags for all meters in the near surface group were short, ranging from 0-5 h, most being 3 h or less. The surface layers were evidently responding very quickly to the wind, and all in the direction of the wind. Five of the seven current meter records in the next group (average depth = 7.4 m) showed a short (≤3 h) lag to the wind, with residual velocity variations in the same direction as the wind. The other two records showed a relatively long lag (18-26 h) and in the opposite direction to the wind.

For the third group (average depth = 12.0 m) two of the residual velocity records showed a short lag (0-1 h) and were positively correlated with axial wind stress, while five of the records showed a long lag (23-31 h) and were negatively correlated with the wind. For the three current meters deployed at depths >18 m, two of the residual velocity records showed an intermediate value for the lag (8-9 h) and the third showed a relatively long lag (20 h) to wind stress. All three showed a negative correlation with wind stress. The records from the deepest meter in each section (H2/25.9 and G4/19.5) had intermediate lags (8-9 h) to the wind, and the residual velocity fluctuations were in the opposite direction to the wind. The records from meters in each section which were the next closest to the bottom (H2/18.3 and G4/13.5) showed increased time lags (20-23 h), and are also negatively correlated with the wind.

TABLE 2. Lag time (h), correlation coefficient (r), coefficient of determination (r^2) and sign of linear regression coefficient (b) for the linear regression of the LLP axial component of the current velocity upon the LLP axial component of the wind stress, for the designated current meter locations.

Station/Depth	*Lag(h)*	*r*	r^2	*Sign b*
A/3.0	0	0.688	0.473	+
B/4.0	2	0.653	0.426	+
C/2.9	4	0.817	0.667	+
G1/2.2	0	0.413	0.171	+
G2/3.0	0	0.722	0.521	+
G4/4.4	5	0.604	0.365	+
G5/4.0	1	0.779	0.607	+
H2/3.3	3	0.813	0.661	+
H3/3.7	2	0.867	0.752	+
A/7.6	0	0.769	0.591	+
B/7.1	2	0.753	0.567	+
C/7.4	3	0.601	0.361	+
G1/6.8	26	0.408	0.166	-
G2/7.6	0	0.794	0.630	+
G3/7.1	1	0.643	0.413	+
H1/8.0	18	0.623	0.388	-
B/11.7	1	0.642	0.412	+
C/11.9	24	0.538	0.289	-
G2/10.6	31	0.068	0.005	-
G3/11.6	0	0.560	0.314	+
G4/13.5	23	0.597	0.356	-
H1/12.5	23	0.605	0.366	-
H2/12.2	24	0.574	0.329	-
G4/19.5	8	0.767	0.588	-
H2/18.3	20	0.753	0.567	-
H2/25.9	9	0.819	0.671	-

It appears then that fluctuations in the axial wind stress produce a rapid response in residual velocity in an upper layer of about 7 m deep in mid-Bay. Relative to the LLP water surface elevation at the Bay mouth, a wind blowing down the Bay would lower the water surface elevation at the head of the Bay, while a wind blowing up the Bay would lead to an increase in the tidal average water surface elevation at the head of the Bay. Consequently, fluctuations in the wind would produce fluctuations in the slope of the Bay's surface, such that there would be a fluctuating barotropic pressure gradient in the Bay directed opposite to the direction of the wind stress. Variations in this slope-induced pressure gradient will be lagged somewhat to the wind fluctuations because it takes some time for the surface elevations in the Bay to respond to a shift in the wind speed and direction. This slope-induced pressure gradient would drive a current in the opposite direction from the wind. The direct drag of the wind dominates flow in the surface layers. The onset of the counterflow then first occurs at the bottom, where the direct drag of the wind has the least impact, and hence the 8 to 9 h lag observed for the deepest meter in each of the cross sections should be close to the lag of the variations in slope relative to the variations in the wind. As the slope builds up under the influence of the wind, the counterflow occurs progressively higher in the water column, thus accounting for the longer lags for records from meters deployed further from the bottom, but still below the directly wind driven surface layer. The pycnocline at intermediate depths (7 to 13 m) inhibits the downward transfer of momentum from the surface layer, and so in the layers below the pycnocline the slope term can be dominant over the direct effect of the drag of the wind.

In the remaining part of this paper, we will examine whether further data analysis, coupled with the use of a diagnostic model, supports the above hypothesis concerning the vertical variations in response of the currents to wind forcing in the middle reaches of Chesapeake Bay.

FURTHER STATISTICAL TREATMENT

An inspection of almost any time series record of the residual current velocity, LLP water surface elevation, or of axial wind stress suggests that the variations in each of these parameters is the sum of individual variations of different frequencies. The lagged correlation function does not provide any information on the response of the Bay to wind fluctuations falling within discrete frequency bands. The use of more sophisticated statistical tools is required.

One such useful tool involves the determination of the rotary coherence squared between the wind stress vector and one of the parameters which serve as indicators of the response of the estuarine system. In the discussion of the time variations in the net volume flux term, $\partial V_\tau/\partial t$, as shown in Fig. 9, it was pointed out that the meteorologically driven motions through each of the cross sections probably accounted for the major part of these time variations. Support for this view is provided by Fig. 11(A), which shows the rotary coherence squared for the $\partial V_\tau/\partial t$ upon the wind stress. Briefly, the procedure used in arriving at this figure involves the resolution of the wind stress vector along axes which are oriented along each 10° of compass heading. The time record of each of

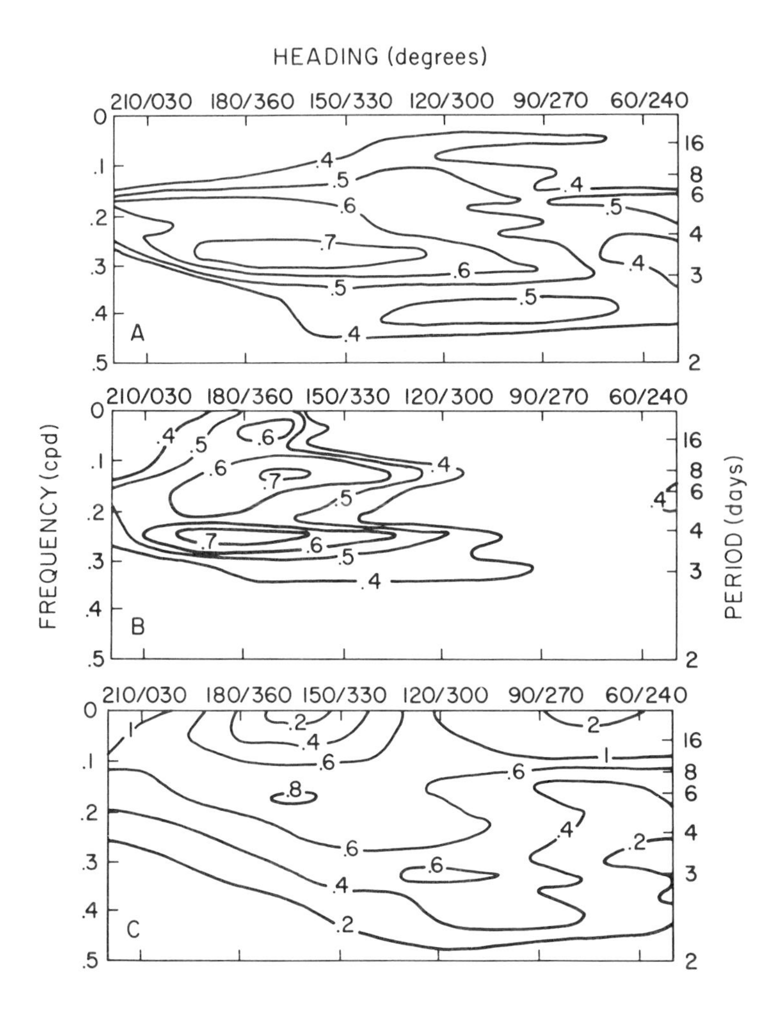

FIGURE 11. (A) Rotary coherence squared between the Patuxent wind stress and $\partial V_\tau/\partial t$ above cross-section H. The 95% significance level is 0.36; (B) Rotary coherence squared between the Patuxent wind stress and the surface slope between Solomons Island and Annapolis. The 95% significance level is 0.36; (C) The directional distribution of the spectral energy density [(dyn cm^{-2})^2/cpd] of the Patuxent wind stress.

these wind stress components, as well as the time record of net volume flux, was then subjected to a Fourier transform, which provided the amplitude and phase of the various harmonic constituents which comprise each of the records. A linear correlation coefficient was then determined for a given harmonic constituent of the wind stress component and of the net volume flux, for each 10° of compass direction. The square of the correlation coefficient was then contoured as a function of the frequency of the harmonic constituent and of the compass direction of the wind stress components used in the analysis. The plotted values are called the rotary coherence squared because of the functional dependence on compass heading.

The longitudinal axis of the Bay in the area of this study is oriented along 140°/320°, but the Bay over its full length is oriented more nearly along 180°/360°. Figure 11(A) shows that there was a maximum coherence between $\partial V_\tau/\partial t$ and the component of wind stress for directions in this range. This figure further shows that the maximum coherences were for an inverse frequency band of 3 to 4 d.

Figure 11(B) shows the rotary coherence squared between the Patuxent wind stress and the slope of the water surface, as determined from the tide gauge records at Annapolis and at Solomons, Maryland. It is clear from this figure that the slope was highly coherent with the longitudinal component of the wind stress at all frequencies < 0.3 per day; that is, for all periods greater than about 3 d. Maximum coherences occurred at about 4 and 7 d.

In order to utilize fully the information contained in Fig. 11(A and B), it is necessary to know the relative amount of energy associated with wind fluctuations at the various frequencies and along the various directions. Figure 11(C) shows the directional distribution of the spectral energy density of the Patuxent wind stress. A band of relatively high energy of wind fluctuations existed for wind stress components along the longitudinal axis of the Bay for frequencies of 3.5 to 8 d.

Values of the residual current velocity for each meter of depth, for each section, computed at 3 h intervals over the period 0900 16 October to 1800 30 October, produce time series records such as those shown in Fig. 12. This diagram shows the time series record of the residual velocity at Section G for layer 4 (3-4 m), layer 8 (7-8 m), layer 12 (11-12 m), and layer 20 (19-20 m).

The time series of axial wind stress and slope forcing functions are also shown in Fig. 12. These functions have the dimensions of acceleration and are, respectively, the axial wind stress divided by the product of density and vertical thickness of the near surface layer, and the slope of the water surface multiplied by gravity. Visual inspection of these figures, particularly (Fig 12a) for the near surface residual velocity record and (Fig 12d) for the near bottom residual velocity record, would appear to support our stated hypothesis that the time fluctuations of the residual velocity in the near surface layers are directly driven by fluctuations in the axial wind stress with little time lag, and that the time fluctuations in the near bottom layers are driven by fluctuations in the surface slope. The fluctuations in the time record of the residual velocity at 7-8 m (Fig. 12b) appear to be similar to the fluctuations in the wind stress term, but the coherence between these two records does not appear to be as strong as for the record for 3-4 m. Fluctuations in the residual velocity at 11-12 m (Fig. 12c) do not appear to be clearly coherent with either the wind stress term or the slope term, a condition to be expected since at this depth neither of the forcing terms are dominant.

Impressions based on visual inspection of complex time series cannot be considered as firm conclusions. Based on the information contained in the plots of rotary coherence squared (Fig. 11A,B), and in the plot of the directional distribution of spectral energy density of the wind (Fig. 11C), we conclude that the primary driving mechanism for fluctuations in the LLP residual velocity, within the frequency band we can resolve with a data set covering just 20 d, is axial wind stress. Figure 11(B) demonstrates that a major fraction of the variance in the surface slope (about 70%) is explained by variations in the axial wind stress over the inverse frequency band of 3 to 8 d. Figure 11(C) shows that there was a significant amount of energy in this frequency band for the axial wind stress.

Therefore, to quantify our visual impression of the depth dependent response to wind forcing, the following computations were made for the data from Section G (the results from Section H are essentially the same): (a) partial coherence squared, phase and gain between axial wind stress and residual velocity at each meter of depth, with the slope effect removed; (b) partial coherence squared, phase and gain between slope and residual velocity at each meter of depth with axial wind stress effect removed; and (c), ordinary coherence squared, phase and gain between axial wind stress and residual velocity at each meter of depth. In this last case, the effects of that portion of the variations in surface slope which is driven by variations in axial wind stress are implicitly included.

For the sake of brevity the results of these computations are not presented in detail here, although they will be provided to interested readers on request. They are summarized below.

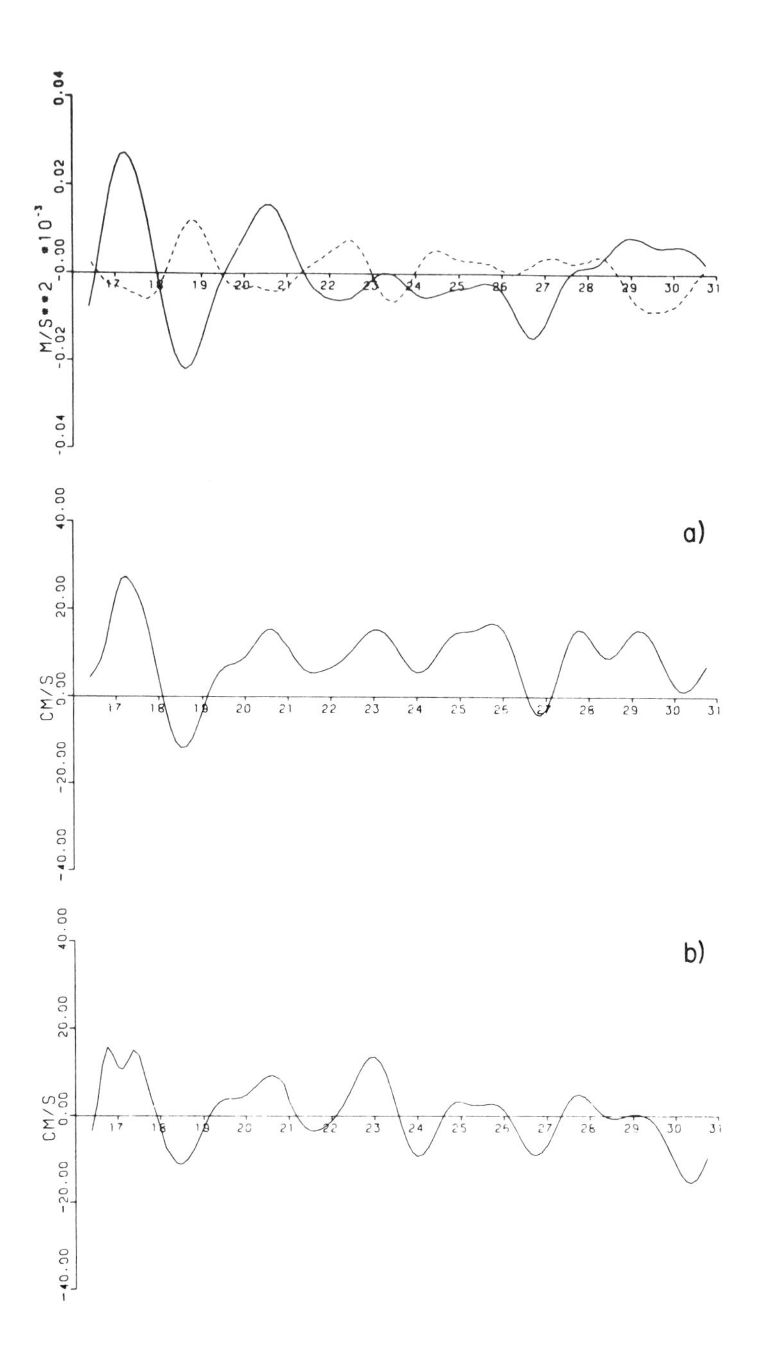
M/S**2 *10^-3
0.04
0.02
0.00
-0.02
-0.04
17 18 19 20 21 22 23 24 25 26 27 28 29 30 31
a)
CM/S
40.00
20.00
0.00
-20.00
-40.00
17 18 19 20 21 22 23 24 25 26 27 28 29 30 31
b)
CM/S
40.00
20.00
0.00
-20.00
-40.00
17 18 19 20 21 22 23 24 25 26 27 28 29 30 31

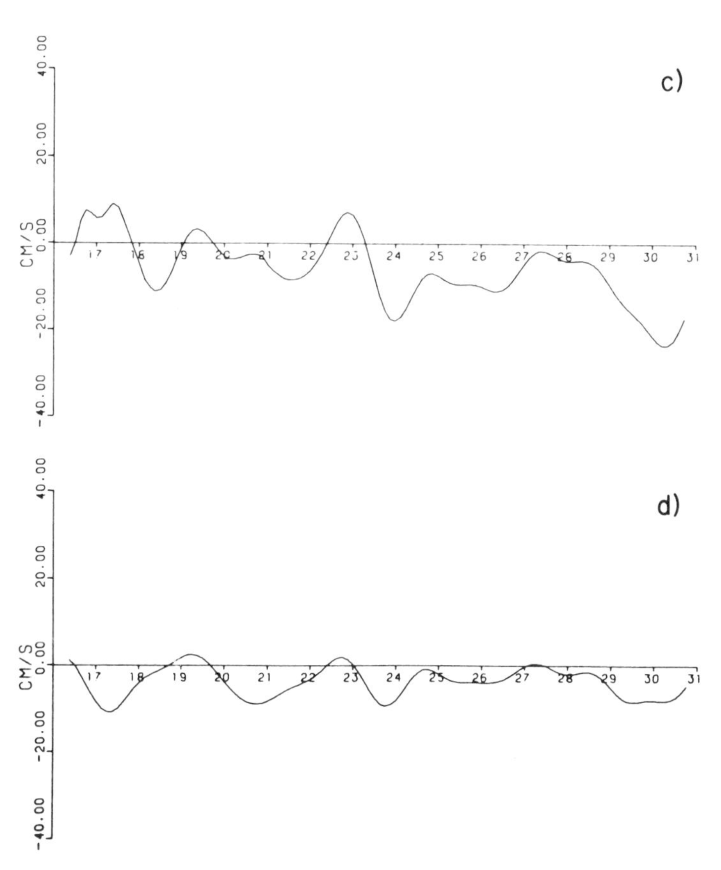

FIGURE 12. Top graph, opposite page: Forcing functions with the mean and linear trend removed, for the period 16-30 October; the Patuxent longitudinal wind stress (solid line) and the surface slope between Solomons Island and Annapolis (dashed line). Remaining graphs: The lateral mean non-tidal flow in cross-section G, during the same time period. (a) Between 3-4 m; (b) Between 7-8 m; (c) Between 11-12 m; (d) Between 19-20 m.

The current and the wind alone were strongly coherent in the upper 8 m or so, with the current lagging the wind by less than 4 h for periods of less than 3.2 days. The gain, i.e., the ratio of the amplitude of the output to that of the input, decreases slowly with depth.

The current and the slope alone were significantly coherent at all depths at periods shorter than 3.2 d, but were significantly coherent only at increasingly deeper depths as the period increased beyond 3.2 d, with the fluctuations in the current lagging those of the slope by 7 to 15 h. The gain was very nearly independent of depth down to 18 m. Below this depth, effects of bottom friction rapidly dampened the amplitude of the velocity fluctuations.

Significant values of the ordinary coherence squared occurred above 10 m for periods ≤ 4 d (Table 3). In both Table 3 and 4, values which are below the 95% significance level are shown in parentheses. For periods > 4 days the vertical thickness of the upper layer having significant coherence decreased with increasing period. This was possibly due to the fact that at these longer periods, our simple method of removal of effects of fluctuations in density driven flow, by removal of the mean and linear trend of the record, is inadequate. A mid-depth region where the ordinary coherence squared between wind and current had values below the 95% significance level separated the upper from the lower significant coherence layers. The lower layer extended from about 18 m to the bottom.

TABLE 3. Ordinary coherence squared between layer mean velocities and axial wind stress. 95% SL = 0.39.

	PERIOD (days)						
Layer	*8.0*	*5.3*	*4.0*	*3.2*	*2.7*	*2.3*	*2.0*
2	.39	.52	.60	.68	.79	.84	.81
4	.40	.53	.61	.69	.80	.83	.80
6	.40	.51	.58	.66	.75	.77	.74
8	(.34)	.43	.51	.57	.61	.61	.62
10	(.25)	(.36)	.42	.44	.42	.43	.49
12	(.19)	(.32)	.40	.39	(.33)	(.35)	.44
14	(.17)	(.28)	(.37)	(.33)	(.27)	(.30)	(.38)
16	(.27)	(.33)	.41	(.36)	(.28)	(.29)	(.34)
18	.49	.52	.58	.53	.43	.42	.40
20	.70	.73	.76	.74	.67	.65	.58
22	.72	.76	.77	.76	.70	.68	.62
24	.73	.76	.78	.76	.70	.67	.61

In the upper layer, fluctuations in the current lead the fluctuations in the wind (Table 4). Other investigators (e.g., Wang 1979b) have stated that current fluctuations lead wind fluctuations at all depths, with the magnitude of this phase difference increasing with depth. Wang's results showed that for a given frequency the phase lead of current to wind increased sharply over a thin middepth layer. When considering two harmonically varying parameters, say, two cosine functions of the same frequency but different phase, it is not possible to state, without considering the physics of the process, which of the two cosine functions is leading, since the two functions may be negatively rather than positively correlated, and since the functions are continuously repeating at intervals of one period. The algorithm for computing the phase difference in a coherence squared computation selects the phase difference which is $< 180°$. It is possible that the physically correct phase difference is the complement of the phase angle computed by the algorithm, with the sign of the correlation reversed. Thus if the phase difference between the time variations in parameter A and in parameter B at, say a period of 2 d, is computed to be 144°, or 19 h, with parameter A leading parameter B, it is possible that in fact parameter B leads parameter A by 36°, or by about 5 h.

Our computations of the partial coherence squared between residual current and slope show that the current lagged the slope in the lower layer where there were significant values of the coherence. Further, the computations of partial coherence

TABLE 4. Phase in hours for ordinary coherence between layer mean velocities and axial wind stress. Positive values are for current lagging wind.

	PERIOD (days)						
Layer	*8.0*	*5.3*	*4.0*	*3.2*	*2.7*	*2.3*	*2.0*
2	1	0	-1	0	0	0	-1
4	-2	-1	-1	-1	-1	-1	-1
6	-6	-4	-3	-2	-2	-2	-2
8	(-17)	-10	-8	-6	-5	-4	-4
10	(-29)	(-18)	-14	-10	-8	-7	-5
12	(-38)	(-23)	-18	-14	(-11)	(-9)	-7
14	(37)	(32)	(25)	(20)	(18)	(16)	(15)
16	(26)	(25)	20	(16)	(15)	(13)	13
18	19	18	15	12	10	9	8
20	15	13	10	8	7	6	5
22	14	13	10	8	6	5	5
24	14	13	10	8	6	5	5

squared between residual current and wind show that the current was not significantly correlated with the wind in this lower layer when the effects of slope were removed. In a separate computation we have shown that variations in slope lag variations in wind. Consequently, in the lower layer where the wind has little direct effect, and can act only through its effect on the fluctuations in the slope, fluctuations in the current must lag those of the wind.

Therefore, below a depth of 13 m, we have taken the complement of the phase produced in the ordinary coherence squared computations between fluctuations in the wind and the current. This produces the result that fluctuations in current lag those of wind in the bottom layer, with the lagged record giving negative residual currents for positive wind, and vice versa (Table 4). This lag increases with distance upward from the bottom. At a period of 4 d, for example, the fluctuations in the current lag those of the wind by 10 h at 24 m, and this lag increases to 20 h at a depth of 16 m. These results are quite consistent with those obtained from the simple lagged correlation analysis.

The negative values for phase shown in the upper layer of Table 4, indicating that fluctuations in the current lead those in the wind by a few hours, can in fact be shown to be physically realistic. As noted above, the wind builds up a slope in the Bay which produces a barotropic pressure force directed opposite to the wind stress, but with a slight lag. Consider now the axial wind stress as a harmonically varying function which alternately varies from being directed down the Bay to being directed up the Bay. The wind would then produce a harmonically varying slope with an associated upwind-directed harmonically varying barotropic pressure force, which is lagged somewhat to the fluctuations in the wind. The surface layers are primarily driven by the wind, and so during most of the time the residual current flows in the direction of the wind. However, during the interval when the wind is decreasing, as the time approaches for a shift in wind direction, the pressure force associated with the slope begins to become a relatively more important forcing term, since its decrease in magnitude is lagged to that of the wind. At some point in time before the wind has changed direction, the oppositely directed slope induced pressure force will overcome the direct drag of wind stress, and the surface layer current will reverse to flow against the weakening wind stress. This phenomenon will be repeated at each reversal of the longitudinal component of the wind, and so an analysis of time variations of wind and current will result in the conclusion that, in the surface layers, the current leads the wind.

The lead of current fluctuations over wind fluctuations increases somewhat with depth in the surface layer (Table 4).

This is consistent with the fact that the fluctuations in the slope-induced pressure force are independent of depth, while the effects of the direct drag of the wind stress decreases with depth. Hence the time at which the effect of the longitudinal pressure term exceeds the effect of the wind stress would occur earlier for currents at deeper depths in the surface layer.

APPLICATION OF A DIAGNOSTIC MODEL TO THIS DATA SET

In this last section we briefly describe some preliminary results of the use of a diagnostic model to investigate vertical variations in response of residual current to fluctuations in axial wind stress. The model is based on a linearization of the tidally averaged longitudinal component of the equation of motion, such that field acceleration terms are removed but local acceleration is retained. Our concern is with residual meteorologically forced flow, and not with density driven flow. We assume that there is no nonlinear interaction between these flows, and so we remove the baroclinic pressure force term from the equation, retaining only the surface slope-induced barotropic term.

The equation is integrated over n layers, where conceptually n may be any integer. To date we have used the model with a maximum of 4 layers. Interlayer friction is simulated by use of a term containing the product of an interlayer frictional coefficient and the difference between adjacent layer mean velocities. A wind stress term is applied at the surface and a linearized bottom frictional term is applied for the solid boundaries.

The Fourier transform is taken of each of these n equations. This converts the set of simultaneous differential equations, the solution of which would require some type of numerical time marching, to a simultaneous set of algebraic equations, the solution of which is straightforward and requires much less computer time than would the solution of the original equation in time space. The transformed equations are defined in the frequency domain, and the solutions exist in that domain. However, the solutions can be transformed back into the time domain, giving the amplitude and phase of a harmonic constituent of the time-varying velocity in each layer at each of a number of discrete frequencies. The combination of these constituents then gives the time record of the computed velocity for each layer.

The observed axial wind stress, and the observed slope of the water surface, are Fourier transformed and used as input to the model. These forcing functions are shown in Fig. 12.

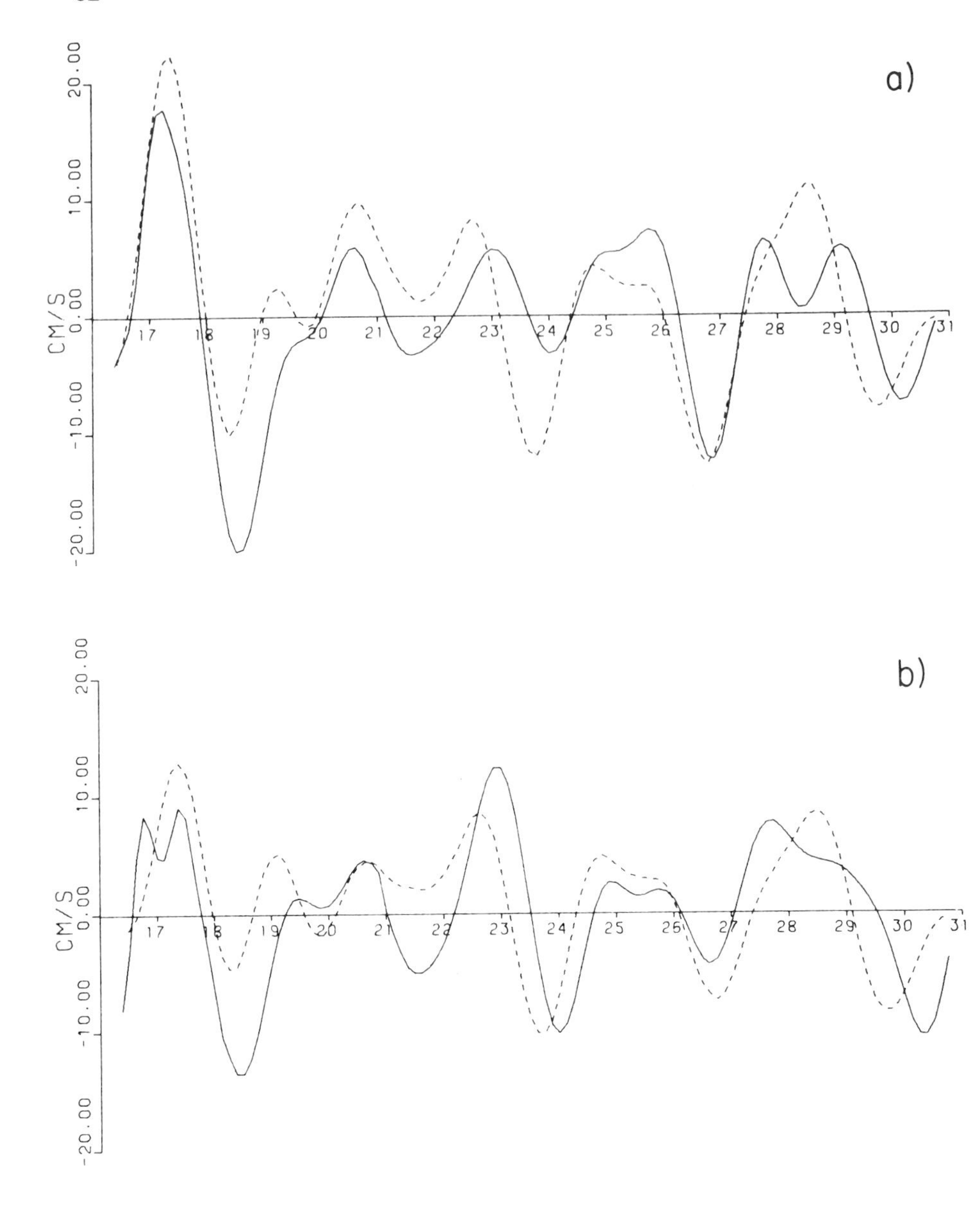

FIGURE 13. Four-layered model, showing mean flow for cross-section G as observed (solid line) and as modelled (dashed line) with K_l = 0.06 cm/s and K_i = 0.1 cm/s. (a) Upper layer (MSL - 6 m); (b) Upper middle layer (6-12 m); (c) Lower middle layer (12-18 m); (d) Lower layer (18-25 m).

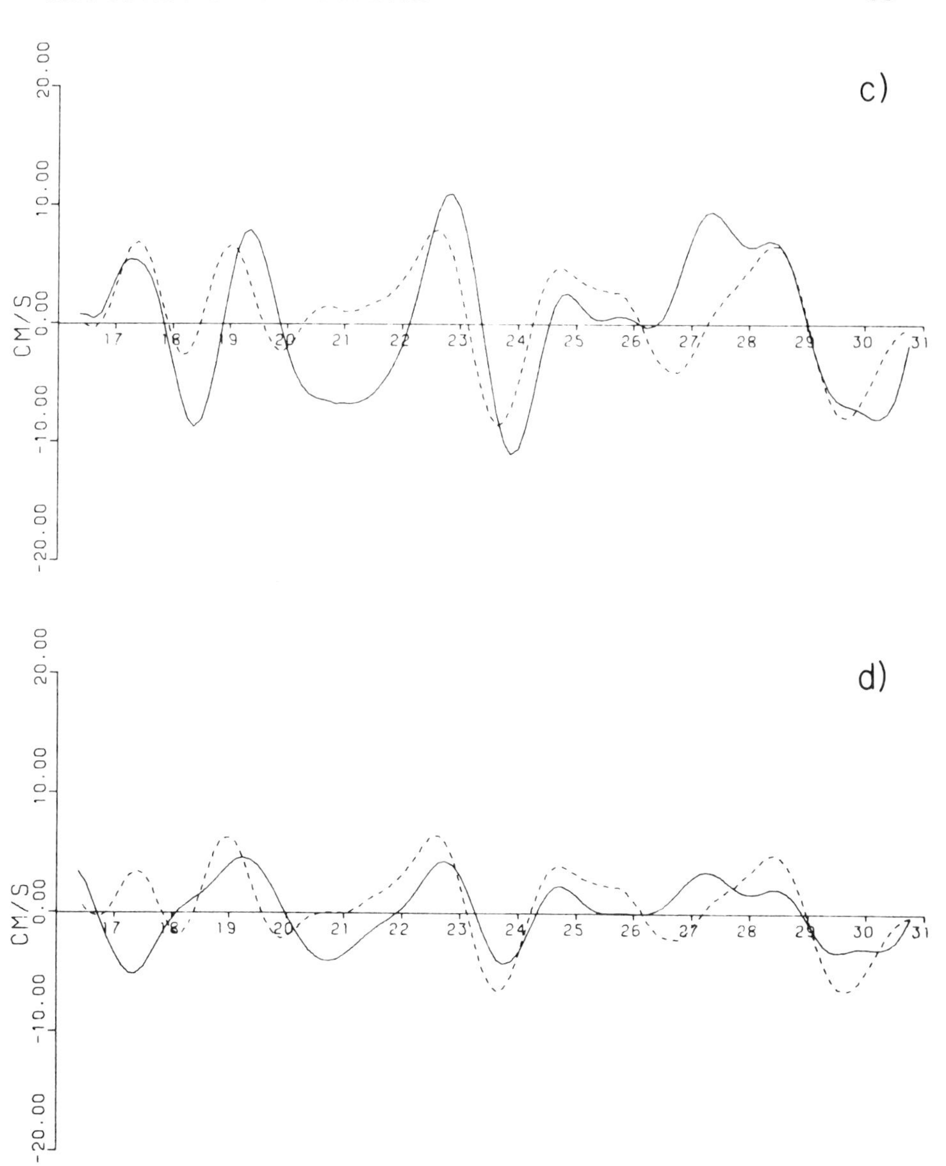

FIGURE 13 (c) and (d).

The bottom frictional coefficient and the interfacial frictional coefficient are not known a priori. The model was run as a one layer model, in which case the bottom frictional coefficient was the only free parameter. This coefficient was adjusted until variance in the computed velocity matched variance in the observed sectionally mean residual velocity. Using this bottom frictional coefficient, the model was run with two layers, and the interlayer frictional coefficient adjusted until the variance in the computed velocity for each layer matched, as nearly as possible, the variance for the mean of the observed residual velocity for each of these two layers. Finally,the model was run with four layers, using the bottom and interlayer frictional coefficients as determined in the one layer and two layer experiments.

The results of the four layer run are shown in Fig. 13. Also shown are the observed mean residual velocity records for each of the modeled layers. There is a curious stretch of about one day duration at the beginning of the graph for the bottom layer when the observed and computed currents were out of phase. This may result from the methods we used in smoothing the initial portion of the input functions. In any case, except for this stretch, the computed and observed velocities for the near surface layer and the near bottom layer were in reasonably good agreement. The agreement for the upper middle layer was fair to good, while that for the lower middle layer was only fair. In all layers the major features of the observed time variation in the current also appeared in the computed record, though with some change in amplitude and time of occurrence. Agreement between the computed and observed time fluctuations in velocity could be improved by allowing the interlayer frictional coefficient to vary with depth.

The results of these model runs indicate that the major features of the vertical response of the residual current to fluctuations in the axial wind stress can be simulated using a relatively simple dynamic model.

ACKNOWLEDGMENTS

The data used in this study were obtained during a field study funded by the Department of Natural Resources, State of Maryland. Funding for the analysis of the data and the modeling effort described here was provided by the National Science Foundation, under Grant ID Number OCN8208363A01.

This paper is Contribution no. 435 of the Marine Sciences Research Center, State University of New York, Stony Brook.

REFERENCES CITED

Elliott, A.J. 1978. Observations of the meteorologically induced circulation in the Potomac Estuary. *Estuar. Coastal Mar. Sci. 6*:285-299.

Elliott, A.J. and D-P Wang. 1978. The effect of meteorological forcing on the Chesapeake Bay: the coupling between an estuarine system and its adjacent coastal waters, pp. 127-145. *In*: J.C.J. Nihoul (ed.), *Hydrodynamics of Estuaries and Fjords*. Elsevier Scientific Publishing Co., Amsterdam.

Elliott, A.J., D-P. Wang and D.W. Pritchard. 1978. The circulation near the head of Chesapeake Bay. *J. Mar. Res. 36*:643-655.

Halpern, D. 1976. Measurements of near-surface wind stress over an upwelling region near the Oregon Coast. *J. Phys. Oceanogr. 6*:108-112.

Pritchard, D.W. 1954. A study of the salt balance in a coastal plain estuary. *J. Mar. Res. 13*:133-144.

Pritchard, D.W. 1980. A note on the Stokes transport in tidal estuaries, pp. 217-226. *In*: B. Patel (ed.), *Management of the Environment*. Health Physics Div., Bhabha Atomic Research Centre, Wiley Eastern Limited, Bombay, India.

Pritchard, D.W. and S.R. Rives. 1979. Physical hydrography and dispersion in a segment of the Chesapeake Bay adjacent to the Calvert Cliffs Nuclear Power Plant. Chesapeake Bay Institute, The Johns Hopkins University, Special Report #74, Baltimore, MD 374 pp.

Vieira, M.E.C. 1983. A study of the non-tidal circulation in a segment in the middle reaches of the Chesapeake Bay. Ph.D. Thesis, The Johns Hopkins University, Baltimore, MD, 154 pp.

Wang, D-P. 1979a. Subtidal sea level variations in the Chesapeake Bay and relations to atmospheric forcing. *J. Phys. Oceanogr. 9*:413-421.

Wang, D-P. 1979b. Wind-driven circulation in the Chesapeake Bay, Winter 1975. *J. Phys. Oceanogr. 9*:564-572.

Wang, D-P. and A.J. Elliott. 1978. Non-tidal variability in the Chesapeake Bay and Potomac River: Evidence for non-local forcing. *J. Phys. Oceanogr. 8*:225-232.

TRANSPORT VARIABILITY IN A FJORD

G. A. Cannon
D. E. Bretschneider
J. R. Holbrook

Pacific Marine Environmental Laboratory
National Oceanic and Atmospheric Administration
Seattle, Washington

Abstract: Bottom water renewal in Puget Sound has been found to occur over short intervals several times during the year and is dependent on tidal and mixing processes over the entrance sill. In addition, the inflow of new bottom water entrains some fraction of seaward flowing upper water at the landward side of the sill. This process acts as a "filter" in that some dissolved and suspended contaminants do not leave the system, but rather make one or more circuits through the basin and possibly contribute to some unknown amount of long-term accumulations.

INTRODUCTION

Fjords are estuaries which have resulted from glacial erosion and which generally occur at higher latitudes. Typically, they are relatively long and deep and possess one or more sills. The oceanography of fjords has been studied extensively in Scandinavia and on the west coast of North America. Much of the early work was descriptive covering a wide variety of fjords. More recently, a broad range of time variations in currents has been elucidated which must be considered when studying particular kinds of processes.

Puget Sound is a fjord-like estuarine system connecting through Admiralty Inlet to the Strait of Juan de Fuca and

ISBN 0-12-405070-0

thence westward to the Pacific Ocean (Fig. 1). It is the southernmost glacially carved waterway in western North America. The main basin off Seattle exceeds 200 m depth and extends about 60 km southward (Fig. 2). It is bounded by an entrance sill region which extends about 31 km through Admiralty Inlet (actually a double sill, shoalest at 64 m) and by a landward sill in The Narrows (44 m) connecting to a southern basin. Recent studies have focused on quantifying the flux of water in the Puget Sound system and on describing spatial and temporal variations (Cannon 1983). Emphasis has been placed on flow interactions at the junction of the main basin and the entrance sill, along the solid line in Fig. 1, where downward entrainment of water is believed to occur. The purpose of this paper is to describe our present understanding of transport variations and how the entrainment may be able to trap contaminants (*i.e.*, act as a filter) in this kind of estuary.

Flow in the estuary is tidal with a superimposed time-varying gravitational circulation (Fig. 2). Vigorous vertical mixing of water occurs in the sill zones. Amplitudes of the tidal currents in the deep basin are about half as large as those shown for the sill, and there is no discernible tidal component in salinity. The larger net flows and higher salinities over the sill generally occur during neap tides, and these are associated with bottom water intrusions into the deep basin. The salinity section shows strong horizontal stratification through Admiralty Inlet similar to some coastal plain estuaries, while that in the deeper basin shows a relatively weaker horizontal stratification more characteristic of fjords.

When the tides are filtered from the observations, the net flow in most of the estuary is seaward near the surface and landward near the bottom (Fig. 2). In the northern half of the main basin (site M), the change from inflow to outflow occurs at about 50 m in a depth of 200 m, or 25% of the total depth. This is intermediate between the deeper level for partially mixed estuaries of 40-50% (shown here in Admiralty Inlet, site A) and the shallower level for more classical fjords of 10-15% (shown here in Saratoga Passage, site S). The net flow in most of the estuary is two-layered. An exception to two-layered flow appears to occur in the southern part of the main basin (sites C and E)

FIGURE 1 (opposite). Puget Sound estuarine system showing station locations and the schematic section used in Fig. 3 (solid line). Mooring sites S, A, M, C and E are used in later figures. A is the middle square, and M is the open square.

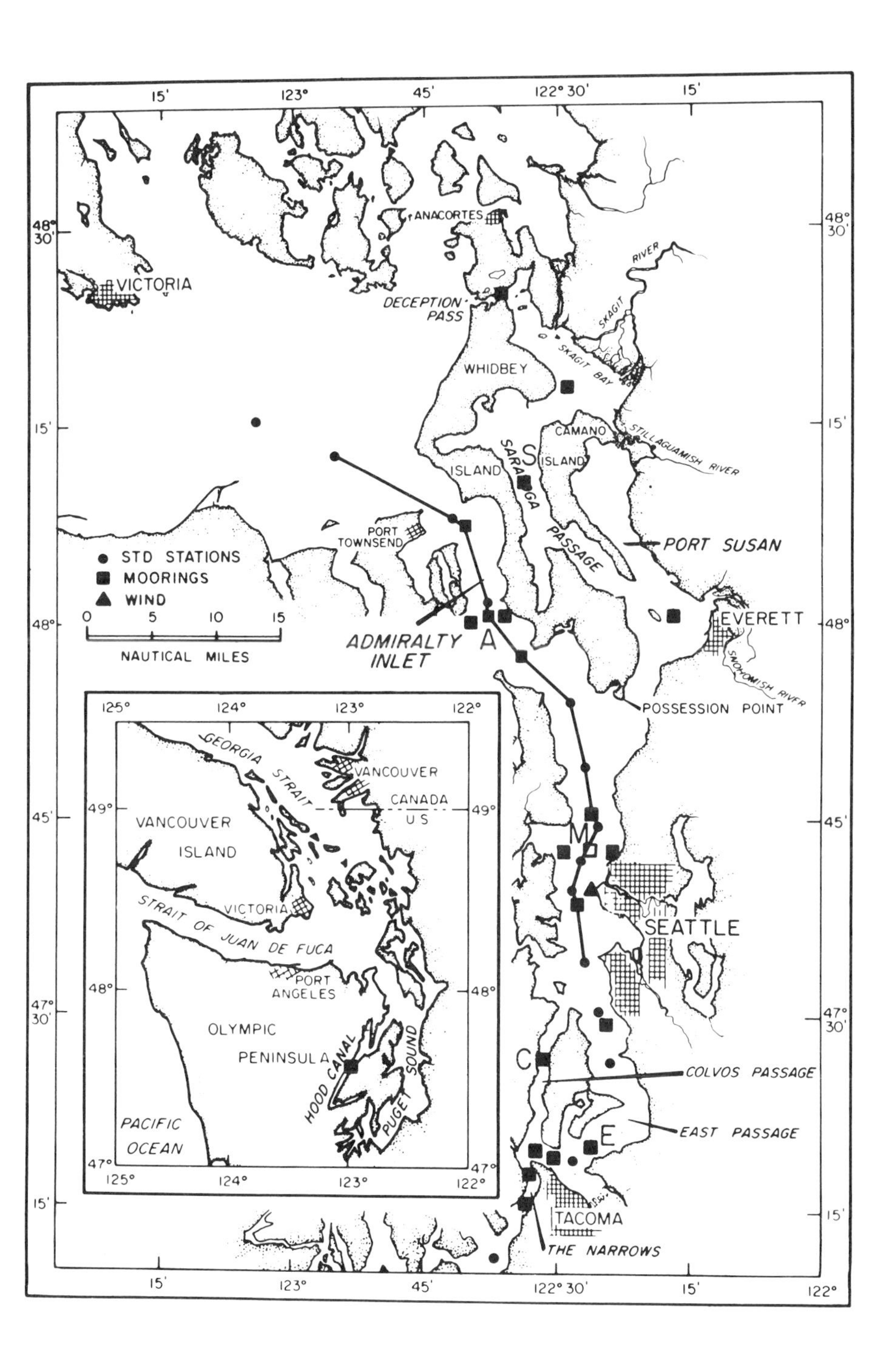

15'
123°
45'
122° 30'
15'
122°
48° 30'
15'
48°
45'
47° 30'
ANACORTES
VICTORIA
DECEPTION PASS
SKAGIT RIVER
SKAGIT BAY
WHIDBEY ISLAND
CAMANO ISLAND
STILLAGUAMISH RIVER
SARATOGA PASSAGE
S
PORT TOWNSEND
PORT SUSAN
STD STATIONS
MOORINGS
WIND
0
5
10
15
NAUTICAL MILES
ADMIRALTY INLET
A
EVERETT
SNOHOMISH RIVER
POSSESSION POINT
M
SEATTLE
C
COLVOS PASSAGE
E
EAST PASSAGE
TACOMA
THE NARROWS
125°
124°
49°
47°
GEORGIA STRAIT
VANCOUVER
CANADA
US
VANCOUVER ISLAND
VICTORIA
STRAIT OF JUAN DE FUCA
PORT ANGELES
OLYMPIC PENINSULA
HOOD CANAL
PUGET SOUND
PACIFIC OCEAN

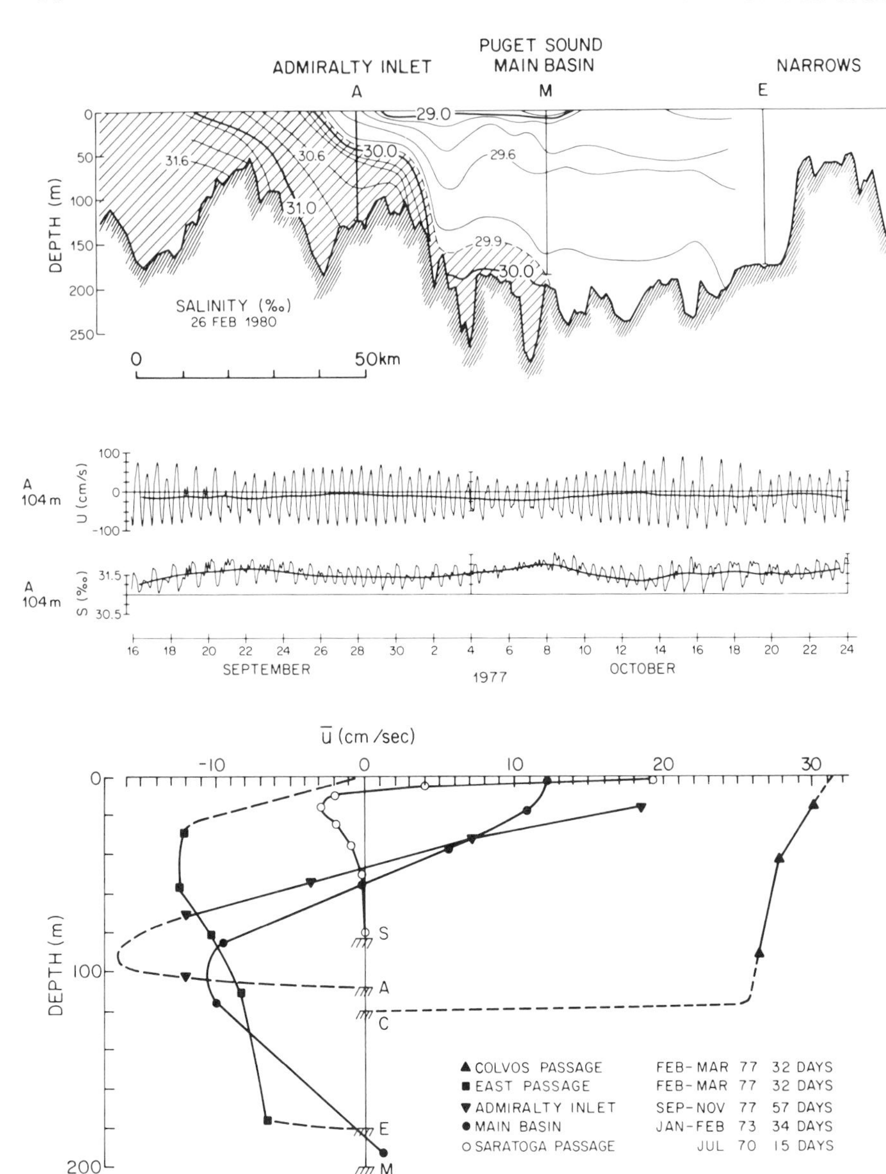

FIGURE 2. Salinity section along Puget Sound (top), time series of flow and salinity on the entrance sill (middle), and long-term average along-channel current profiles at five locations (S, A, C, E and M) in the Puget Sound system (bottom). Positive flow is seaward. Mooring locations are shown in Fig. 1. Bold lines (middle) are time series with tides removed.

where geometry has a significant effect. The net flow at all depths appears to be southward in East Passage and northward in Colvos Passage due to the residual tidal circulation around the intervening island.

CONCEPTUAL MODEL

A schematic representation of the circulation in Puget Sound at the junction of the entrance sill and the deep basin has evolved based on extensive water property observations and some current measurements (Fig. 3). This conceptual model suggests that considerable seaward-flowing surface water is mixed downward (or entrained) into new water entering the deep basin at the southern end of the Admiralty Inlet entrance sill, thus refluxing the entrained surface water back into the Sound in the deep water below sill depth. The downward mixing near the sill of seaward-flowing water is probably unique to the fjord-like Puget Sound estuary compared to other estuaries in the lower 48 states. Most other U.S. estuaries mix incoming water upward into the seaward-flowing surface water continuously along the estuary. This refluxing of part of the seaward-flowing water has significant implications regarding flushing of the Puget Sound main basin. Some fraction of dissolved and suspended contaminants will not leave the basin immediately but will make additional trips through the Sound, and this can lead to some unknown amount of long-term accumulations of these contaminants.

Observations of currents from several years of research (Cannon 1983) have been used to obtain first estimates of the refluxing and to show the magnitudes of spatial and temporal variations in water transport in the Sound. The observations have been used with a simple box model of the Sound developed elsewhere (Hamilton et al. 1984) to obtain a first estimate of possible seasonal variations in transport.

TRANSPORT ESTIMATES

Estimates of the flux of water (Fig. 3) have concentrated on the lower layer or inward flow, partly because the observations rarely were extended to the surface because of shipping. The upper layer or seaward flow must be the same plus the amount of freshwater being transported seaward. Cannon and Ebbesmeyer (1978) calculated an average southward flux near Seattle of 4.3×10^4 m^3/sec using

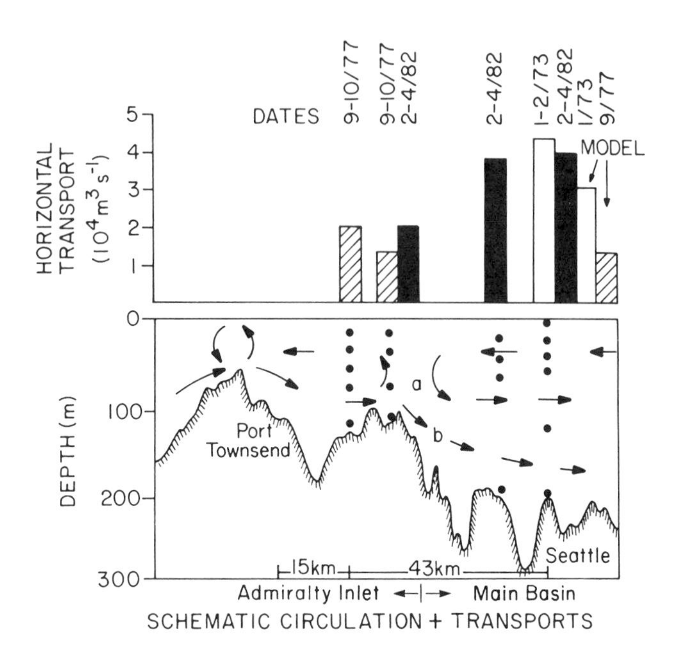

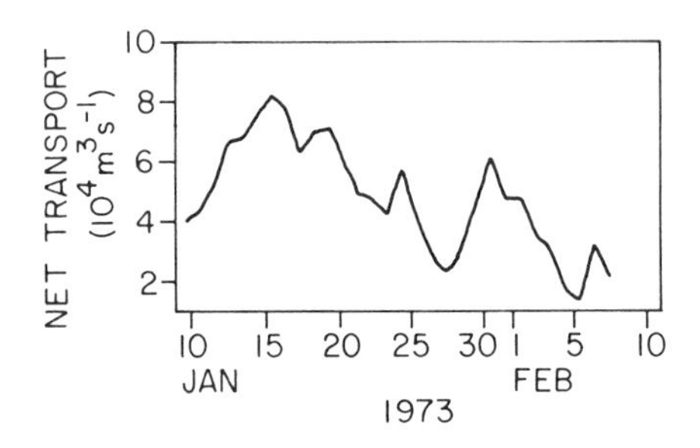

FIGURE 3. Schematic representation of recirculation in Puget Sound (middle) and estimates of lower layer transport (top) from observations at the locations shown by dots. Dates of estimates are given, and simultaneous estimates are indicated by similar hatching. Model estimates from Hamilton et al. (1984). Lower figure shows variations in transport at site M for one interval using daily average currents.

mid-channel month-long average currents (assumed uniform across channel) for winter 1973 (Fig. 1, at mooring M, open square). Daily average transports ranged from 2-8 x 10^4 m^3/sec. Ebbesmeyer et al. (1984) estimated that those values should be multiplied by 0.75 to account for cross section variations. Thus, the 1973 winter average would be 3.2 x 10^4 m^3/sec. Cannon (1983) estimated the inflow midway over the Admiralty Inlet sill using a three station cross section at 2.0 x 10^4 m^3/sec (Fig. 1, at mooring A), about 1/2 - 2/3 of the estimate off Seattle, thus supporting the refluxing concept that some seaward flowing water is entrained into the bottom inflowing water. However, the sill observations were made in autumn 1977, and there may have been seasonal or interannual variations in the flow. A decrease in inflow calculated at an inner station in Admiralty Inlet (Fig. 1, 7 km SE of A), but with less data, is an indication of some flow into Hood Canal, and this implies a larger reflux in the Sound.

Hamilton et al. (1984) used a simple two-layered box model of the Sound to calculate an average transport of about 3.0 x 10^4 m^3/sec for winter observations and about 1.4 x 10^4 m^3/sec for autumn. The winter calculation agrees with the adjusted measured estimate. Observations in Admiralty Inlet south of the Hood Canal entrance, however, are about the same as the model in the main basin, implying a much smaller recirculation, if any, in late summer to early fall. A reflux box model, currently under development (E. Cokelet and R. Stewart, Pacific Marine Environmental Laboratory, personal communication), estimates the average annual transport off Seattle at about 2.0 x 10^4 m^3/sec, which is about midway between the box model seasonal extremes.

Observations in winter 1982 also show an increased inflow between the sill (7 km SE of site A) and two sections in the main basin (site M and 9 km north). Considerable variations exist over the sill, as was shown for the main basin in Fig. 3, depending whether there is a specific inflow event (Fig. 4). However, the average inflow over the two months was about 2 x 10^4 m^3/sec. The average inflow during the event on 14-20 February was 3.5 x 10^4 m^3/sec and during the subsequent low inflow on 21-27 February was 0.13 x 10^4 m^3/sec .

Longer term average current profiles in addition to the one shown in Fig. 2 for the main basin show characteristics somewhat between those of classical fjords and partially mixed estuaries. However, significant variations in these profiles have been observed when averaging over shorter intervals of about a day. The residual currents with the tides removed from the record are obviously a function of time, and the concept of "a" mean current is questionable.

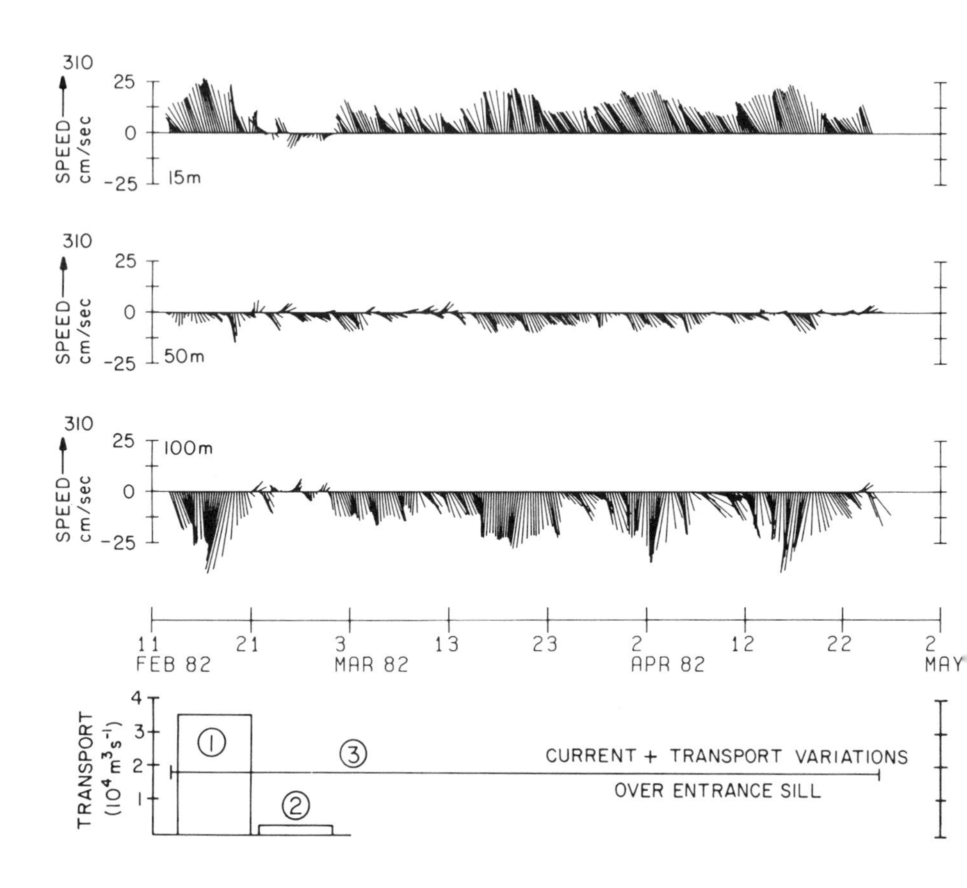

FIGURE 4. Low-pass filtered currents at three depths on the entrance sill (7 km SE of site A) and estimates of landward transport for the indicated intervals. The vertical axis of the currents is along 310° T.

There appears to be a delicate balance of forces along the main basin because of relatively small horizontal gradients. Thus, the wind contribution to the total horizontal pressure gradient may result in major changes in flow throughout the water column (Cannon 1983). Ebbesmeyer et al. (1984) indicate that wind effects may be correlated to variations in southward flow at deeper depths in the main basin. Similar flow variations have been observed in the more classical fjord flow in Saratoga Passage, but little work has been carried out on this aspect of fjord circulation.

SPATIAL VARIATIONS

The current profiles observed at our long-term study site (M) are probably representative for the northern part of the main basin at mid-channel (Fig. 2). However, the few long-term observations at other main basin locations and in other parts of the system show considerable spatial variations. Concurrent month-long observations in East Passage (E) and Colvos Passage (C) indicate clockwise circulation around the intervening island as was previously deduced from hydraulic model studies. A discrepancy in volume fluxes (assuming profiles are uniform across channel) of 3.8 and 2.7 x 10^4 m^3/sec southward and northward, respectively, has not been resolved, but it may indicate that some near-surface water is going northward in East Passage.

Current profiles in Saratoga Passage (S) show what has been considered a classical fjord current profile (Fig. 2). The location of the station can be thought of as being in the middle reach of a fjord, downstream from the Skagit River at the head of the fjord and upstream from the Admiralty Inlet sill. Observations (not shown) from the middle reach of Hood Canal show two-layer flow which differs from that of Saratoga Passage, possibly because Hood Canal lacks a significant or single source of freshwater.

Current profiles in Admiralty Inlet (A) resemble those of a coastal plain estuary (Fig. 2). When new bottom water enters the main basin, it shows here as increased inflow in the bottom layer, and because of continuity, it is accompanied by increased outflow in the surface layer (Geyer and Cannon 1982). However, when this increased inflow reaches the fjord-like main basin, it results in a flow adjustment at mid-depth, but below the sill depth, and there is negligible response near the surface (Fig. 5).

IMPLICATIONS

New deep water has been observed to enter the Sound at about fortnightly intervals usually associated with the largest flood tides through Admiralty Inlet, but also associated with the smallest neap tides (Geyer and Cannon 1982). In the first case, incoming bottom water can transit Admiralty Inlet in one flood cycle and is least mixed. In the second, during the lowest flood tide ranges during the equinoxes in late winter and late summer, incoming bottom water transits most of Admiralty Inlet on the larger flood

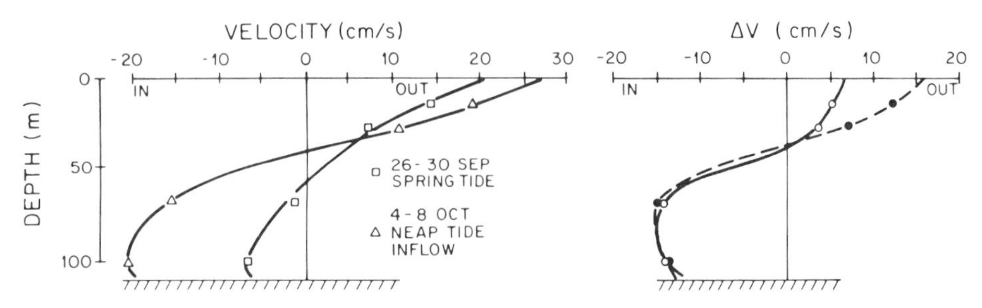

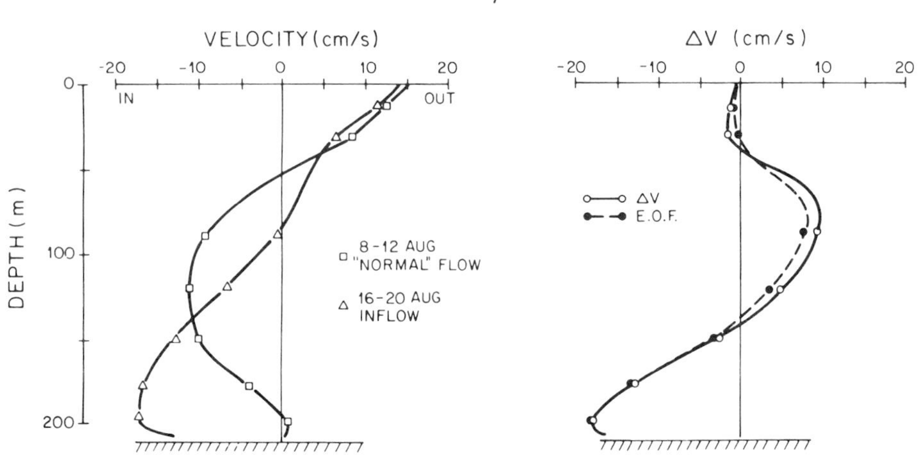

FIGURE 5. *Vertical profiles of tidally averaged velocities (at sites A and M) during periods with bottom inflow and with more normal flow (adapted from Geyer and Cannon 1982).*

cycle. The subsequent small ebb is almost nonexistent at the bottom, and the following smaller flood is capable of completing the transit of the sill with minimal mixing. However, denser water is always more readily available at sill depth in the Strait than it is in the bottom of the Sound, and the intrusions do not occur every fortnight. The efficiency of mixing processes over the sill appears to be the controlling mechanism.

Through use of the transport estimates, an overall rate of replacement of intermediate and deep water south of the main basin mooring (M) has been approximated (Cannon and Ebbesmeyer 1978; Cannon 1983). The average replacement time was less than two weeks when an intrusion of new bottom water was occurring, and observations of water properties along the Sound showed the same transit time. The first conclusion one might draw is that the main basin water below sill depth could be renewed in about two weeks, but not during every two-week interval. However, in light of the conceptual model and observations discussed here, the question now being addressed in the reflux model mentioned above (Cokelet and Stewart, personal communication) is, how much water entering the deep basin at the sill is new and how much is refluxed from the seaward-flowing surface water by downward entrainment? Understanding this phenomenon and its episodic and seasonal variations is of utmost importance in determining the fate of contaminants in this kind of estuarine system. Present efforts are focusing on better estimates of transport through a section which include both balancing the water and salt fluxes and obtaining more extensive observations across a section.

ACKNOWLEDGMENTS

This work was partially funded first by NOAA's MESA Puget Sound Project and later by the Long-Range Effects Research Program. We are particularly grateful to D. Pashinski and D. Kachel who have helped with field work and data processing over many years, and to H. Mofjeld and R. Stewart for critically reviewing the manuscript. Dr. William Biocourt kindly presented the paper at the ERF biennial meeting when the authors could not attend. Figure 5 was adapted from Geyer and Cannon (1982) as published by the American Geophysical Union. Contribution number 681 from Pacific Marine Environmental Laboratory.

REFERENCES CITED

Cannon, G.A. 1983. An overview of circulation in the Puget Sound estuarine system. *NOAA Technical Memorandum ERL PMEL-48*, 30 pp., Seattle, WA.

Cannon, G.A. and C.C. Ebbesmeyer. 1978. Winter replacement of bottom water in Puget Sound, pp. 229-238. *In: B. Kjerfve (ed.), Transport Processes in Estuarine Environments*, Univ. South Carolina Press, Columbia, SC.

Ebbesmeyer, C.C., C.A. Coomes, J.M. Cox, J.M. Helseth, L.R. Hinchey, G.A. Cannon and C.A. Barnes. 1984. Synthesis of current measurements in Puget Sound, Washington. Vol. 3. Circulation in Puget Sound: an interpretation based on historic records of currents and water properties. *NOAA Technical Memorandum NOS OAD*, Rockville, MD. (in press).

Geyer, W.R. and G.A. Cannon. 1982. Sill processes related to deep water renewal in a fjord. *J. Geophys. Res. 87:* 7985-7996.

Hamilton, P., J. T. Gunn and G. A. Cannon. 1984. A box model of Puget Sound. *Estuarine, Coastal and Shelf Science*. (in press).

GEOLOGICAL PROCESSES

THE ESTUARY AS A FILTER FOR FINE-GRAINED SUSPENDED SEDIMENT

J.R. Schubel
H.H. Carter

Marine Sciences Research Center
State University of New York
Stony Brook, New York

Abstract: Estuaries function as "filters" for the signals they receive from the land and the sea. The nontidal residual circulation, tidal mixing, and estuarine geometry determine the efficiency of this filter for suspended sediment. Tidal currents provide energy for mixing saltwater from the ocean with freshwater from the river. The resulting salinity distribution drives that part of the nontidal circulation caused by density differences, which in turn influences the salinity patterns and resulting density gradients. This feedback between salinity redistribution and gravitational circulation places a constraint (filter) on the range of variations in flow and salt concentration in the estuary. The nontidal circulation also controls the distribution and transportation of suspended sediment, and the deposition of fine particles within the estuary. The distributions of salt and fine suspended matter control the behavior of many non-conservative constituents such as nutrients, radionuclides and some metals, their modes of occurrence and transport, and their reservoirs of accumulation. The result of these processes is that the estuary modifies significantly the strength and the form of the signals it receives from land and sea. Alteration of an estuary's gravitational circulation pattern by changing its freshwater input or its geometry modifies its filtering efficiency. In general, as an estuary changes from being highly stratified toward being well-mixed, its filtering efficiency for land-derived constituents first increases to a maximum (in the partially-mixed region) and then decreases. A simple kinematic model is used to

ISBN 0-12-405070-0

demonstrate the relationship between estuarine type (gravitational circulation pattern) and filtering efficiency for Chesapeake Bay, USA.

INTRODUCTION

In this paper we use Pritchard's (1967) definition of an estuary: An estuary is a semi-enclosed coastal body of water which has a free connection with the open sea and within which seawater is measurably diluted with freshwater derived from land drainage. A number of factors concerning the origin and development of estuaries are well documented. The following are among the major ones:

- The origin of estuaries is determined primarily by those climatic events that control eustatic sea level changes--glaciations and deglaciations.
- The distribution of estuaries is controlled by the interplay among climatic events and regional and more local geological processes.
- Estuaries are relatively large and abundant following periods of rising sea level, particularly where continental margins are broad and relatively flat such as along the Atlantic and Gulf coasts of the United States.
- Estuaries are smaller and less abundant during lowstands of sea level and where continental margins are narrow and the coast has high relief such as along the west coast of the United States.
- Once formed, estuaries are destroyed rapidly. On geological time scales they are ephemeral features having life spans measured in thousands of years to perhaps a few tens of thousands of years; they fill rapidly with sediments.
- Characteristically, sedimentation rates in estuaries are highest near their heads where an estuarine delta forms in the upper reaches of the estuary near the new river mouth. The delta grows progressively seaward within the estuary to extend the realm of the river and force the intruding sea out of the semi-enclosed coastal basin.
- Estuaries in various stages of geological evolution can be found around the world.

As a result of the last three of these "facts", estuaries often are characterized as being effective "filters" for the fine-grained sediments that they receive from their tributary rivers. Webster's Third New International Dictionary of the English Language (1976) defines a filter as a device through which fluids are passed to separate out matter in suspension.

Estuaries appear to be effective filters, although their effectiveness--their filtering efficiency--differs for different estuarine circulation types (Pritchard 1955), and for any given estuary it changes as the estuary evolves.

In this paper we discuss some of the physical characteristics of estuaries that cause them to function as filters for fine-grained sediments, we assess the relative importance of these characteristics in determining the filtering efficiency of different estuaries, and we evaluate how an estuary's filtering efficiency changes as the estuary evolves.

The filtering efficiency--the filter factor--of an estuary for suspended matter is determined by a variety of physical oceanographic, geochemical and biological processes, and by the geometry of the estuarine basin. The most important physical oceanographic characteristic is the nontidal circulation pattern. The most important biological processes are the activities of filter feeding organisms on the bottom and in the water column (Biggs and Howell 1984). Geochemical processes that affect suspended and dissolved materials include flocculation, absorption, and adsorption (Kranck 1984). Here we concentrate on the effects of physical processes and geometry on estuarine filtering of fine-grained suspended sediment. The importance of biological and geochemical processes are discussed elsewhere in this volume.

FACTORS THAT AFFECT FILTERING EFFICIENCY

For the purposes of this paper, we define the filtering efficiency of an estuary as that fraction of the total sediment introduced to an estuary through its open boundaries--by its rivers and the ocean--that is retained within the estuary. Filtering efficiency is determined primarily by the strength of the estuary's nontidal circulation and by the size of the estuary, particularly its length. We define the length of the estuary as the distance from the mouth of the estuary to the last measurable traces of sea salt. The length of an estuary varies with river discharge and can fluctuate by many tens of kilometers. In all estuaries, except sectionally homogeneous estuaries, the landward limit of sea salt penetration coincides roughly with the "null zone"--the zone within which the net upstream flow of the lower layer dissipates until the net flow is downstream at all depths. The maximum filtering efficiency is achieved in partially mixed (Type B) estuaries (Pritchard 1955)--which have the strongest gravitational flows[1]--when the null zone is well within the estuary and cannot be displaced from the basin even during periods of very high river flow. Two estuaries that fit these criteria are Chesapeake Bay and Long Island Sound, USA.

For any circulation pattern, the geometry of the estuarine basin plays a role. The larger the basin, the greater the estuary's filtering efficiency for any given circulation pattern. As an estuary evolves and the basin is filled, its depth, volume, and surface area decrease. Consequently, intertidal volume decreases and the strength of the river flow relative to the tidal flow increases, causing the circulation pattern to shift in the direction of a river-dominated estuary. Since most estuaries fill from their heads, there is a progressive shortening of the length of the estuary. An increase in river discharge relative to tidal flow also shortens the estuary, increases the longitudinal density gradient and hence the gravitational circulation. The increases in circulation, however, are buffered by the increased stratification that also results from increased river flow. Depending upon where an estuary is in the sequence of estuarine circulation types (Pritchard 1955), altering the gravitational circulation could either increase or decrease the estuary's filtering efficiency. Ultimately, however, the estuary reaches a size beyond which further ageing results in a decrease in filtering efficiency.

Any reduction in the length of an estuary leads to a reduction in its filtering efficiency. This effect is particularly important in stratified and partially mixed estuaries. When the estuary is shortened below some threshold length, the entire "estuary"--the penetration of sea salt--may be expelled from its basin during periods of high riverflow and its filtering efficiency will approach zero during these times. Since periods of high riverflow are usually periods of high fluvial suspended sediment input, the fraction of this sediment discharged directly to the ocean increases.

In salt wedge (Type A) and partially mixed (Type B) estuaries, settling of particles into the lower layer reverses the direction of net sediment transport. This increases the retention time within the estuary, the time available for settling, the probability of deposition, and as a result, the filtering efficiency. In vertically homogeneous (Type C) and

---

[1]Partially mixed estuaries have the strongest gravitational flows in terms of multiples of river flow, Q_R. Fjords, according to Hansen and Rattray (1966), have the strongest gravitational flows as measured by the ratio, u_s/U_f, where u_s is the longitudinal mean velocity at the surface and U_f is the integral mean velocity (the river discharge rate/cross-sectional area of the estuary).

sectionally homogeneous (Type D) estuaries, settling does not lead directly to a reversal in the direction of sediment transport. Therefore, for particles with low settling velocities, settling has less effect on retention time within the estuary and hence little effect on filtering efficiency. Estuaries that approach sectional homogeneity throughout the tidal cycle must be short in length, at most a few tidal excursions long, because the potential mechanisms for transporting salt upstream are limited in number and in efficacy. In estuaries that approach sectional homogeneity, the diffusion and advection of salt are in opposite directions--the diffusion of salt is into the estuary and the advection of salt is out of the estuary. On the other hand, the diffusion and advection of fine-grained suspended sediment are in the same direction--seaward. The addition of these two processes decreases the retention time for fine particles and the filtering efficiency of sectionally homogeneous estuaries.

Estuaries with the lowest filtering efficiencies share one or more of the following features. Either they are highly stratified (salt wedge) estuaries with restricted vertical mixing across the halocline or they approach sectional homogeneity. In either case, the average length of the estuary as indicated by the landward limit of sea salt penetration and the null zone is limited to the vicinity of the mouth of the estuary.

The most dramatic manifestation of the filtering action of estuaries is the well-known "turbidity maximum"--a zone in the upper reaches of estuaries within which concentrations of suspended sediment are greater than those either farther upstream in the source river or farther seaward in the estuary. Zones of turbidity maxima have been reported in a large number of estuaries throughout the world (Glangeaud 1938; Postma and Kalle 1955; Nichols and Poor 1967; Postma 1967; Schubel 1968a,b; Conomos and Peterson 1976). Their origin and maintenance have been attributed to flocculation (Lüneburg 1939; Ippen 1966) and to deflocculation (Nelson 1959), but it is now generally accepted that they are produced and maintained primarily by the nontidal circulation (Glangeaud 1938; Postma and Kalle 1955; Inglis and Allen 1957; Nichols and Poor 1967; Postma 1967; Schubel 1968a,b; Allen 1973; Gallene 1974; Conomos and Peterson 1976). A comprehensive descriptive and theoretical treatment of the characteristic circulation pattern in partially mixed estuaries was provided first by Pritchard (1952, 1954, 1956). Our theoretical understanding of the full sequence of the several types of estuaries was expanded by Rattray and Hansen (1962), Hansen and Rattray (1966), and more recently and in somewhat simpler form by Officer (1976).

Postma (1967), Schubel (1968a,b, 1969, 1971) and others have provided the conceptual basis for the coupling between the gravitational circulation and the origin and maintenance of the turbidity maximum. Suspended particles moving seaward in the upper layer and which sink into the lower layer are carried back upstream in the lower layer to produce accumulations of particles on the bottom (rapid sedimentation) and within the water column (the turbidity maximum). The process provides an effective sorting mechanism. Particles with settling velocities comparable to the mean vertical mixing velocities are concentrated within the turbidity maximum, those with greater settling velocities are deposited, and those with smaller settling velocities are carried out of the zone of the turbidity maximum and not returned. Schubel (1969, 1971) demonstrated that the population of suspended particles in Chesapeake Bay's turbidity maximum is composed of two subpopulations: those particles alternately resuspended and deposited by waxing and waning of tidal currents and those particles in more-or-less continuous suspension. Particles in the first subpopulation had mean Stokes' diameters ranging from 2-64 μm; particles in the second subpopulation had a much narrower size and more stable size distribution with a mean size of approximately 2-3 μm.

Festa and Hansen (1976) developed and applied a two-dimensional, numerical model of estuarine gravitational circulation to investigate the effects of changes in basin geometry and freshwater inflow on circulation and salinity distributions in estuaries. Later, Festa and Hansen (1978) used this model to investigate the relationships between the suspended sediment turbidity maximum and the gravitational circulation. The model assumed no net deposition or erosion of bottom sediments.

Festa and Hansen (1978) found that the turbidity maximum was associated with salinity values of approximately 0.10 S_o where S_o is the salinity of the ocean or the embayment into which the estuary discharges. This value of salinity, 0.10 S_o, also corresponded closely to the value associated with the landward flowing lower layer in the null zone.

Both the vertical stratification and the strength of the gravitational circulation increase with increased freshwater discharge. Festa and Hansen (1978) demonstrated that the principal response to changes in the volume of freshwater discharge occurred within the upper layer and that the total transport in the lower layer was almost independent of the amount of freshwater discharged. The length of the estuary is a strong function of freshwater discharge. Festa and Hansen (1978) related salinity intrusion to U_f, the vertically averaged riverflow per unit width, and showed that the relationship is nonlinear. In most partially mixed estuaries a characteristic value of U_f is about 1 cms^{-1}. Within the range of conditions considered by them, the length of salinity intrusion changes in behavior in the vicinity of U_f = 1 cms^{-1}.

Increases of U_f from 1 cms^{-1} led to a modest retreat of salinity intrusion, but reductions of U_f resulted in greatly increased salinity intrusion. The location of the stagnation point, the mean position of the sediment filter, is related closely to the extent of salinity intrusion. Their model indicated it was bracketed between salinity values of 1.5 and 0.15 $^o/oo$ at the bottom.

Whereas changes in river discharge affect horizontal velocity primarily in the top half of the water column, depth changes primarily influence landward transport in the bottom half of the water column. Festa and Hansen's model indicated that the salinity intrusion increased with increasing depth as expected and that "salinity stratification decreased with increasing depth." This is true only at the mouth of the estuary; elsewhere in the estuary an increase in depth increases the salinity stratification.

Using a simple two-dimensional box model (Officer 1980a,b) was able to reproduce many of the features of turbidity maxima predicted by Festa and Hansen (1978) with their hydrodynamic numerical model. The simpler box model predicted the same approximate location and magnitude (strength) for the turbidity maximum, particularly for small ($\cong 1 \times 10^{-3}$ cms^{-1}) settling velocities. Officer (1981) pointed out that while the box model approach has limitations, it can be applied easily to actual conditions and can be extended to include sediment exchange with the bottom to estimate first-order values for net erosion or deposition.

Using a simple box model, Schubel and Carter (1976) concluded that there is a net transport of suspended sediment from the ocean into Chesapeake Bay and a net transport of suspended sediment from the Bay into its major tributary estuaries. Officer and Nichols (1980) used their box model approximation for gravitational circulation for two of these estuaries, the James and the Rappahannock, and showed that during conditions of moderate river flow the fluxes of suspended sediment into these estuaries from their rivers and from the Bay were comparable. Their model indicated that the principal depositional zone in each estuary was seaward of the turbidity maximum.

DEPARTURES FROM AVERAGE CONDITIONS AND THEIR EFFECTS ON FILTERING

Two assumptions implicit in the discussion to this point are that wind is of relatively little importance in determining the net estuarine (gravitational) circulation pattern and that lateral variations in the velocity and density fields are small. One of the most extensive and most thoroughly and thoughtfully analyzed sets of observations for any estuary was that made in the James River estuary (Virginia, USA) and described by Pritchard (1952, 1954, 1956). The data consisted of three separate subsets of observations of 4, 5, and 11 days, all taken during relatively calm weather. This set of measurements and their analysis laid the foundation for much of our understanding of the physics of estuaries, but as Carter et al. (1979) point out, it so influenced our thinking of how estuaries should operate--particularly partially mixed estuaries--that observations of circulation patterns which were at variance with the classical two-layered circulation pattern were considered to be measurement artifacts or to represent anomalous conditions. Most investigators clung to the belief that observations over several tidal cycles provided a sufficiently long averaging period for representative nontidal flow estimates (Dyer 1973). Later, Weisberg (1976) and Weisberg and Sturges (1976) pointed out that the nontidal flow also was highly variable and that the customary rule of thumb of averaging over a few tidal cycles could lead to serious errors.

Prompted by significant advances during the 1960's in instruments to sense and internally record and store large quantities of current speeds and directions, conductivities and temperatures, Pritchard and co-workers set out in 1974 to repeat the "James River experiment" but over a longer period and in the lower Potomac River estuary (another tributary to Chesapeake Bay). As part of this study, they maintained a reference mooring of three current meters for a full year. Elliott (1976, 1978) analyzed the records from this vertical array. The records were filtered to remove the major tidal components and the residuals were averaged over each calendar day to produce daily estimates of the mean residual flows. The frequencies of occurrence of the six circulation patterns observed are summarized in Table 1. The two-layered circulation pattern occurred a surprisingly small 43% of the time. However, for averaging intervals of ten days or longer, the two-layered pattern always emerged. Analysis of a three month record from a two-current-meter mooring in the upper reaches of the main body of Chesapeake Bay revealed similar variability in the residual circulation (Elliott, Wang and Pritchard 1978).

The variability was produced in part by local wind forcing associated with passage of storms along the U.S. east coast and in part by far-field wind forcing on the continental shelf (Elliott 1978; Elliott et al. 1978; Wang and Elliott 1978). Their results showed that it was possible to separate variations in current and sea level forced by local wind stress which respond at approximately the seiche period of the Bay from the longer period variations forced nonlocally by coastal winds which produce large sea level fluctuations

TABLE 1. Circulation patterns observed in lower Potomac estuary (USA) over one year and their frequencies of occurrence (Data from Elliott 1978).

Estuarine circulation pattern	*Frequency of occurrence (% of time)*
Classical	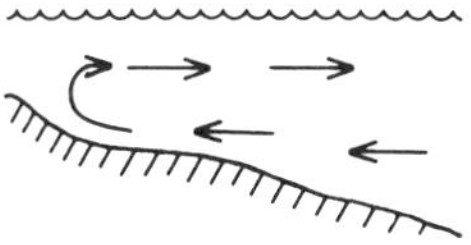 43%
Reverse	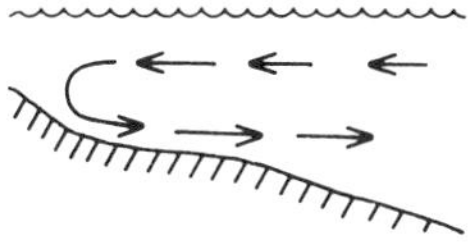 21%
Three-layered	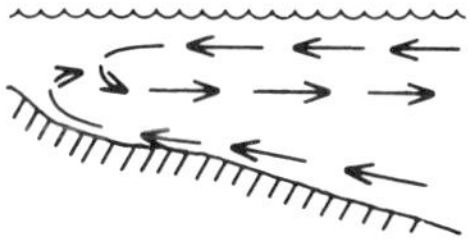 1%
Reverse three-layered	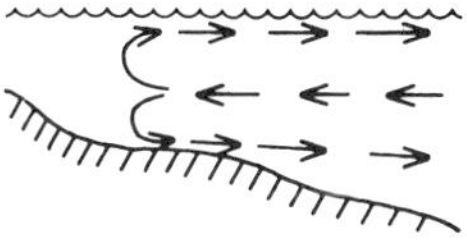 7%
Discharge: flow out at all depths	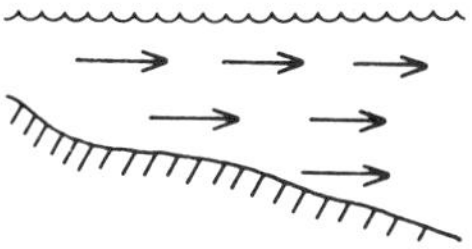 6%
Storage: flow in at all depths	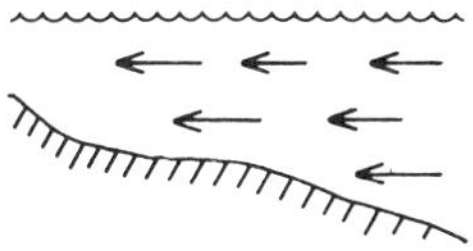 22%
	100%

at the mouth of the Bay and which have a period of about 20 days. These large sea level fluctuations were produced by an onshore or offshore Ekman transport associated with longshore winds.

Since wind-driven flows can at times be much larger than the gravitational circulation, the effects of atmospheric forcing cannot be neglected. According to Elliott (1978), ..."the concept of mean nontidal velocity may have little physical significance for many processes, in particular sediment transport, and effort should be directed toward modeling the variability of the nontidal flow."

The significance of the variability of estuarine circulation patterns to estuarine sedimentation has not been established. The effects of these fluctuations on sediment transport, on patterns of sediment accumulation, and on the filtering efficiency of an estuary will depend strongly upon when they occur and upon the size of the estuarine basin. A wind-driven discharge at all depths coupled with a high riverflow event could lead to significant sediment discharge from an estuary to the ocean if the estuarine basin were relatively small. For a large estuary like Chesapeake Bay proper, it would have little effect.

The discharge of the Susquehanna River associated with Tropical Storm Agnes (June 1972) had a recurrence interval of approximately 200 years. The extremely high riverflow resulting from that storm displaced the landward limit of sea salt penetration and the null zone more than 20 km farther seaward than had ever been recorded, but this was still more than 240 km from the mouth of the estuary (Schubel 1974). While the Chesapeake Bay discharged a significant amount of suspended sediment to the ocean following Tropical Storm Agnes, the filtering efficiency during and following that event was very high, greater than 90% (Schubel 1974). No wind event we can imagine would have had a significant effect on the Chesapeake Bay's filtering efficiency for fine-grained suspended fluvial sediment even if coupled with a flood with a recurrence interval of 200 years.

There are a variety of approaches one can use to estimate the filtering efficiency of estuaries for fine-grained sediment. All require two common pieces of information: (a) estimates of the total amount of suspended sediment introduced by the river(s), by the ocean, by shore erosion, by primary productivity, and by other sources, and (b) an estimate of the fraction of that mass which is retained within the estuary. The input term for the fluvial contribution requires serial measurements of the concentration of suspended sediment in the mouth of the river and of the riverflow. The fraction of an equivalent mass that is retained within the estuary can be estimated from (a) calculations of the fluxes of material in and out of the estuary through its mouth by combining measurements of current velocity and concentrations of suspended sediment, (b) calculations of the rate of accumulation of fine-grained sediment on an estuary-wide basis by combining measurements of local sedimentation rates with areas of accumulation for which each of the rates is representative, and (c) calculations using a variety of modeling techniques. Schubel, Bokuniewicz and Gordon (1978), Schubel and Hirschberg (1980), and Schubel (1983) have discussed the relative merits of these various approaches.

In establishing the filtering efficiency of an estuary, direct calculations of fluxes through its mouth by combining measurements of current velocity and suspended sediment concentration and the subtraction of ebb and flood fluxes to estimate net flux of sediment are of questionable value. The problem is the familiar one of correctly estimating the small difference between two large numbers. Not only is the resulting magnitude of dubious precision and accuracy, but often even its sign is equivocal. Still, these difficulties often have not deterred the use of this superficially attractive, simple and direct method of estimating net flux (Inglis and Allen 1957; Cassie et al. 1962; Terwindt 1967; Flemming 1970; Oostdam and Jordan 1972; Allen and Castaing 1973; Nichols 1974; and many others).

While the determination of an estuary's filtering efficiency for suspended sediment is difficult, determination of the filtering efficiency for reactive components is far more difficult. These calculations require measurements of the concentrations of these components in solution as well as in suspension. Few budgets exist for such components.

A MODEL FOR ESTIMATING ESTUARINE FILTERING EFFICIENCY

We constructed a simple model to assess the relationship between estuarine circulation and the filtering efficiency of estuaries for fine-grained fluvial suspended sediment. In our model we consider only two sources of sediment: river and ocean. The elements of the model are shown schematically in Fig. 1. We define the following terms:

- $\bar{u}_u$, $\bar{s}_u$, $\bar{c}_u$ are monthly averages of velocity, salinity and concentration of suspended inorganic sediment in the upper layer at the mouth of the estuary;
- $\bar{u}_\ell$, $\bar{s}_\ell$, $\bar{c}_\ell$ are monthly averages of the velocity, salinity and concentration of suspended inorganic sediment in the lower layer at the mouth of the estuary;
- ν is the fraction of the net seaward flux of salt, $\bar{u}_u\bar{s}_u$, that is balanced by an upstream

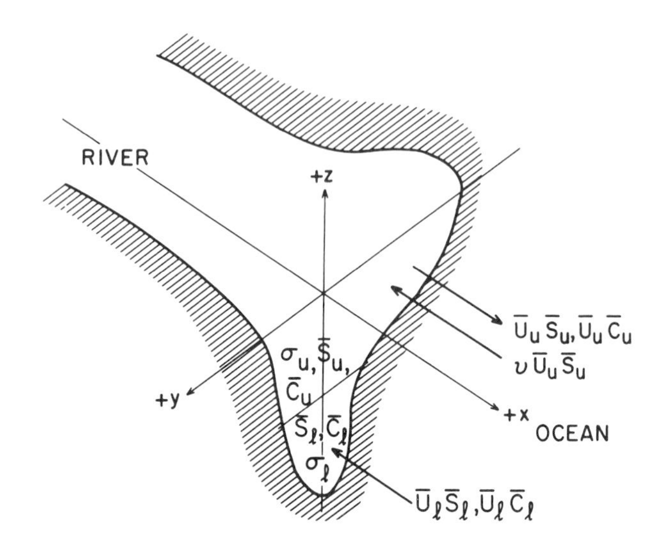

FIGURE 1. Presentation of the simple model used to estimate the exchanges of water, salt, and suspended sediment between an estuary and the ocean.

diffusive flux of salt, $\nu\bar{u}_u\bar{s}_u$;

ν^* is a similar coefficient for inorganic suspended sediment;

x,y,z are the axes of a left-handed coordinate system with the positive x-direction toward the ocean;

σ_u, σ_ℓ are the cross-sectional areas of the upper and lower layers at the mouth of the estuary, respectively;

f is the ratio σ/σ_ℓ where $\sigma \equiv \sigma_\ell + \sigma_u$ at the mouth of the estuary;

δc is the difference in the concentration of inorganic suspended sediment in the lower and upper layers, $\bar{c}_\ell - \bar{c}_u$, at the mouth of the estuary;

δs is the difference in salinity in the lower and upper layers, $\bar{s}_\ell - \bar{s}_u$, at the mouth of the estuary;

U_f is the monthly cross-sectional average of the velocity in the mouth of the estuary;

Q_R is the flux of freshwater through the cross-section at the mouth of the estuary and is equal to $U_f\sigma$, again a monthly average; and
C_R is the concentration of inorganic suspended sediment in the mouth of the river.

In our model we assume that water, salt, and suspended sediment are transported in the upper and lower layers through a vertical cross-section separating the estuary from the ocean. We assume that the transports of these materials are by longitudinal advection and longitudinal turbulent diffusion. Vertical mixing and vertical advection are not included explicitly in the model. Since turbulent diffusion is an exchange process with no net flux of water, salt and suspended sediment will be transported in the direction of decreasing concentration of each of these materials. In the case of salt, this will always be upstream. For suspended sediment, the diffusive transport will almost always be downstream (seaward) since the gradient usually is directed toward the ocean. Thus, ν will always be > 0 but ν^* will be > 0 only if the suspended sediment gradient is directed downstream but < 0 if directed upstream. In addition, $|\nu| \neq |\nu^*|$ since the gradients will differ in magnitude as well as in direction. The absolute value in each case will be proportional to the gradient. It should be noted that our ν differs from that of Hansen and Rattray (1966); it is smaller by the factor

$$\frac{U_f}{\bar{u}_u} \cdot \frac{S_o}{\bar{s}_u} \approx \frac{U_f}{\bar{u}_u}$$

where S_o is the cross-sectional mean salinity.

Water Budget

Integrating the velocity, u, over the cross-section, σ, we have

$$\int_o^{\sigma} u d\sigma = \int_o^{\sigma_\ell} u d\sigma + \int_o^{\sigma_u} u d\sigma = U_f \sigma \qquad \text{or}$$

$$U_f = \frac{\bar{u}_\ell}{f} + \frac{(f-1)}{f}\bar{u}_u \quad \text{where } f \equiv \frac{\sigma}{\sigma_\ell} \tag{1}$$

Salt Budget

The flux of salt is given by

$$\overline{\text{Flux}}_{salt} = \bar{u}_u(\sigma-\sigma_\ell)\bar{s}_u + \bar{u}_\ell\sigma_\ell\bar{s}_\ell - \nu(\sigma-\sigma_\ell)\bar{u}_u\bar{s}_u \tag{2}$$

Assuming there is no upstream storage or depletion of salt, $\overline{\text{Flux}}_{salt}$ is zero and equations (1) and (2) give

$$\frac{\bar{u}_u}{U_f} = \left[\left(\frac{f}{f-1}\right) \right] \frac{1}{\frac{\delta s}{\bar{s}_\ell}(1-\nu) + \nu} \tag{3}$$

In our model there are two extremes: when $\nu = 1$ and when $\nu = 0$. When $\nu = 1$, all of the salt advected downstream in the upper layer is balanced by diffusion upstream; when $\nu = 1$ there can be no advective flux of salt upstream. That is, $\sigma_\ell = 0$ and the estuary is sectionally homogeneous with only one layer and $\bar{u}_u/U_f = 1$. At the other extreme $\nu = 0$. When $\nu = 0$, downstream longitudinal advection in the upper layer is balanced by longitudinal advection (and vertical advection and mixing) upstream in the lower layer. In other words, when $\nu = 0$

$$\frac{\bar{u}_u}{U_f} = \left(\frac{f}{f-1}\right) \frac{1}{\frac{\delta s}{s_\ell}}$$

Suspended Sediment Budget

The flux of suspended sediment is given by

$$\overline{\text{Flux}}_{ss} = \bar{u}_u(\sigma-\sigma_\ell)\bar{c}_u + \bar{u}_\ell\sigma_\ell c_\ell + \nu*(\sigma-\sigma_\ell)\bar{u}_u\bar{c}_u \tag{4}$$

From equation (4) we have that

$$\overline{\text{Flux}}_{ss} = \sigma U_f \bar{c}_\ell \left[1 - \frac{\bar{u}_u}{U_f} \left(\frac{f-1}{f} \right) \frac{\delta c}{\bar{c}_\ell} (1+\nu^*) - \nu^* \right] \tag{5}$$

Substituting for $\bar{u}_u/U_f$ from equation (3) we obtain

$$\overline{\text{Flux}}_{ss} = Q_R \, \bar{c}_\ell \left[1 - \frac{\frac{\delta c}{\bar{c}_\ell}(1+\nu^*) - \nu^*}{\frac{\delta s}{\bar{s}_\ell}(1-\nu) + \nu} \right] \gtrless 0 \tag{6}$$

It can be seen from equation (6) that the relative strengths of $\delta c/\bar{c}_\ell$ and $\delta s/\bar{s}_\ell$ determine whether there will be an import or export of suspended sediment through the mouth of the estuary. If the vertical gradient of the concentration of suspended sediment is stronger than the vertical gradient in salinity, there will be an import of sediment to the estuary from the ocean. If the vertical gradient of suspended sediment is weaker than the vertical gradient of salinity, suspended sediment will be exported from the estuary to the ocean. In Fig. 2 we have plotted the monthly mean estimates of $\delta c/\bar{c}_\ell$, $\delta s/\bar{s}_\ell$ for a station near the entrance to Chesapeake Bay. The salinity and suspended sediment data are from Schubel et al. (1970). The upper and lower layer values of $\bar{s}$ and $\bar{c}$ have been estimated from single vertical casts for temperature, conductivity, and suspended sediment, repeated at approximately monthly intervals. Monthly mean values of Q_R for the entire Bay are from the U.S. Geological Survey.

Figure 2 indicates that over the period from August 1969 through July 1970, Chesapeake Bay imported sediment from the Atlantic Ocean over more than 80% of the time. Only during the period of high riverflow in the spring of 1970 did the Bay export sediment to the Ocean.

There is some evidence from our data that suspended sediment concentrations at station 0659W in the mouth of Chesapeake Bay are influenced by local resuspension. Because resuspension increases concentrations of suspended sediment more in the lower layer than in the upper layer, any resuspensions would appear in our model as an increase in the importation of sediment to the Bay from the Ocean. We used

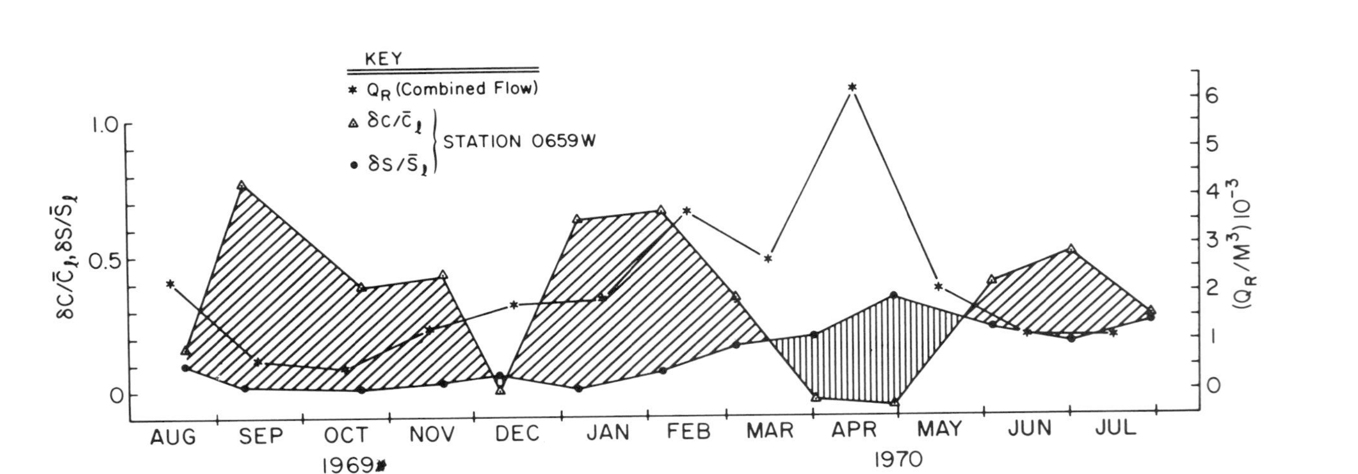

FIGURE 2. The annual cycle of $\delta s/\bar{s}_\ell$, $\delta c/\bar{c}_\ell$, and Q_R for Station 0659W, midway between Cape Charles and Cape Henry in the mouth of Chesapeake Bay. $\delta s/\bar{s}_\ell$ and $\delta c/\bar{c}_\ell$ are one-time estimates from a single vertical cast for temperature, conductivity, and suspended sediment for the day plotted; Q_R are monthly averages from the US Geological Survey.

▥ *are periods when the flux of suspended sediment is to the Ocean (>0) and*

▨ *are periods when the flux is into the estuary (<0).*

data from a station farther up the Bay to assess the sensitivity of the output of our model to resuspension near the mouth. This analysis indicated that magnitude of the net transport of suspended sediment through the mouth decreased, but that the direction of net transport was unaffected. In other words, during periods of low to moderate riverflow the Bay is a sink for fine-grained suspended sediment, and only during periods of high riverflow is it a source of sediment to the Ocean.

Filtering Efficiency

We can define the filtering efficiency of an estuary for fine-grained, fluvial suspended sediment, $Q_R C_R$, as

$$\text{Filter Efficiency} = 1 - \frac{\overline{\text{Flux}}_{ss}}{C_R Q_R} \tag{7}$$

A filtering efficiency of 0 implies that an estuary passes to the ocean an amount (mass) of suspended sediment equivalent to the total suspended sediment input from the river(s). An efficiency of 0 does not mean that all of the sediment that is discharged by the estuary to the ocean is fluvial sediment. Nor does it mean that no sediment is deposited within the estuary. There may be other marginal and internal sources of suspended sediment. A filtering efficiency of 0 implies only that the amount of sediment that leaves is equivalent to the amount the rivers bring in.

A filter efficiency of 1 indicates only that an estuary filters out and retains a quantity of sediment equal to the total amount of suspended sediment introduced by its rivers. Since there can be some substitution of sediment brought in from the shelf and since there are other sources of suspended sediment to the estuary, it does not follow that no sediment is discharged to the ocean or that none of the sediment discharged to the ocean is fluvial sediment. A filter efficiency of > 1 indicates that the estuary receives and retains a quantity of sediment greater than the total amount of suspended sediment introduced by its rivers. In the case of our model, a filtering efficiency > 1 indicates a net import of sediment to the estuary from the ocean since the only sources considered are the rivers and the ocean.

The data in Fig. 2 indicate that for the 12 month period from August 1969 through July 1970, Chesapeake Bay had a filtering efficiency of > 1 except for approximately two months during the freshet in the spring of 1970 and, except perhaps, for a brief period in early December 1969. During the 1970 spring freshet when the filtering efficiency was < 1, it was still high (0.92).

The data set that approaches most closely the requirements of our model was reported by Schubel et al. (1970) and is summarized in Fig. 2. While published data sets appropriate for calculating filtering efficiencies with our model are limited to a very small number of estuaries, we can make some estimates of the relationship between filtering efficiency and estuarine circulation type. We have summarized these estimates in Table 2, and in Fig. 3 we have plotted both filter efficiency and a measure of the strength of gravitational circulation as a function of estuarine circulation type. The similarity of the shapes of the two curves are what one would expect based on the discussion presented earlier in this paper; the agreement is striking in view of the paucity of data--or perhaps the agreement is striking because of the paucity of data. The filter efficiency curve reaches a maximum value of > 1 for partially mixed estuaries because of the importation of sediment to the estuary from the ocean. The large volume rates of flow associated with partially mixed estuaries and the relatively steep vertical gradient (increasing with depth) in suspended sediment--relative to salinity--account for the importation of sediment during periods of low to moderate river flow. The strength of gravitational circulation as represented by u_s/U_f reaches a maximum for fjord estuaries. If fjords were excluded from the analysis, the strength of gravitational circulation would be greatest for partially mixed estuaries.

SOME CLOSING OBSERVATIONS

The report prepared for UNESCO's Intergovernmental Oceanographic Commission (1983) highlighted the need to improve our understanding of estuarine filtering processes.

> "Estuaries and coastal areas trap significant quantities of material and thus act as filters between land and the oceans. It is important to quantify the capacity of these sedimentation basins as filters for different materials and elements. Biological processes may also play an important role in trapping and mobilization of material carried by rivers and land runoff."

The ability of estuaries to trap significant quantities of fine-grained sediment and associated particle-bound contaminants is recognized widely. The physical and geological processes that contribute to the filtering efficiency of estuaries are understood qualitatively, but as Officer (1981)

TABLE 2. Filter efficiency as a function of estuarine circulation type.

Type[a]	*Description*	$\frac{\delta s}{s_\ell}$	$\frac{\delta c}{c_\ell}$	ν	ν^*	$\overline{Flux}_{ss}$	*Filter efficiency*[b]
4	Salt wedge[c]	1	0.3	0	0[d]	$+0.7\ \bar{c}_\ell Q_R$	$1-0.7\ \bar{c}_\ell/C_R \cong 0.65$[e]
3	Fjord	1	?	0	0[d]	?	1[f]
2	Partially-mixed[g] annual average	0.14	0.35	0.1	0[d]	$-0.55\ \bar{c}_\ell Q_R$	$1-0.55\ \bar{c}_\ell/C_R \cong 1.19$
1	Well-mixed	0	0	1	0[d]	$+\bar{c}_\ell Q_R$	0[h]

a Hansen and Rattray (1966)

b *Filter Efficiency* $\equiv 1-(\overline{\text{Flux}}_{ss}/C_R Q_R)$

c Meade (1972)

d Assumes $\partial c/\partial x \cong 0$

e Assumes $\bar{c}_\ell/C_R \cong 0.5$

f Estimate

g $\bar{c}_\ell$ is estimated at 11.6 mg/ℓ from Schubel et al. (1970); C_R is estimated at 34 mg/ℓ from Schubel and Carter (1975)

h Assumes $\bar{c}_\ell \cong C_R$

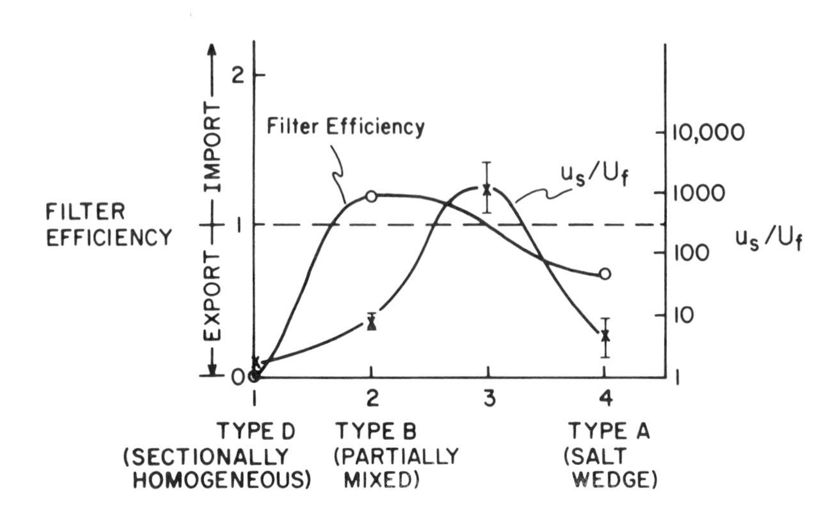

FIGURE 3. Filter efficiency and gravitational circulation plotted as a function of estuarine circulation type. Filter efficiency = 1- ($\overline{Flux_{ss}}/C_R Q_R$). The term u_s/U_f is a measure of the strength of the gravitational circulation and has been estimated from Fig. 2 of Hansen and Rattray (1966). Estuarine types 1-4 from Hansey and Rattray (1966); A,B, and D from Pritchard (1955).

points out ..."in many cases we do not understand the processes, or the critical parameters affecting them sufficiently well to be able to give adequate quantitative descriptions." Officer (1981) summarized the needs for information and knowledge regarding estuarine sedimentation when he stated:

> "There is a need for theoretical developments and scientific numerical modeling of some of the more complex defining equations. There is a corresponding need for field observations aimed at understanding specific processes such as bottom source conditions or tidal transport."

There are other more mundane, but no less important data needs. These include routine observations of the concentrations and composition of suspended sediment in the mouths of the world's major rivers, and monthly, or at least seasonal, observations of temperature, salinity, and concentration of suspended sediment throughout the water column at a series of axial stations extending over the entire length of important estuaries.

The filtering efficiency of estuaries for fine-grained sediments and for associated particle-bound contaminants has major implications for management of these important coastal environments to ensure the multiple uses that society wishes to make of them. Since estuaries typically have lifetimes of a few thousands of years, however, estuarine filtering probably has little impact on long-term sedimentation patterns in the ocean or on long-term oceanic geochemical cycles.

ACKNOWLEDGMENTS

We thank James Liu for assistance in obtaining suspended sediment concentration data, and Connie Hof for typing the manuscript. Contribution No. 415 of the Marine Sciences Research Center of the State University of New York.

REFERENCES CITED

Allen, G.P. 1973. Etude des processus sédimentaires dans l'estuaire de la Gironde. *Mémoires de l'Institute de Géologie du Bassin d'Aquitaine*, No. 5.

Allen, G.P. and P. Castaing. 1973. Suspended sediment transport from the Gironde estuary (France) onto the adjacent continental shelf. *Mar. Geol.* *14*:47-53.

Biggs, R.B. and B.A. Howell. 1984. The estuary as a sediment trap: Alternate approaches to estimating its filtering efficiency,pp. 107-129. *In*: V.S. Kennedy (ed.), *The Estuary as a Filter*. Academic Press, New York.

Carter, H.H., T.O. Najarian, D.W. Pritchard and R.E. Wilson. 1979. The dynamics of motion in estuaries and other coastal water bodies. *Reviews of Geophysics and Space Physics* *17*:1585-1590.

Cassie, W.F., J.R. Simpson, J.H. Allen and D.G. Hall. 1962. Final Report. Hydraulic and Sediment Survey of the Estuary of the River Tyne. Univ. of Durham, Kings College, Newcastle upon Tyne, Bull. 24.

Conomos, T.J. and D.H. Peterson. 1976. Suspended-particle transport and circulation in San Francisco Bay: An overview, pp.48-62. *In*: M.L. Wiley (ed.), *Estuarine Processes, Vol. 2*, Academic Press, New York.

Dyer, K.R. 1973. *Estuaries: A Physical Introduction*. John Wiley & Sons, New York. 140 pp.

Elliott, A.J. 1976. A study of the effect of meteorological forcing on the circulation in the Potomac estuary. Chesapeake Bay Inst., The Johns Hopkins Univ., Special Report No. 55, Ref. 76-8, 35 pp.

Elliott, A.J. 1978. Observations of the meteorologically induced circulation in the Potomac estuary. *Estuar. Coastal Mar. Sci., 6*:285-299.

Elliott, A.J., D-P. Wang and D.W. Pritchard. 1978. The circulation near the head of Chesapeake Bay. *J. Mar. Res. 36*:643-655.

Festa, J.A. and D.V. Hansen. 1976. A two-dimensional numerical model of estuarine circulation: The effects of altering depth and river discharge. *Estuar. Coastal Mar. Sci. 4*:309-323.

Festa, J.A. and D.V. Hansen. 1978. Turbidity maxima in partially mixed estuaries: A two-dimensional numerical model. *Estuar. Coastal Mar. Sci. 7*:347-359.

Flemming, G. 1970. Sediment balance of Clyde estuary. *J. Hydraulics Div. Proc. Amer. Soc. Civil Eng. 96*:2219-2230.

Gallene, B. 1974. Study of fine material in suspension in the estuary of the Loire and its dynamic grading. *Estuar. Coastal Mar. Sci. 2*:261-272.

Glangeaud, L. 1938. Transport et sédimentation dans l'estuaire et á l'embouchure de la Gironde. Caracteres petrographiques des formations fluviatiles, saumatres, littorales, et neritiques: *Soc. Géol. France Bull., ser. 5., 8*:599-630.

Hansen, D.V. and M. Rattray. 1966. New dimensions in estuary classification. *Limnol. Oceanogr. 11*:319-326.

Inglis, C.C. and F.H. Allen. 1957. The regimen of the Thames estuary as affected by currents, salinities and river flow. *Proc. Inst. Civil Eng. 7*:827-878.

Intergovernmental Oceanographic Commission. 1983. Ocean science for the year 2000. Report of an Expert Consultation Organized by SCOR/ACMRR with the support of IOC and the Division of Marine Sciences of UNESCO. UNESCO, Paris, France, 124 pp.

Ippen, A.T. 1966. Sedimentation in estuaries, pp.648-672. *In*: A.T. Ippen (ed.), *Estuary and Coastline Hydrodynamics*. McGraw Hill, New York.

Kranck, K. 1984. The role of flocculation in the filtering of particulate matter in estuaries, pp.159-175. *In*: V.S. Kennedy (ed.), *The Estuary as a Filter*. Academic Press, New York.

Lüneburg, H. 1939. Hydrochemische Untersuchungen in der Elbmundung Mittels Elektrokolorimeter. *Arch. Deutsch Seewarte 59*:1-27.

Meade, R.H. 1972. Transport and deposition of sediments in estuaries, pp.91-120. *In*: B.W. Nelson (ed.) *Environmental Framework of Coastal Plain Estuaries*, Memoir 133, Geol. Soc. of America, Boulder, CO.

Nelson, B.W. 1959. Transportation of colloidal sediment in the fresh water marine transition zone (abstract), pp.640-641. *In*: Mary Sears (ed.), *1st Internat. Oceanog. Cong. preprints*. Am. Assoc. Adv. Sci., Washington, DC.

Nichols, M.M. 1974. Development of the turbidity maximum in the Rappahannock estuary, Summary. *Mémoires de l'Institut de Géologie du Bassin d'Aquitaine 7*:19-25.

Nichols, M.M. and G. Poor. 1967. Sediment transport in a coastal plain estuary. *Proc. Amer. Soc. Civil Eng. 93(WW4)*:83-95.

Officer, C.B. 1976. *Physical Oceanography of Estuaries (and Associated Coastal Waters)*. John Wiley & Sons, New York. 465 pp.

Officer, C.B. 1980a. Discussion of the turbidity maximum in partially mixed estuaries. *Estuar. Coastal Mar. Sci. 10*:239-246.

Officer, C.B. 1980b. Box models revisited, pp.65-114. *In*: P. Hamilton and K.B. MacDonald (eds.), *Estuarine and Wetland Processes*. Plenum Press, New York.

Officer, C.B. 1981. Physical dynamics of estuarine suspended sediments. *Mar. Geol. 40*:1-14.

Officer, C.B. and M.M. Nichols. 1980. Box-model application to a study of suspended sediment distributions and fluxes in partially mixed estuaries, pp.329-340. *In*: V.S. Kennedy (ed.), *Estuarine Perspectives*. Academic Press, New York.

Oostdam, B.L. and R.R. Jordan. 1972. Suspended sediment transport in Delaware Bay, pp.143-150. *In*: B.W. Nelson (ed.), *Environmental Framework of Coastal Plain Estuaries*. Memoir 133, Geol. Soc. of America, Boulder, CO.

Postma, H. 1967. Sediment transport and sedimentation in the estuarine environment, pp.158-179. *In*: G.H. Lauff (ed.), *Estuaries*. Amer. Assoc. Adv. Sci. Pub. No. 83, Washington, DC.

Postma, H. and K. Kalle. 1955. Die Entstehung von Trübungszonen im Unterlauf der Flüsse, Speziell im Hinblick auf die Verhältnisse in der Unterelbe. *Dtsch. Hydrogr. Z.* *8*:137-144.

Pritchard, D.W. 1952. Salinity distribution and circulation in the Chesapeake Bay. *J. Mar. Res.* *11*:106-123.

Pritchard, D.W. 1954. A study of the salt balance in a coastal plain estuary. *J. Mar. Res.* *13*:133-144.

Pritchard, D.W. 1955. Estuarine circulation patterns. *Proc. Amer. Soc. Civil Engrs.*, Separate 717, Vol. *81*:717/1-717/11.

Pritchard, D.W. 1956. The dynamic structure of a coastal plain estuary. *J. Mar. Res.* *15*:33-42.

Pritchard, D.W. 1967. What is an estuary: Physical viewpoint, pp.3-5. *In*: G.H. Lauff (ed.), *Estuaries*, Amer. Asso. Adv. Sci. Pub. No. 83, Washington, DC.

Rattray, M., Jr. and D.V. Hansen. 1962. A similarity solution for circulation in an estuary. *J. Mar. Res.*, *20*:121-123.

Schubel, J.R. 1968a. Suspended sediment of the northern Chesapeake Bay. Chesapeake Bay Inst., The Johns Hopkins Univ., Tech. Rept. 35, Ref. 68-2, 264 pp.

Schubel, J.R. 1968b. Turbidity maximum of the northern Chesapeake Bay. *Science* *161*:1013-1015.

Schubel, J.R. 1969. Size distributions of the suspended particles of the Chesapeake Bay turbidity maximum. *Neth. J. Sea Res.* *4*:283-309.

Schubel, J.R. 1971. Tidal variations of the size variation of suspended sediment at a station in the Chesapeake Bay turbidity maximum. *Neth. J. Sea Res.* *5*:252-266.

Schubel, J.R. 1974. Effects of Tropical Storm Agnes on the suspended solids of the northern Chesapeake Bay, pp.113-132. *In*: R.J. Gibbs (ed.), *Suspended Solids in Water*, Marine Science, Vol. 4. Plenum Press, New York.

Schubel, J.R. 1983. Estuarine fine particle sediment systems: The need to know. Marine Sciences Research Center Working Paper No. 10, SUNY, Stony Brook, NY. Ref. 83-4, 33 pp.

Schubel, J.R., H.J. Bokuniewicz and R.B. Gordon. 1978. Transportation and accumulation of fine-grained sediments in the estuarine environment: Recommendations for research. Marine Sciences Research Center Spec. Rept. 14, SUNY, Stony Brook, NY. Ref. 78-2, 13 pp.

Schubel, J.R. and H.H. Carter. 1976. Suspended sediment budget for Chesapeake Bay, pp.48-62. *In*: M.L. Wiley (ed.), *Estuarine Processes, Vol. 2*. Academic Press, New York.

Schubel, J.R. and D.J. Hirschberg. 1980. Accumulation of fine-grained sediment in estuaries, pp.77-83. *In*: *River Inputs to Ocean Systems*. Proc. of SCOR Workshop, 26-30 March 1979, Rome, Italy, UNESCO, Paris. 384 pp.

Schubel, J.R., C.H. Merrow, W.B. Cronin and A. Mason. 1970. Suspended sediment data summary, August 1969 to July 1970. CBI Spec. Rept. Ref. No. 70-10, 39 pp.

Terwindt, J.H.J. 1967. Mud transport in the Dutch delta area along the adjacent coastline. *Neth. J. Sea Res.* *3*:305-331.

Wang, D-P. and A.J. Elliott. 1978. Non-tidal variability in the Chesapeake Bay and Potomac River: Evidence for non-local forcing. *J. Phys. Oceanogr.* *8*:225-232.

Weisberg, R.H. 1976. A note on estuarine mean flow estimation. *J. Mar. Res.* *34*:387-394.

Weisberg, R.H. and W. Sturges. 1976. Velocity observations in the West Passage of Narragansett Bay: A partially mixed estuary. *J. Phys. Oceanogr.* *6*:345-354.

THE ESTUARY AS A SEDIMENT TRAP: ALTERNATE APPROACHES TO ESTIMATING ITS FILTERING EFFICIENCY

Robert B. Biggs
Barbara A. Howell

College of Marine Studies
University of Delaware
Newark, Delaware

Abstract: The trapping efficiency of estuaries for particulate matter is reviewed using box models, evaluation of historical changes in bathymetry, geochronologic data, and the capacity-inflow ratio. Most of the open water estuaries of the world are more or less efficient sediment "filters". Those which are not have evolved to estuarine deltaic environments. For estuarine systems that we have evaluated, it appears that biologically mediated sedimentation processes have the capability to overwhelm all others in the deposition of fine sediments. For example, the major filter-feeding species in Delaware Bay are capable of depositing 200 times the annual fluvial input of suspended sediment.

INTRODUCTION

Emery and Uchupi (1972) have estimated that, if all of the suspended sediment discharged by rivers of the Atlantic and Gulf Coasts of the United States were deposited in the associated estuaries, then these basins would fill in 9500 years, on the average, assuming constant sea level and excluding sediments from the Mississippi River. Meade (1982) concluded that only 10% of the sediments eroded from hillslopes in the Atlantic drainage of the U.S. has actually

ISBN 0-12-405070-0

reached estuaries. The remainder is stored, more or less temporarily, in river valleys, impoundments, and flood plains. The "wave" of sediment produced in part by poor soil conservation practices in the 1920's and which is now progressing down fluvial river valleys of the eastern U.S. may increase the rate of infill of our estuaries.

Geologists sometimes state relatively simple "rules". One, not stated elsewhere, is "There must be space (volume) in which a sedimentary lithosome may accumulate." Contemporaneous estuarine sediments may accumulate vertically to highest high tide; above that elevation, fluvial or aeolian lithosomes occur. Estuarine sediments may occur longitudinally to the mouth of the estuary; beyond it, contemporaneous sediments are marine, though, as with fluvial deposits, the gradation may be difficult to recognize. As rivers and the ocean deliver materials to the estuary, particles can accumulate and biochemical processes can create solids from dissolved components. These materials may be deposited in the bottom of the estuary, further reducing the space available for deposition (Roy, Thom and Wright 1980). At maturity, the estuary can be filled with intertidal deposits with little remaining space for estuarine materials. Finally, at old age, the former estuary has been overlain with flood deposits and fluvial materials are discharged directly to the sea.

When the quantity of material discharged exceeds the competence of coastal energy processes to distribute it, a delta may form and prograde across a continental shelf extending the life of the estuary by projecting the new land and enclosed estuarine waters longitudinally out to sea. Alternately, the estuarine environment may extend vertically upward, through a relative sea level rise. As long as the rate of sea level rise exceeds the average rate of sedimentation in an estuary, then the estuary will deepen (that is, provide a net increase in space with time); when sedimentation and relative sea level rise are equal, the estuary will be maintained; and when sedimentation exceeds sea level rise, the estuary will age.

We are at a time in Holocene history when the long term rate of relative sea level rise has decreased or approached zero, even though coastlines are still retreating. Sediment yield from some watersheds has intensified, in part due to man's activities. Rates of sedimentation frequently exceed the rate of sea level rise, causing generalized estuarine shoaling. Simultaneously, in commercially important waterways, deeper channels are required to handle modern ships.

It is the purpose of this paper to review the present approaches taken to estimate the "filtering" efficiency of estuarine systems for particulate matter, to review the processes that have been linked to the filtering capacity, and to suggest both a new approach to and a quantification of a potentially important process, namely, biologically mediated sedimentation, that occurs in estuaries.

SEDIMENT TRAPPING EFFICIENCY

Various methods have been used to estimate the trapping efficiency of estuarine systems. Box models have been developed to measure or estimate all of the inputs and outputs of an estuarine system. Internal sedimentation rates are computed using radiometric, pollen, geochemical or other event markers or chronologic indicators.

Box Models

Officer (1980) has reviewed box model methodology in considerable detail. The estuary is segmented longitudinally, or longitudinally and vertically if a strong pycnocline is present. The number of segments is usually limited by the availability of salinity data, because salinity is used as a natural tracer and the salt continuity equation is used to determine both advective and nonadvective exchange coefficients between adjacent boxes. In estuaries, box models have been used to estimate ammonium and nitrate fluxes (Taft, Elliott and Taylor 1978), silica fluxes (Peterson, Festa and Conomos 1978; Rattray and Officer 1979) and suspended sediment fluxes (Schubel and Carter 1976; Officer and Nichols 1980).

Box models have been used to compute the sediment trapping efficiency of estuaries by measurement or estimation of sources (inputs) and sinks (outputs) of suspended material, usually over an annual cycle. Biggs (1970) used a box model to estimate the trapping efficiency of Chesapeake Bay for sediment generated from the Susquehanna River, from shore erosion and primary production in the system, and from seaward sources during the three year interval 1966-68. He concluded that 90% of the sediment input is trapped in the northern third of the estuary. Through box modelling, he could not ascertain where the sediment is deposited in the estuary, or predict the trapping efficiency of the system for conditions outside of the observed range.

Low frequency, high intensity storms can affect estuaries because runoff from extreme precipitation changes estuarine circulation patterns, reduces salinity, and is capable of transporting materials long stored in floodplains (Meade 1982). Gross et al. (1978), examining the sediment discharge from the Susquehanna to the Chesapeake during the period 1966-1976, found that about 50 x 10^6 tons of suspended sediment were discharged, and that 80% of that total was discharged in ten days (0.3% of the decade-long record), during two extreme events, Tropical Storm Agnes (1972) and Hurricane Eloise (1975). Because his research was conducted before these extreme events occurred, Biggs (1970) underestimated the long term river input of sediment to northern Chesapeake Bay by a factor of five and therefore misestimated the mass sedimentation rate or trapping efficiency, or both (Biggs and Cronin 1981).

Boon (1978) and Pritchard and Schubel (1981) have demonstrated the technical difficulties of attempting flux measurements through a given estuarine cross-section. The uncertainty, even for salinity flux, arises primarily because of the large temporal and spatial variance of concentration, not from analytical inaccuracy. Box modellers, as a rule, avoid measurements at the estuary mouth because of the complex circulation patterns, the wide cross section that must be sampled, and the resulting large error potential. Sediment entering the estuary from the sea is usually coarse grained material transported by longshore and coastal currents. This coarse grained material moves close to the bottom, usually as bed load, and its flux is extremely difficult to quantify. We are aware of no box models that address the flux of bed load materials at estuarine mouths.

Bathymetric Change

While the use of box models is an attempt to predict mass and volume of sediment accumulated by measuring inputs and outputs, the measurement of change in bathymetry of an estuary involves an attempt to estimate mass and volume of accumulated sediment, comparing the amount accumulated with the mass of the inputs. The ratio of accumulated mass to input mass is the trapping efficiency (expressed as a decimal fraction).

Bathymetric surveys, separated in time by years, decades or centuries, are compared by contrasting depth contours, cross-sectional areas along longitudinal sections, or mean cell depths on a user-defined grid system (Sallinger, Goldsmith and Sutton 1975). The use of this technique can have several advantages. When the interval between surveys is large (50-100 years), the effects of extreme events, involving either erosion or deposition, are included in the net rate of change which has occurred. Second, because bathymetric surveys almost always have a high density of sampling points, patterns of erosion or deposition throughout the region under study can be estimated.

Considerable data manipulation is required to rectify two bathymetric data sets separated in time by decades. Both must be brought to the same mean low water datum by correcting for eustatic sea level changes, crustal (tectonic) changes, and seasonal tidal variations. If possible, one must correct the bathymetry for muddy sediment compaction, especially if the time interval is long (50-100 years). Finally, one has to hope that the sounding and location methods are comparable, or at least quantifiable, in terms of errors.

Bathymetric comparisons have been used on scales as small as 1 km^2 (Demarest 1978) and as large as Chesapeake Bay (Kerhim et al. 1982; Byrne, Hobbs and Carron 1982). Most authors assume an error estimate of 0.3 m for two soundings from the same location and for water depths less than 20 m (Sallinger et al. 1975). Byrne et al. (1982) have estimated the standard deviation of individual soundings at a given location for their study of the southern Chesapeake as ±0.57 m. The 95% confidence interval between co-located individual depths, on separate surveys, is ±1.12 m. Thus, when individual co-located points are compared, the bathymetric change for the southern Chesapeake must be greater than 1 m to sense, statistically, a depth change. By pooling a number of depth measurements into an average for a cell, Byrne et al. (1982) were able to infer a reduction in error by some undetermined amount.

The major problem with using change in bathymetry as a tool for measuring time rate of sediment accumulation is that the error term, if it is generally as large (±1.1 m) as that found by Byrne et al. (1982), is of the same magnitude as the expected average sedimentation rate. Rusnak (1967) has reasoned that the average rate of net sedimentation in estuaries cannot exceed the rate of sea level rise over the last 10,000 to 15,000 years. That rate of sea level rise, 0.6 m per 100 years, plus or minus subsidence or uplift, represents the maximum sustainable sedimentation rate.

Sediment Geochronology

Sediment cores from estuaries may contain datable chronologic horizons. The presence of anthropogenic materials, like coal from mining, (Ryan 1953), anthropogenic lead (Murozumi, Chow and Patterson 1969; Edgington and Robbins 1976), and man-made organic materials like Kepone (Nichols and Cutshall 1981) has been used as time horizons in sediments. Pollen assemblages, particularly oak and ragweed ratios, have been found to be valuable event markers (Newman and Munsart 1968; Brush et al. 1978). All of these techniques can establish chronology in sediments deposited in the last 300 years or less. Some (pollen and ^{14}C) can be extended back in time to the pre-Holocene.

Some geochemists have argued that chronologies established by any one of these techniques are ambiguous, principally because of potential bioturbation of the sediment layers. A plot of concentration (or a derivative of concentration) vs. depth may be interpreted as reflecting sedimentation or sedimentation plus bioturbation. Core radiographs and/or some radioisotope (^{137}Cs and others) analyses can be useful to assess the degree to which the sediments in a core have been turbated (Benninger et al. 1979). Officer and Lynch (1982) have developed a formulation for the determination of sedimentation rate, diffusion coefficient(s), and mixed layer thickness from observed ^{137}Cs and ^{210}Pb measurements. The problem with establishing a core chronology is its high cost. Most investigators have taken a few cores and have interpreted sediment fluxes and changes in other preserved parameters for the core sites (Goldberg et al. 1978). Seldom have there been data from an adequate number of core sites so that the mass rate of accumulation for the estuary could be computed. Officer et al. (1984) have used the radiochronological data from 23 cores to estimate the rate of accumulation of Chesapeake Bay sediments. They have found that the northern Chesapeake accumulates slightly more than 100% of the material entering the Bay, including extreme events.

In Table 1, we have summarized the sediment trapping characteristics of several major estuaries where investigators have utilized one or more of the methods described previously. From all of the estimates on open water estuaries of even moderate depth and length, it appears that the estuary behaves as a more or less efficient sediment trap, especially under average conditions.

Capacity Inflow Ratio

A considerable body of empirical knowledge has been developed for man-made freshwater impoundments and their ability to trap sediments (Linsley and Franzini 1972). Brune (1953) defined a parameter, the capacity-inflow ratio, to predict the useful life of a reservoir before it filled with sediment. The fraction of inflowing sediment that is trapped in a reservoir is a function of the ratio of reservoir water volume or capacity (C) to total water inflow (I). Figure 1 illustrates the C/I ratio for over forty impoundments,

TABLE 1: Selected estuarine trapping efficiency in eastern Atlantic and Gulf coasts of the United States

Location	*Sediment Trapped*	*Method*	*Notes*	*References*
Chesapeake Bay				
Northern third	90%	Box model	Normal conditions	Biggs (1970)
Entire bay	100%	Box model	Normal conditions, import from ocean	Schubel and Carter (1976)
Northern third	100%	Radiochronology	Includes extreme events	Officer et al. (1984)
Maryland portion	100%	Bathymetry	Includes extreme events	Kerhin et al. (1982)
Rappahannock River	90%	Box model	Includes extreme events	Nichols (1977)
James River	100%	Box model	Normal conditions	Officer and Nichols (1980)
James River	60%	Box model	Extreme event	Officer and Nichols (1980)
Virginia portion	100%	Bathymetry	Includes extreme events, import from ocean	Byrne et al. (1982)
Savannah Estuary	100%	Bathymetry and Box model	Includes extreme events, import from ocean	Meade (1976)
Mobile Bay	70%	Bathymetry	Includes extreme events	Ryan and Goodell (1972)

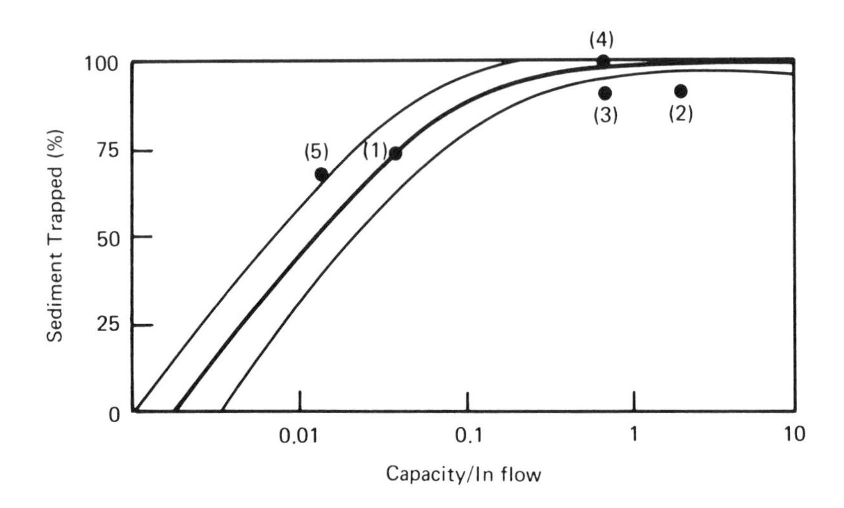

FIGURE 1. The capacity (C) volume: inflow (I) ratio of Brune (1953). The heavy line represents the best fit and the lighter lines represent the envelope that encloses the C/I ratio of 40 impoundments whose trapping efficiency was measured. Similar data, using MLW volume for C and potential runoff for I, along with measured trapping efficiency, are presented for Chesapeake Bay (1), Rappahannock River (2), Choptank River (3), James River (4), and Mobile Bay (5).

ranging in drainage area from 0.1 Km^2 to 4.8 x 10^6 Km^2 (Brune 1953). Also plotted on the illustration is the modified C/I ratio and five independently measured trapped inflowing sediment fractions for (1) the Rappahannock estuary (90% sediment trapped; Nichols 1977), (2) Choptank estuary (88% sediment trapped; Yarbro et al. 1981), (3) the northern third of Chesapeake Bay (75% of Susquehanna River sediment trapped; Biggs 1970), (4) James River (90% sediment trapped; Nichols 1977) and (5) Mobile Bay (70% sediment trapped; Ryan and Goodell 1972). It appears as though the C/I index may grossly predict the sediment trapping ability of estuaries. In Table 2, we have computed the C/I index for a number of estuaries to illustrate that the index can vary over at least 4 orders of magnitude. Our use of mean low water (MLW) volume minimizes the effect of tidal variations between systems, and total potential freshwater inflow (annual precipitation x watershed area), instead of actual discharge, makes the ratio a conservative predictor. We expect that the ratios represent the minimum sediment trapping efficiency of estuaries.

TABLE 2. C (estuary volume): I (total potential freshwater inflow) ratio for selected U.S. estuaries. C computed from estuary area x mean depth; I computed from watershed area x annual precipitation; precipitation data (Ppt) from Raffner (1978); estuary dimensions from Nixon 1982; percent sediment trapped estimated from Brune (1953).

Estuary	*Watershed area* (Km^2)	*Ppt* (*M*)	*Low tide volume* (Km^3)	*C/I*	*Predicted trapping efficiency* (%)
Narragansett Bay, RI	4.8×10^3	1.3	2.4	0.4	95±5
Long Island Sound, NY	4.2×10^4	1.1	60.8	1.3	97±3
New York Bay	3.8×10^4	1.1	2.3	0.05	77±10
Delaware Bay	3.3×10^4	1.1	19.4	0.5	98±3
Chesapeake Bay	1.1×10^5	1.1	80.5	0.7	98±3
James River	3.7×10^4	1.1	2.5	0.7	98±3
Patuxent River	2.2×10^3	1.1	6.1	2.5	>99
Potomac River	3.8×10^4	1.1	7.5	1.8	>99
Rappahannock River	9.6×10^3	1.1	1.8	0.7	98±3
Chester River	1.6×10^3	1.1	0.5	2.0	>99
Choptank River	2.5×10^3	1.1	1.5	2.0	>99
Patapsco River	2×10^3	1.1	0.5	0.8	98±3
Pamlico Sound, NC	1.1×10^4	1.3	0.9	0.06	80±10
Apalachicola Bay, FL	4.4×10^4	1.4	0.4	0.006	30±15
Mobile Bay, AL	1.0×10^5	1.5	3.2	0.02	61±12
Barataria Bay, LA	4×10^3	1.5	0.3	0.05	76±10
San Francisco Bay, CA	1.6×10^5	0.7	2.5	0.02	61±12
Kaneohe Bay, HA	97	1.9	0.4	0.001	0

It is important to recognize that the C/I ratio is based on annual conditions but may not include effects of extreme events. Hydraulic engineers use sluicing and venting techniques to scour sediments from impoundments during floods. We suspect that the same or similar processes occur naturally during extreme events in estuarine systems. Further, regarding estuarine evolution, trapping efficiency decreases as the estuary fills with sediment over time and estuarine volume decreases, until the estuary passes essentially all of its sediment to the sea and becomes a delta.

PROCESSES THAT CAUSE PARTICLE TRAPPING

Suspended sediment processes are complex in estuarine systems. Erosion, transportation, and deposition are produced by river discharge, tidal currents, meteorological events, biological processes, and chemical reactions. Two important processes, electrochemical flocculation and gravitational circulation, have received the most attention in determining the estuarine characteristic of trapping suspended matter.

Estuarine circulation results when seaward-moving fresh water flows over salt water, creating a residual upstream flow in the bottom water. In partially mixed estuaries, the landward-moving bottom flow is sufficiently intense to move sediment up the estuary, often causing the concentration of suspended sediment to be greater at the tip of the salt intrusion, and creating a turbidity maximum in which the turbidity levels exceed those of the proximal river or estuary. In this manner, estuarine circulation acts as a sediment trap for particulates. Intimately associated with turbidity maxima is the process of flocculation. Field observations reveal striking changes in suspended particle size as river water is slightly diluted with salt water (Whitehouse, Jeffery and Debgrecht 1960; Kranck 1981; Gibbs, Konwar and Terchunian 1982). However, there is still considerable controversy over whether flocculation is the principal cause (Schubel 1982). Here we focus on biologically mediated sedimentation.

Biologically Mediated Sedimentation

A wide range of biological sediment interactions occur in estuarine and coastal environments. The scale of sediment-animal interaction ranges from microscopic to macroscopic and includes both chemical and physical changes in the sediment, both in suspension and on the bottom. The extent of interaction depends on several factors, including the population density of organisms, size of the organisms, taxonomic peculiarities such as feeding, filtering and egestion rates, seasonal variations in activity, sediment composition, and in the case of suspended material, sediment concentration (Aller 1982; Rhoads and Boyer 1982).

A number of investigators have quantified the interactions between estuarine organisms and their surrounding environment. Rhoads (1974) outlined the modification of sedimentary characteristics by benthic organisms including the effects they have on sediment transport, chemical reactions, particle aggregation, and nutrient recycling. Haven and Morales-Alamo (1966) discussed the rates of, and factors influencing, bio-deposition by oysters and other filter feeders. Pryor (1975) investigated the role of filter feeders in the rate of sedimentation and their relationship to the process of glauconization. Glauconization involves the pelletization of argillaceous material which, upon deposition, undergoes ionic fixation and leads to the adsorption of Fe and K and eventually to the formation of a hydrous K, Fe, silicate (gluconite) pellet. Zabawa (1982) showed that bacteria affect the rate of deposition by attaching to suspended solids and secreting a mucal slime webbing of a sticky polysaccharide that not only holds particles together, but traps isolated mineral grains as they collide with the agglomerated particle. Aller (1982) reported on physical and chemical changes taking place in the water overlying sediments due to the biogenic activities of benthic organisms.

The purpose of this section is to evaluate the hypothesis that biologically mediated sedimentation processes are as important, if not more so, than the mechanical or physical processes which lead to the deposition of fine sediments.

TABLE 3. Average biological populations, by numbers, for typical northeast Atlantic estuaries

	Average zooplankton[a] m^{-3}	*Average benthos*[b] m^{-2}
Delaware Bay	1.0×10^2	200
Chesapeake Bay	1.0×10^4	14,000
Narragansett Bay	3.0×10^4	25,000
Long Island Sound	3.0×10^4	16,000

[a]*Data from Nixon (1982);* [b]*Data from Watling and Maurer (1975).*

Delaware Bay will be used as an example to show that, despite averaging 3 - 335 times fewer organisms (Watling and Maurer 1975) than most other northeast Atlantic estuaries (see Table 3), the Bay's sparse biotic populations can affect sedimentation processes.

Suspension feeders actively or passively entrap suspended seston on ciliated tentacles, mucus nets or on ciliated and mucus covered respiratory surfaces (Jørgensen 1966). A major and well documented effect of filter feeding organisms is the packaging or pelletization of fine-grained material into agglomerated fecal pellets and pseudofeces which exhibit substantially greater settling rates than their composite particles (Haven and Morales-Alamo 1966, 1968; Schubel 1971; Johnson 1974; Rhoads 1974; Pryor 1975; Honjo and Roman 1978; Prahl and Carpenter 1979; Black 1980). An obvious accompaniment to this effect is the deposition of fine particles in areas which would not otherwise experience accumulations of fine-grained material. There are a variety of organisms which remove sediment from suspension by the process of filter feeding, including tunicates, molluscs, barnacles, and copepods (Haven and Morales-Alamo 1968; Rhoads 1974).

Temperate estuaries are subjected to hot summers and cold winters requiring their inhabitant organisms to undergo seasonal metabolic adjustments which, in turn, directly affect the organisms' feeding and filtration rates. During periods of extreme cold or heat, ±10-15^{0}C of their optimum temperature, organisms will reduce their filtering rates to a minimum, or totally "shut down". In general, filtration rates are dependent upon temperature, species, type of food available, concentration of food and suspended material in the water column, and in some cases, sex and life cycle stage (Jørgensen 1966; Mullin 1969). Obviously, to consider all of these factors for each of the Bay's populations would be tedious and almost impossible. Therefore, we have used ranges of filtration values and averaged the rate of fecal pellet deposition to give an estimate, but not a quantitative value, of biological influence on sedimentation processes in estuarine systems.

In Delaware Bay, an average yearly zooplankton standing crop consists primarily of *Acartia tonsa*, *Temora longicornus*, *Centrophages hatamus* and *Pseudocalanus minutus* (Watling and Maurer 1975). By determining the annual weighted populations of these zooplankters in the Bay and multiplying those numbers by the minimum and maximum filtration rates recorded for each group, we have estimated that the total zooplankton population has the capability of filtering the entire volume of Delaware Bay between three to five times a year and depositing about 4 x 10^{11} Kg of fecal material (see Table 4).

TABLE 4. Estimated filtration rate and fecal pellet production for Delaware Bay[a]

ZOOPLANKTON *Species*	*Abundance*[b] *(No. m^{-3})*	*Filtration*[c] *rate (ml d^{-1})*	*Population filtration rate (L y^{-1})* *(min)*	*(max)*	*Fecal pellet*[d] *production (Kg y^{-1})*
Acartia tonsa	2.3×10^{2}	5-10	0.9×10^{13}	1.7×10^{13}	1.0×10^{11}
Temora longicornis	-	12-20			
Centrophages hatamus	8.0×10^{2}	8-10	5.2×10^{13}	7.6×10^{13}	3.0×10^{11}
Pseudocalanus minutus	-	9			
Combined effect of zooplankton	-	-	6.1×10^{13}	9.3×10^{13}	4.0×10^{11}
BIVALVES	*Population size*	*(L h^{-1})*	*(L y^{-1})* *(min)*	*(max)*	*(Kg y^{-1})*
Crassostrea virginica	6.9×10^{7}[e]	0.4-0.8[c]	2.8×10^{11}	5.5×10^{11}	5.3×10^{6}[f]
Mytilus edulis	2.0×10^{4}[a]	0.5-1.5[c]	8.8×10^{7}	2.6×10^{8}	-

[a] *Population filtration rates were calculated by multiplying the average number of organisms times the low (min) and high (max) estimates of the filtration rate. Fecal pellet production was determined similarly by multiplying average pellet production rates times number of organisms to derive a yearly production value.* [b] *Watling and Maurer (1975);* [c] *Jorgensen (1966);* [d] *Honjo and Roman (1978);* [e] *Maurer et al. (1971);* [f] *Haven and Morales-Alamo (1966).*

These estimates are impressive; however, Rhoads (1974) states that a large population of filter feeding zooplankton has the capability of filtering the entire volume of its estuary every few weeks, and Schubel (1971) estimated that the zooplankton assemblages of Chesapeake Bay, dominated by copepods, could filter a volume of water equivalent to the entire Bay in a few days, or almost 24 times faster than the rate computed for the Delaware Bay organisms. Narragansett Bay and Long Island Sound have similar zooplankton populations in kind and slightly greater in number as Chesapeake Bay (Nixon 1982), therefore, it would be expected that these estuaries would be subjected to similar filtration rates as is the Chesapeake. These rates and subsequent deposition have remarkable implications when considering the volume of material deposited.

The American oyster, *Crassostrea virginica*, has been the subject of a number of filtration rate and biodeposition studies (Lund 1957; Haven and Morales-Alamo 1966; Jørgensen 1966). It is an efficient filter feeder and feeds between 10-14 h a day at a rate of 0.4-0.8 $L\ h^{-1}$ (Jørgensen 1966), and deposits between 0.98-1.56 grams of fecal material per animal per week (Haven and Morales-Alamo 1966). In Delaware Bay, the oyster population has been severely depleted in the last 50 years due to overfishing and disease (Maurer, Watling and Keck 1971). However, even in its present reduced status, the oyster population, consisting of about 6.9×10^7 individuals (Maurer et al. 1971) filtering at the above mentioned rates, can filter between $2.8\text{-}5.5 \times 10^{11}\ L\ yr^{-1}$ and deposit more than 5.2×10^6 Kg of fecal material.

Another common filter feeding bivalve of northeast Atlantic estuaries is the blue mussel, *Mytilus edulis*, with densities as great as several hundred organisms per m^2 (Haven and Morales-Alamo 1966). Under normal conditions (12-20^{0}C) they will filter 97-99% of the time at a rate of 0.5-1.5 $L\ h^{-1}$ (Jørgensen 1966). Delaware Bay's total mussel population is estimated to be 20,000 individuals (Watling and Maurer 1975) and filters about $3.5 \times 10^7\ L\ yr^{-1}$. The mussels' filtrating capabilities are enormous and can have a substantial effect on sedimentation rates in bays where their numbers are greater than those seen in Delaware Bay.

Delaware Bay represents the low end of the scale of biologically affected estuaries due to its low density of organisms. It is safe to assume that any process of sedimentation which is a result of biological intervention can be occurring at least three-fold in any other North American northeast Atlantic estuary, due to the substantially greater number of organisms compared to Delaware Bay. Therefore, when investigating the processes of sedimentation in any given estuary, one must consider the effects of biological activities on suspended matter. The annual rate of input of fluvial sediment to Delaware Bay is about 1.4×10^9 Kg and between the zooplankton, oysters and mussels they fix about 4×10^{11} Kg yr^{-1} or 200 times the annual input of fluvial material. Clearly, most of this material has been resuspended and both gravitational circulation and flocculation are important, or even dominant, processes occurring in some estuarine reaches. Nevertheless, this analysis provides convincing evidence that biologically mediated sedimentation processes are a principal mechanism for trapping sedimentary materials in the estuary.

A CONCEPTUAL MODEL FOR SEDIMENT TRAPPING IN ESTUARIES

Figure 2 illustrates a conceptual model for sediment trapping in an estuary. Dominance by the fluvial endmember to the extent that net fluvial transport extends beyond the estuary mouth results in incomplete trapping of land-derived sediment and escape of that sediment to sea, with or without the formation of a prograding delta outside of the estuary mouth. The Mississippi is an example of a system in which there is a large net annual loss of fluvial material from the estuary to the shelf and the creation of a delta (Fig. 2A). Mobile Bay (Fig. 2B) exemplifies a system in which periodic freshets transport fluvial material to the shelf, yet during other times fluvial material is wholly trapped in the estuary and marine derived materials enter the estuary mouth. So long as relative sea level rises, space is created for additional sediment and the estuarine sedimentary environments can move landward and upward. Flocculation and gravitational circulation trap suspended material, especially near the head of the estuary. Biologically mediated sedimentation occurs as an overprint along the entire estuarine gradient. Bedload transport in and out of the estuary mouth results in a net input of sediment from the sea. The Savannah and Chesapeake are examples of this estuarine type (Fig. 2B).

Finally, there are inverse estuaries (Rusnak 1967) in which most sediment transport is from the sea and fluvial transport is small (Fig. 2C). Coastal embayment estuaries that have relatively small fresh water discharge fit into that category.

While the conceptual model represents the axial distribution of materials and rates, the reader is cautioned that the lateral complexities of circulation and sediment sources may modify the generalized distribution presented here. For example, bottom sediment distribution in the main channel of

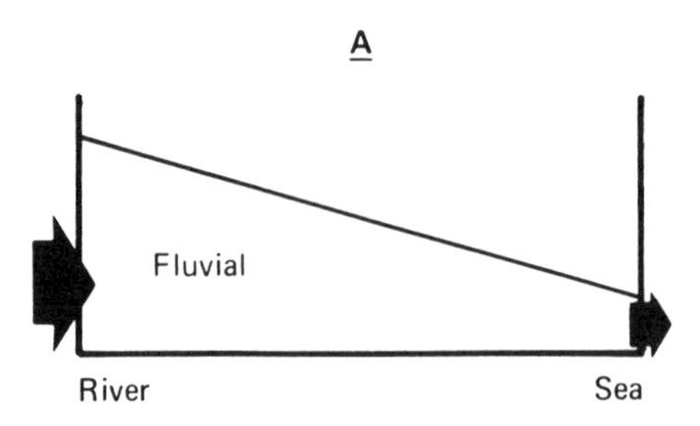

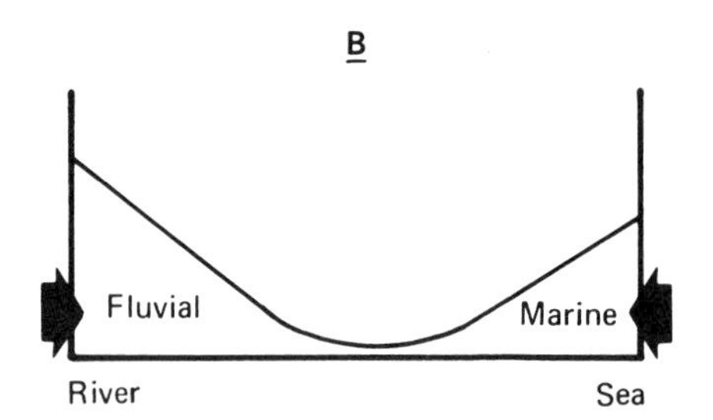

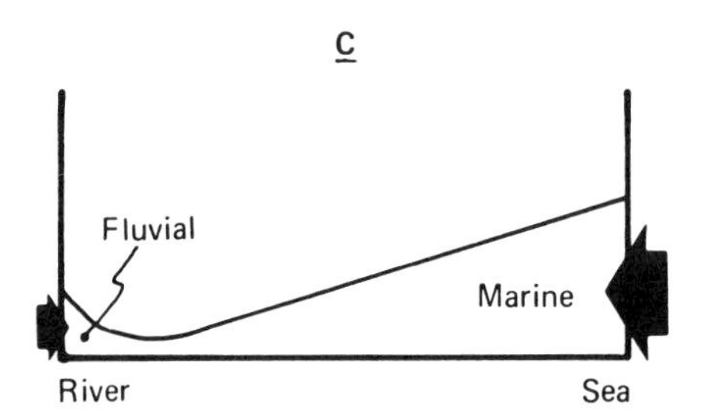

FIGURE 2. A conceptual model for the source and intensity of axial depositional processes in estuaries. (A.) A system in which fluvial material escapes seaward of the estuary, like the Mississippi. (B.) A system in which fluvial and marine sediments are both trapped in the estuary, like Chesapeake Bay. (C.) A system with relatively small fluvial input, like most coastal lagoons. Note that a system may change its character in response to periodic and aperiodic events, and that channel parallel sections on opposite sides of an individual estuary may vary from A (on the right side in the northern hemisphere) to C (on the left side in the northern hemisphere).

Delaware Bay resembles Fig. 2B, the west (State of Delaware) side is similar to Fig. 2A, while the east (State of New Jersey) side most nearly matches Fig. 2C.

CONCLUSIONS

In summary, the use of C/I ratios along with the conceptual model outlined above provide easy and fairly accurate assessments of estuarine trapping efficiency but, as in all models, these estimates should be calculated by employing dependable sampling techniques and in consideration to storm events, biological influence and estuarine circulation and hydrographical patterns.

There is sufficient evidence presented in this paper, and others (Schubel 1971; Rhoads 1974), that biologically mediated sedimentation processes are of primary importance in estuaries with substantial inhabitant organism populations and therefore must be included in predictions of sediment trapping efficiencies.

ACKNOWLEDGMENTS

This work was partially supported by grants from the U.S. Environmental Protection Agency, Cooperative Agreement #X-003256-01, the Delaware River and Bay Authority, and the Office of Sea Grant (NA80-AA-D-00106) of the U.S. National Oceanic and Atmospheric Administration.

REFERENCES CITED

Aller, R. C. 1982. The effects of macrobenthos on chemical properties of marine sediment and overlying water, pp. 53-102. *In:* P. L. McCall and M. J. Tevesz (eds.), *Animal -Sediment Relations*. Plenum Press, New York.

Anderson, F. E.; L. Black, L. M. Mayer, W. Mook and L. E. Watling. 1981. A temporal and spatial study of mudflat erosion and deposition. *J. Sed. Pet. 51*: 729-736.

Benninger, L. K., R. C. Aller, J. K. Cochran and K. K. Turekian. 1979. Effects of biological sediment mixing on the ^{210}Pb chronology and trace metal distribution in a Long Island Sound core. *Earth Planet. Sci. Lett. 43:* 241-259.

Biggs, R. B. 1970. Sources and distribution of suspended sediment in northern Chesapeake Bay. *Mar. Geol. 9:* 187-201.

Biggs, R. B. and L. E. Cronin. 1981. Special characteristics of estuaries, pp. 3-24. *In:* B. J. Neilson and L. E. Cronin (eds.), *Estuaries And Nutrients*. Humana Press, Clifton, NJ.

Black, L. 1980. The biodeposition cycle of a surface deposit-feeding bivalve, *Macoma balthia* (L.), pp. 389-402. *In*: V. S. Kennedy (ed.) *Estuarine Perspectives*. Academic Press, New York.

Boon, J. D. 1978. Suspended solids transport in a salt marsh creek-an analysis of errors, pp. 147-160. *In:* B. Kjerfve (ed.), *Estuarine Transport Processes*. University of South Carolina Press, Columbia, SC.

Brune, G. M. 1953. Trap efficiency of reservoirs. *Trans. American Geophysical Union. 34:* 407-418.

Brush, G. S., E. A. Martin, R. S. DeFries and C. A. Rice. 1982. Comparison of ^{210}Pb and pollen methods for determining rates of estuarine sediment accumulation. *Quaternary Research 18:* 196-217.

Byrne, R. J., C. H. Hobbs and M. J. Carron. 1982. Baseline sediment studies to determine distribution, physical properties, sedimentation budgets and rates in the Virginia portion of Chesapeake Bay. Draft Final Report to U.S. Environmental Protection Agency, Grant #R806001010. Virginia Inst. Mar. Sci., College of William and Mary, Gloucester Pt., VA 194 p.

Demarest, J. M. 1978. The shoaling of Breakwater Harbor, Cape Henlopen area, Delaware Bay, 1842-1971. Delaware Sea Grant Tech. Report. DEL-SG-1-78, College of Marine Studies, U. of Delaware, Newark, DE 169p.

Edgington, D.M. and J. A. Robbins. 1976. Records of lead deposition in Lake Washington sediments since 1800. *Envir. Sci. Technology 10:* 266-274.

Emery, K.O. and E. Uchupi. 1972. Western North Atlantic Ocean: Topography, rocks, structure, water, life and sediments. Amer. Assoc. Pet. Geologists Memoir 17, Tulsa, OK 532 p.

Gibbs, R.J., L. Konwar and A. Terchunian. 1983. Size of flocs suspended in Delaware Bay. *Can. J. Fish. Aquatic Sci. 40 Supp. (1):* 102-104.

Goldberg, E. D. , V. Hodge, M. Koide, J. Griffin, E. Gamble, O.P. Bricker, G. Matisoff, G. R. Holdren and R. Braun. 1978. A pollution history of Chesapeake Bay. *Geochim. Cosmochim. Acta 42:* 1413-1425.

Gross, M. G., M. Karweit, W. B. Cronin and J. R. Schubel. 1978. Suspended sediment discharge from the Susquehanna River to northern Chesapeake Bay, 1966-1976. *Estuaries 1:*106-110.

Haven, D. and R. Morales-Alamo. 1966. Aspects of biodeposition by oysters and other invertebrate filter-feeders. *Limnol. Oceanogr. 11:* 487-498.

Haven, D. S. and R. Morales-Alamo. 1968. Occurrence and transport of faecal pellets in suspension. *Sed. Geo. 2:* 141-151.

Honjo, S. and M. R. Roman. 1978. Marine copepod fecal pellets: production, presentation and sedimentation. *J. Mar. Res. 36:* 45-57.

Johnson, R. G. 1974. Particulate matter at the sediment-water interface in coastal environments. *J. Mar. Res. 32*: 313-330.

Jørgensen, C. B. 1966. *Biology of Suspension Feeding.* Pergamon Press, London. 357 pp.

Kerhin, R. T., J. P. Halka, E. L. Hennessee, P. J. Blakeslee, D. V. Wells, N. Zoltan and R. H. Cuthbertson. 1982. Physical characteristics and sediment budget for bottom sediments in the Maryland portion of Chesapeake Bay. Draft Final Report to U.S. Environmental Protection Agency, Coop. Agreement #R805965. Md. Geol. Survey, Baltimore, MD 190pp.

Kranck, K. 1981. Particulate matter, grain-size characteristics, and flocculation in a partially mixed estuary. *Sedimentology 28:* 107-114.

Linsley, R. K. and J. B. Franzini. 1972. *Water Resources Engineering.* McGraw-Hill, New York. 464 pp.

Lund, E. J. 1957. A quantitative study of clearance of a turbid medium and feeding by the oyster. *Publ. Inst. Marine Sci. Texas 4:* 296-312.

Maurer, D., L. Watling and R. Keck. 1971. The Delaware oyster industry: A reality? *Trans. Amer. Fish. Soc. 100:* 101-111.

Meade, R.H. 1976. Sediment problems in the Savannah River basin, pp. 105-129. *In:* J. R. Dillman and M. M. Stepp (eds.), *The Future of the Savannah River.* Water Resources Res. Inst., Clemson University, Clemson, SC.

Meade, R. H. 1982. Sources, sinks and storage of river sediments in the Atlantic drainage of the United States. *J. Geol. 90:* 235-252.

Murozumi, M., T. J. Chow and C. Patterson. 1969. Chemical concentration of pollutants, lead aerosols, terrestrial dusts, and sea salts in Greenland and Antarctic snow strata. *Geochim. Cosmoschim. Acta 33:* 1247-1294.

Newman, W. S. and C. A. Munsart. 1968. Holocene geology of the Wachapreague Lagoon, Eastern Shore Peninsula, Virginia. *Mar. Geol. 6:* 81-105.

Nichols, M. M. 1977. Response and recovery of an estuary following a river flood. *J. Sed. Petrol. 47:* 1171-1186.

Nichols, M. M. and N. Cutshall. 1981. Tracing Kepone in James Estuary sediments. *Rapp. P.V. Reun. Cons. Int. Explor. Mer. 181:* 102-110.

Nixon, S. W. 1982. Comparative ecology of some estuaries in the United States. Final Report to U.S.E.P.A., Chesapeake Bay Program Grant X-003259-01, Grad. School of Oceangr., University of Rhode Island, Kingston, R. I. 60 p.

Nowell, A. R. M., P. A. Jumars and J. E. Eckman. 1981. Effects of biological activity on the entrainment of marine sediments. *Mar. Geol. 42:* 133-153.

Officer, C. B. 1980. Box models revisited, pp. 65-114. *In:* P. Hamilton (ed.), *Wetlands and Estuarine Processes in Water Quality Modelling.* Plenum Press, New York.

Officer, C. B. and D. R. Lynch. 1982. Interpretation procedures for the determination of sediment parameters from time-dependent flux inputs. *Earth Planet. Sci. Lett. 61:*55-62.

Officer, C. B., D. R. Lynch, G. H. Setlock and G. R. Helz. 1984. Recent sedimentation rates in Chesapeake Bay, pp. 131-157. *In:* V. S. Kennedy (ed.), *The Estuary as a Filter*. Academic Press, New York.

Officer, C. B. and M. M. Nichols. 1980. Box model application to a study of suspended sediment distributions and fluxes in partially mixed estuaries, pp. 329-340. *In:* V. S. Kennedy (ed.), *Estuarine Perspectives*. Academic Press, New York.

Olsen, C. R., H. J. Simpson, R. F. Bopp, S. C. Williams, T. H. Peng and B. C. Deck. 1978. A geochemical analysis of the sediments and sedimentation in the Hudson estuary. *Jour. Sed. Petrol 48:* 401-418.

Peterson, D. H., J. F. Festa and T. J. Conomos. 1978. Numerical simulation of dissolved silica in San Francisco Bay. *Est. Coastal Mar. Sci. 7:* 99-116.

Prahl, F. G. and R. Carpenter. 1979. The role of zooplankton fecal pellets in the sedimentation of polycyclic aromatic hydrocarbons in Dabob Bay, Washington. *Geochim. Cosmochim. Acta. 43:* 1959-1972.

Pritchard, D. W. and J. R. Schubel. 1981. Physical and geological processes controlling nutrient levels in estuaries, pp. 47-69. *In:* B. Neilson and L. E. Cronin (eds.), *Estuaries and Nutrients*. Humana Press, Clifton, NJ.

Pryor, W. A. 1975. Biogenic sedimentation and alternation of argilaceous sediments in shallow marine environments. *Geol. Soc. Amer. Bull. 86:* 1244-54.

Rattray, M. and C. B. Officer. 1979. Distribution of a nonconservative constituent in an estuary with application to the numerical simulation of dissolved silica in the San Francisco Bay. *Est. Coastal Mar. Sci., 8:* 489-494.

Rhoads, D. C. 1974. Organism-sediment relations on the muddy sea floor. *Oceanogr. Mar. Biol. Ann. Rev. 12:* 263-300.

Rhoads, D. C. and L. F. Boyer. 1982. The effects of marine benthos on physical properties of sediments, a successional perspective, pp. 3-52. *In:* P. L. McCall and M. J. S. Tevesz (eds.) *Animal-Sediment Relations*. Plenum Press, New York.

Roy, P.S., B.G. Thom and L.D. Wright. 1980. Holocene sequences on an embayed high-energy coast: An evolutionary model. *Sedimentary Geol. 26:* 1-19.

Raffner, J. A. 1978. *Climates of the States.* Gale Co., Detroit, MI 1184pp.

Rusnak, G. A. 1967. Rates of sediment accumulation in modern estuaries, pp. 180-184. *In:* G. H. Lauff (ed.), *Estuaries.* AAAS Public. 83, Wash. D. C.

Ryan, J. D. 1953. The sediments of Chesapeake Bay. Dept. of Geol., Mines, and Water Resources Bull. 12; Md. Geol. Survey, Baltimore, MD. 120pp.

Ryan, J. J. and H. G. Goodell. 1972. Marine geology and estuarine history of Mobile Bay, Alabama. *Geol. Soc. Amer. Memoir 43:* 116-138.

Sallinger, A. H., V. Goldsmith and C. H. Sutton. 1975. Bathymetric comparisons: A manual of methodology error criteria and techniques. Special Report 66, Applied Marine Science and Engineering, Virginia Inst. Mar. Sci., Gloucester Pt., VA 34pp.

Schubel, J. R. 1971. *The Estuarine Environment.* pp. 1-29. Amer. Geol. Inst. Short Course Lecture Notes. Wash., D. C.

Schubel, J. R. 1982. An eclectic look at fine particles in the coastal ocean, pp. 53-142. *In:* J. J. Kimrey and R. M. Burns (eds.), *Pollutant Transfer by Particulates.* Proceedings of a Workshop. Seattle, WA.

Schubel, J. R. and H. H. Carter. 1976. Suspended sediment budget for Chesapeake Bay, pp. 48-62. *In:* M. L. Wiley (ed.), *Estuarine Processes, Vol. 2.* Academic Press, New York.

Schubel, J. R. and D. J. Hirschberg. 1977. Pb^{210} determined sedimentation rate and accumulation of metals in sediments at a station in Chesapeake Bay. *Chesapeake Sci. 19:* 379-382.

Taft, J.L., A. J. Elliott and W. R. Taylor. 1978. Box model analysis of Chesapeake Bay ammonium and nitrate fluxes, pp. 115-130. *In:* M. L. Wiley (ed.), *Estuarine Interactions.* Academic Press, New York.

Watling, L. and D. Maurer. 1975. Ecological studies on benthic and planktonic assemblages in lower Delaware Bay. NSF Report. College of Marine Studies, Lewes, DE. 630 pp.

Whitehouse, U.C., L. M. Jeffery and J. D. Debgrecht. 1960. Differential settling tendencies of clay minerals in saline waters, pp. 1-79. *In:* A. Swineford (ed.) *Clays and Clay Minerals, 7th Conf.* Pergamon Press, New York.

Yarbro, L. A., P. R. Carlson, R. Crump, J. Chanton, T. R. Fisher, N. Burger and W. M. Kemp. 1981. Seston dynamics and a seston budget for the Choptank River estuary in Maryland. Horn Point Environmental Lab., University of Maryland, Cambridge, MD. 223 pp.

Zabawa, C. F. 1982. Estuarine sediments and sedimentary processes in Winyah Bay, South Carolina. *Geol. Notes 23:*38pp.

RECENT SEDIMENTATION RATES IN CHESAPEAKE BAY

Charles B. Officer

Earth Sciences Department
Dartmouth College
Hanover, New Hampshire

Daniel R. Lynch

Thayer School of Engineering
Dartmouth College
Hanover, New Hampshire

George H. Setlock
George R. Helz

Department of Chemistry
University of Maryland
College Park, Maryland

Abstract: Nonlinear parameter estimation techniques were applied to obtain optimum values and confidence intervals for mass sedimentation rates in Chesapeake Bay from ^{210}Pb, ^{137}Cs and $^{239,240}Pu$ geochemical tracer profiles. The results show high sedimentation rates of around 0.3-1.2 gm $cm^{-2}yr^{-1}$ for the upper bay, modest sedimentation rates of 0.1-0.3 gm cm^{-2} yr^{-1} for the middle bay, and modest to high sedimentation rates of 0.1-0.8 gm $cm^{-2}yr^{-1}$ for the lower bay. The principal source for the upper bay is the Susquehanna River discharge and that for the lower bay is the ocean and lower bay environment. In terms of total sediment budget the results are in

ISBN 0-12-405070-0

agreement with published evidence that episodic sedimentation events are an important feature of the bay system.

INTRODUCTION

The principal purpose of this investigation was to arrive at definitive mass sedimentation rates for Chesapeake Bay from the available ^{210}Pb, ^{137}Cs and $^{239,240}Pu$ geochemical tracer data. Nonlinear parameter estimation techniques were applied to determine the optimum values and their confidence intervals.

The location of the sediment cores used in the analysis are shown in Fig. 1. Cores 914S, 858C, 856E, 834G and 747A are from Goldberg et al. (1978); core C is from Schubel and Hirschberg (1977); core E is from Hirschberg and Schubel (1979); and all other cores are from Helz et al. (1981). The ^{137}Cs and $^{239,240}Pu$ data for the Helz et al. (1981) cores were supplied by G.R. Helz. All other data used in our analysis were as listed in the respective publications.

ANALYSIS PROCEDURES

A summary of the analytic solutions for a steady state flux input of a tracer, applicable to ^{210}Pb observations, has been given by Officer (1982). The analytic solutions and numerical calculation procedures for a time dependent flux input, applicable to ^{137}Cs and $^{239,240}Pu$ observations related to the atmospheric nuclear weapons testing period, have been given by Officer and Lynch (1982). The physical model for these solutions consists of a system undergoing uniform sedimentation with an upper layer of finite thickness subject to bioturbation.

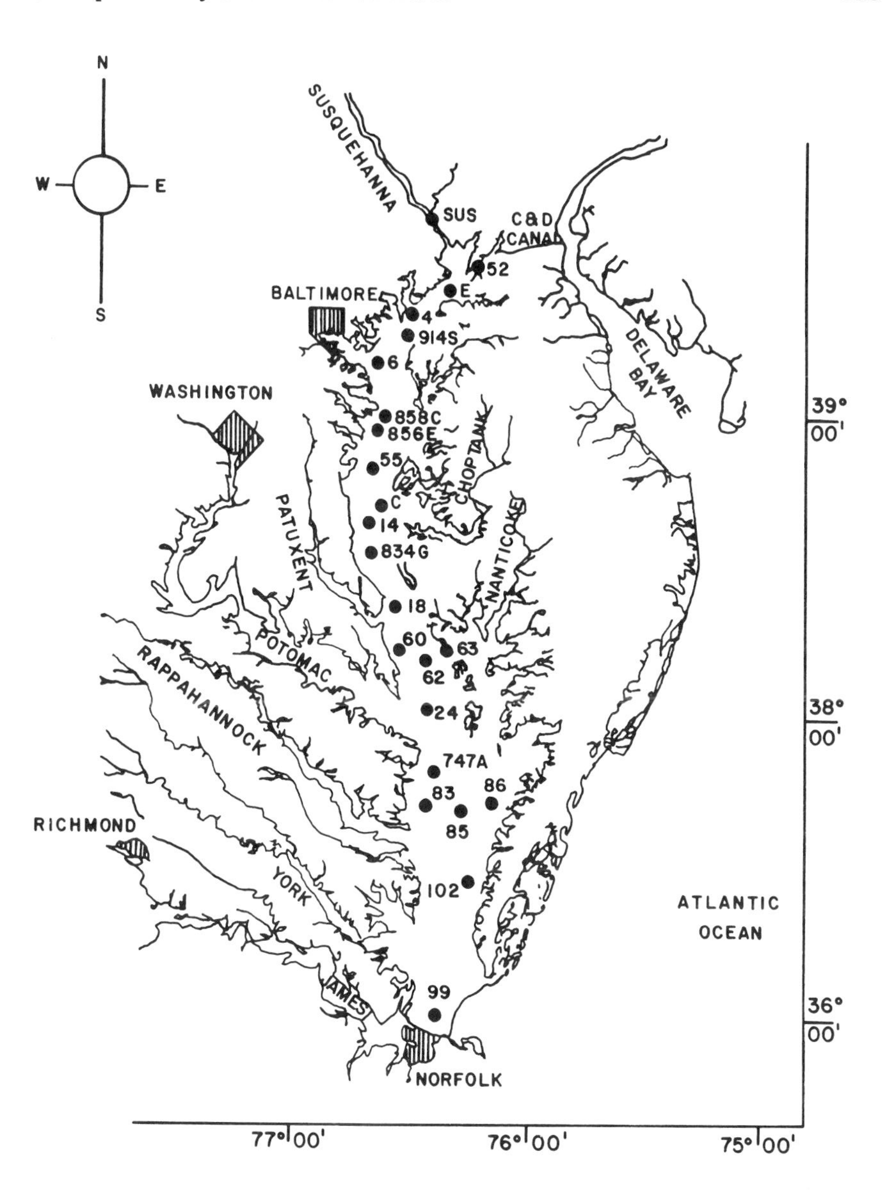

FIGURE 1. Location of Chesapeake Bay cores.

The nonlinear parameter estimation techniques for both the steady state and time dependent flux input problems have been given by Lynch and Officer (1984). In essence the techniques amount to a computer search procedure for the optimum values of sedimentation rate, bioturbation coefficient, and mixed layer depth in their three dimensional parameter space. The optimum corresponds to the parameter values for which the sum of the squared errors of theoretical minus observed concentrations is a minimum. The procedures also provide the confidence intervals for each parameter. A 90% confidence index has been used in these calculations.

In specific application to a given sediment core for either a steady state or time dependent flux input, it is possible to obtain estimates for the sedimentation rate, bioturbation coefficient, and mixed layer depth from simple interpretation procedures and to arrive at some intuitive sense of how well each of the three unknown parameters can be determined. This intuitive sense is important and depending on the problem may be sufficient in itself. For other purposes and specifically for our purpose of intercomparing results from one location to the next it is imperative to be able to arrive at the global optimum in the three dimensional parameter space and to be able to estimate the confidence intervals applicable to the optimum values. These optimization procedures have been applied by Officer and Lynch (1983) to a similar problem of determining the bioturbation coefficients and mixed layer depths appropriate to deep sea sediment cores from observations of microtektites, volcanic ash, and pumice.

As discussed by Robbins and Edgington (1975) as well as others, it is appropriate to use a mass based coordinate system rather than core depth in order to eliminate the effects of porosity variations as a function of depth. The analytic solutions given by Officer (1982) and Officer and Lynch (1982) are applicable to these conditions. The interrelations between the parameters in the two coordinate systems are simply

$$x = \int_0^z \rho_s(1 - \phi)\, dz \tag{1}$$

$$\omega = \rho_s(1 - \phi)\, v \tag{2}$$

$$E = [\rho_s(1 - \phi)]^2\, D \tag{3}$$

$$\delta = \int_0^d \rho_s(1 - \phi)\, dz \tag{4}$$

where x = total sediment particle accumulation in gm cm^{-2}, z = core depth in cm, ω = mass sedimentation rate in gm cm^{-2} yr^{-1}, v = sedimentation rate in cm yr^{-1}, E = mixing parameter in $gm^2 cm^{-4} yr^{-1}$, D = mixing coefficient in $cm^2 yr^{-1}$, δ = total sediment particle accumulation for the mixed layer in gm cm^{-2}, d = mixed layer depth in cm, ρ_s = sediment particle density in gm cm^{-3}, and ϕ = porosity. Care has to be taken in the interpretation of any calculated value for v because it includes both sedimentation and compaction effects. The true sedimentation rate, as referred to bathymetric depth changes, is given by

$$v' = \frac{\omega}{\rho_s(1 - \phi')} \tag{5}$$

where ϕ' is the ultimate, compacted porosity at depth.

Most of the tabulated data have all the information needed for our analysis, viz., observed ^{210}Pb, ^{137}Cs or $^{239,240}Pu$ concentrations, supported ^{210}Pb, and percent water content by weight as a function of core depth. The Goldberg et al. (1978) core tabulations do not include water content; values from adjacent Helz et al. (1981) cores were used. The

Hirschberg and Schubel (1979) core tabulation does not include supported ^{210}Pb levels; the supported level from the adjacent Helz et al. (1981) core 4 was used. Porosity was determined from the water content fraction, ν, by the usual relation

$$\phi = \frac{\rho_s \nu}{\rho_s \nu + \rho_w (1-\nu)} \tag{6}$$

using $\rho_s = 2.6$ gm cm^{-3} and $\rho_w = 1.0$ gm cm^{-3}.

As discussed by Officer (1982) and Carpenter, Peterson and Bennett (1982) as well as others, ^{210}Pb sedimentation rate determinations can be particularly sensitive to mixed layer effects. For example, a straight line plot on a semilogarithmic graph of excess ^{210}Pb concentration versus depth can be interpreted as either rapid sedimentation or a combination of slower sedimentation and mixing of the entire column to the depths observed. Resolution of such questions can be obtained from two information sources. First, the interpretation of ^{137}Cs or $^{239,240}Pu$ profiles does not have this indeterminancy. Second, X-radiograph examination of the cores can assess in a qualitative manner the degree of bioturbation. Fortunately, for the Chesapeake Bay cores a substantial number have either ^{137}Cs or $^{239,240}Pu$ observations in addition to the ^{210}Pb observations. Further, Reinharz et al. (1982) have made a radiographic examination of the physical and biogenic sedimentary structure of the bay cores. They find that the degree of bioturbation follows the salinity gradient with the least amount of biological activity at the head of the bay and the greatest near its mouth. For the upper bay there is little to no bioturbation; individual laminae are well preserved. Near the mouth there is often complete mixing; the sediment is relatively homogeneous as a function of core depth.

RESULTS

Mass Sedimentation Rate

A summary of the optimum solution values for the mass sedimentation rate, ω, mixing coefficient, D, and mixed layer depth, d, as well as the corresponding values for the estimated flux, F, of ^{210}Pb and total mass, M, of ^{137}Cs or $^{239,240}Pu$ is given in Table 1. Figure 2 is a plot of the mass sedimentation rates and their 90% confidence intervals versus latitude and longitudinal distance along Chesapeake Bay. Some of the ω values are tightly defined whereas others have an extended confidence interval. Also included in the figure is a sketch of the bathymetry along the central axis of the bay.

There is a clear trend to the data with high sedimentation rates of around 0.3-1.2 gm $cm^{-2}yr^{-1}$ for the upper bay, modest sedimentation rates of 0.1-0.3 gm $cm^{-2}yr^{-1}$ for the middle bay, and modest to high sedimentation rates of 0.1-0.8 gm $cm^{-2}yr^{-1}$ for the lower bay. The curve in Fig. 2 is included as an estimate of this trend. It is a weighted, least square fit to the data using a second order polynomial. Following a standard procedure the weighting for each data point was taken to be proportional to the reciprocal of the square of the confidence interval. This procedure emphasizes the well defined data points, which generally correspond to the lower sedimentation rate values. A third order polynomial was also fit to the data with no significant change in the resultant sum of the weighted, least square errors.

It is important to emphasize, however, that the sedimentation rate near the bay mouth is poorly delineated. It is defined by only two data points, cores 109 and 99, both of which have large associated confidence intervals. Shideler

TABLE 1 - Summary of Chesapeake Bay core data and analyses. h - water depth, m; ω - mass sedimentation rate, gm $cm^{-2}yr^{-1}$; D - mixing coefficient, cm^2yr^{-1}; d - mixed layer depth, cm; F - flux of ^{210}Pb, dpm $cm^{-2}yr^{-1}$; M - mass of ^{137}Cs or $^{239,240}Pu$ in cores, dpm cm^{-2}.

Core	*Tracer*	*Lat. (N)*	*Long. (W)*	*h*	ω	*D*	*d*	*F*	*M*
SUS	^{210}Pb	39°57'	76°23'		1.89	0		3.94	
52	^{210}Pb	39°29'	75°57'		1.24	0		1.91	
E	^{210}Pb	39°23'	76°06'	5	0.87	0		2.06	
E	^{137}Cs				0.49	0.4	7		18.8
4	^{210}Pb	39°19'	76°14'	4	0.39	0		0.95	
4	^{137}Cs				0.41	0.6	24		22.4
914S	^{210}Pb	39°14'	76°14'	7	5.55	0		12.59	
6	^{210}Pb	39°09'	76°23'	5	0.63	0		2.66	
6	$^{239,240}Pu$				0.38	1.3	7		4.17
858C	^{210}Pb	38°58'	76°23'	32	1.23	0		5.75	
856E1	^{210}Pb	38°57'	76°25'	12	0.35	0		1.54	
856E2	^{210}Pb				0.36	0		1.93	
55	^{210}Pb	38°49'	76°24'	15	0.69	0		3.23	
C	^{210}Pb	38°42'	76°24'	13	0.06	5.1	6	0.75	

14	^{210}Pb	38°39'	76°26'		0.05	0		0.27	
834G	^{210}Pb	38°34'	76°27'	12	0.30	25	10	2.73	
18	^{210}Pb	38°19'	76°20'	15	0.10	0.9	4	1.09	
18	$^{239,240}Pu$				0.16	0.1	7		1.95
60	^{210}Pb	38°12'	76°20'	11	0.08	0		0.62	
62	^{210}Pb	38°11'	76°14'	34	0.28	1.4	9	2.29	
62	$^{239,240}Pu$				0.21	5.4	13		4.03
63	^{210}Pb	38°12'	76°08'	8	0.23	3.0	14	0.99	
24	^{210}Pb	38°00'	76°13'	15	0.68	4.1	20	3.92	
24	^{137}Cs				0.37	3.5	27		25.4
747A	^{210}Pb	37°47'	76°11'	33	0.18	100	30	3.64	
747A	$^{239,240}Pu$				0.49	100	24		
83	^{210}Pb	37°42'	76°15'	13	0.09	0		0.68	
85	^{210}Pb	37°41'	76°05'	10	0.25	0		1.33	
85	$^{239,240}Pu$				0.23	0.6	5		2.89
86	^{210}Pb	37°43'	75°55'	4	0.30	∞	22	3.32	
102	^{210}Pb	37°24'	76°04'	12	0.48	∞	10	1.23	
99	^{210}Pb	34°01'	76°14'	5	0.85	∞	12	1.55	

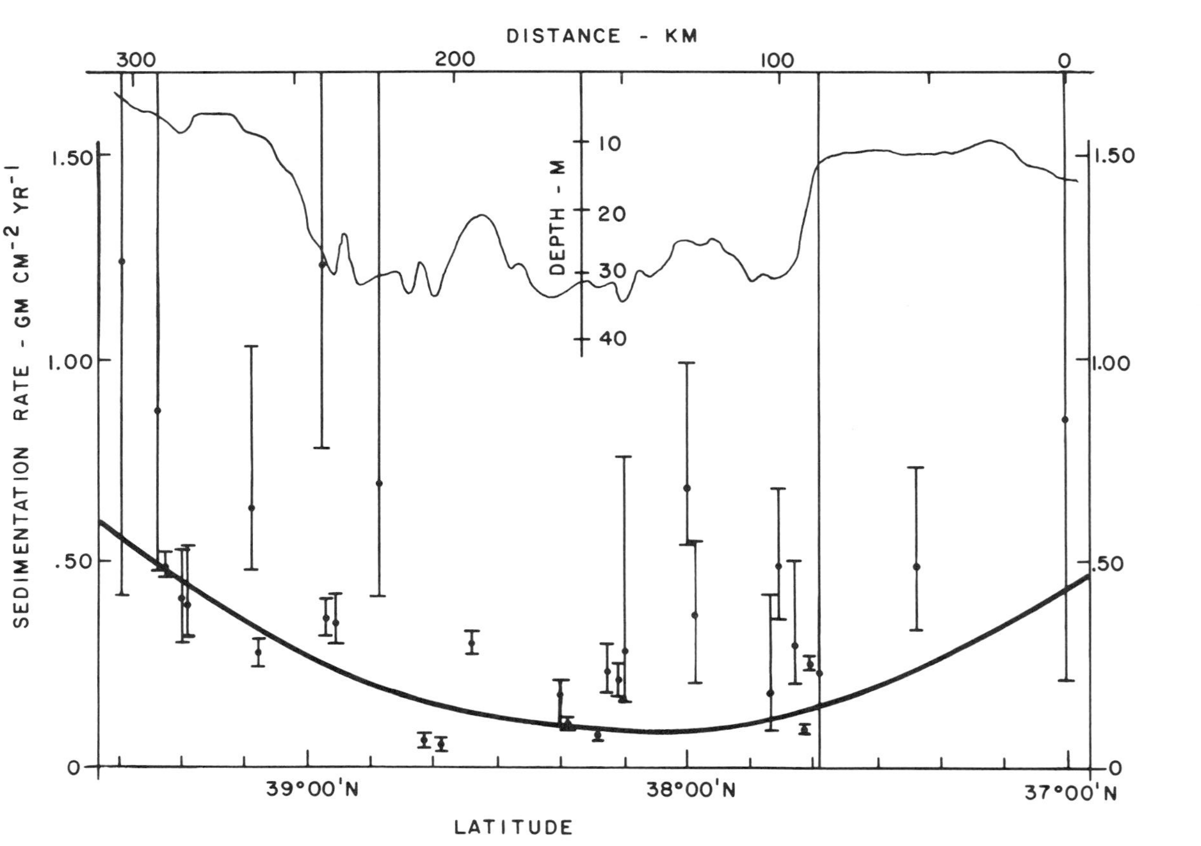

FIGURE 2. Plot of optimum values of mass sedimentation rate and confidence interval versus distance along Chesapeake Bay.

(1975) suggests that sediments covering a substantial portion of the lower bay are relict whereas Ludwig (1981) found high sedimentation rates near the mouth of the James estuary. Clearly, more work is needed to delineate the lower bay sedimentation regime.

The sedimentation trend provides an explanation for the existence of the deep, midbay channel of Fig. 2. There has been high sedimentation in the northern portion of the bay with the Susquehanna River as source and a lesser but substantial sedimentation in the southern portion with the ocean as source. From both sources there has been minimal sedimentation in the central portion of the midbay channel.

The very high sedimentation rate for core 914S requires attention. On a semilogarithmic plot of excess ^{210}Pb versus total sediment particle accumulation the data points follow a straight line trend from the surface down to the maximum depth of 70 cm. Without additional information this could be interpreted as either a high sedimentation rate or mixing over an extended depth interval. There is additional information which definitively indicates that the high sedimentation rate interpretation, originally given by Goldberg et al. (1978), is correct. First, neither the radiographic examination by Reinharz et al. (1982) nor the results from the nearby cores of Fig. 1 and Table 1 indicate any extensive bioturbation. Second, Goldberg et al. (1978) have two determinations of $^{239,240}Pu$ for this core with values of 0.030 dpm gm^{-1} at 1 cm and 0.123 dpm gm^{-1} at 68 cm, indicating again no extensive bioturbation but rather deep burial of the 1963 atmospheric nuclear weapons testing peak of radioactivity. This core, however, is located in or near the dredged shipping channel. We attribute this high sedimentation rate to slumping or rapid infilling of sediments related to dredging.

In other words, it is an anthropogenic rather than a natural sedimentation effect, and we have not included it in Fig. 2 or subsequent discussion.

It is instructive to compare the mass sedimentation rate and mixing parameter values for the ^{210}Pb and ^{137}Cs or $^{239,240}Pu$ methods as a check on the reliability of each method. This comparison is given in Table 2 for those eight cores which have both sets of observations. For most of the value sets the comparison is excellent, and for all except the ω values for core 6 the agreement is within the 90% confidence interval.

It is also of interest as a reliability check to compare these results with those from the tributary estuaries. Knebel et al. (1981) and Brush et al. (1982) report ^{210}Pb sedimentation rates for the Potomac estuary. There are two stations near the mouth of the estuary with values of ω = 0.07-0.09 gm $cm^{-2}yr^{-1}$ in agreement with the results of Table 1 and Fig. 2. The sedimentation rates increase from the mouth up the estuary to values in the range of 0.30-0.70 gm $cm^{-2}yr^{-1}$. For the James estuary, Cutshall, Larsen and Nichols (1981) indicate depth change rates of less than 1 to 5 cm yr^{-1} near the mouth increasing to very high rates of 10-20 cm yr^{-1} upestuary. Upon use of the porosity values of core 99 near the mouth of the James estuary, a value of v = 1.0 cm yr^{-1} converts to ω = 1.2 gm $cm^{-2}yr^{-1}$. In addition, Ludwig (1981) found bathymetric depth change rates of 0.2-1.0 cm yr^{-1} in the Thimble shoals region adjacent to the mouth of the James. The high sedimentation rates at the mouth of the James are also in agreement with the results of Table 1 and Fig. 2.

Bioturbation

As indicated by the values in Table 2, the mixing coefficient, D, is determined at best to first figure accuracy. At the extremes the mixing coefficient values are indistinguishable from D = 0, no bioturbation, or D = ∞, well mixed. The mixed layer depth, d, values are generally determined to within a few centimeters. As indicated by the values in Table 2, the ^{137}Cs and $^{239,240}Pu$ method is somewhat more sensitive to the determination of low D coefficient values. The mixing coefficient values of Table 1 follow a trend from D = 0 at the bay head to D = ∞ near the mouth in agreement with the radiographic results of Reinharz et al. (1982), discussed previously.

Figures 3 and 4 are comparisons of the optimum solution curves and data points for two cores in which there was both ^{210}Pb and ^{137}Cs or $^{239,240}Pu$ information. They were also chosen to illustrate the significant changes that occur for both tracer distributions depending on the degree of bioturbation mixing. In the case of core 4 there is little to no bioturbation; the ^{137}Cs peak is well defined and the ^{210}Pb distribution follows an exponential curve. For core 747A there is extensive bioturbation down to depths of 24-30 cm; the $^{239,240}Pu$ peak is displaced downward in the core and is less pronounced and the ^{210}Pb distribution is essentially at a constant level through the mixed layer followed by an exponential curve.

Tracer Fluxes and Core Masses

As summarized by Robbins (1978), one might expect the atmospheric input flux of ^{210}Pb to produce relatively constant flux levels from one location to the next in Chesapeake Bay in the absence of horizontal sediment transfer effects. As

TABLE 2 - Comparison of ^{210}Pb optimization values and confidence intervals for mass sedimentation rate, ω, gm $cm^{-2}yr^{-1}$; mixing coefficient, D, cm^2yr^{-1}; and mixed layer depth, d, with the corresponding values for the ^{137}Cs and $^{239,240}Pu$ analyses.

Core	*Tracer*	$-\ \omega\ +$	$-\ D\ +$	$-\ d\ +$
E	^{210}Pb	0.48<0.87<5.3		
E	^{137}Cs	0.46<0.49<0.52	0.2 < 0.4 < 1.0	3 < 7<10
4	^{210}Pb	0.31<0.39<0.54		
4	^{137}Cs	0.30<0.41<0.53	0.4 < 0.6 < 3.1	8<24 < -
6	^{210}Pb	0.48<0.63<0.93		
6	$^{239,240}Pu$	0.34<0.38<0.41	0.7 < 1.3 < 3.2	5 < 7<10

18	^{210}Pb	$0.09<0.10<0.12$	$0.2 < 0.9 < -$	$2 < 4 < 8$
18	$^{239,240}Pu$	$0.10<0.16<0.21$	$0.04 < 0.1 < 0.9$	$2 < 7 < -$
62	^{210}Pb	$0.16<0.28<0.76$	$0 < 1.4 < -$	$0 < 9 < -$
62	$^{239,240}Pu$	$0.17<0.21<0.25$	$2.2 < 5.4<20$	$8<13<15$
24	^{210}Pb	$0.54<0.68<0.89$	$0.1 < 4.1<12$	$0<20 < -$
24	^{137}Cs	$0.20<0.37<0.55$	$1.3 < 3.5<12$	$10<27 < -$
747A	^{210}Pb	$0.09<0.18<0.42$	$32 < 100 < -$	$25<30<35$
747A	$^{239,240}Pu$	$0.35<0.49<0.68$	$24 < 100 < -$	$21<24<29$
85	^{210}Pb	$0.24<0.25<0.27$		
85	$^{239,240}Pu$	$0 < 0.23 < -$	$0 < 0.6<-$	$0 < 5 < -$

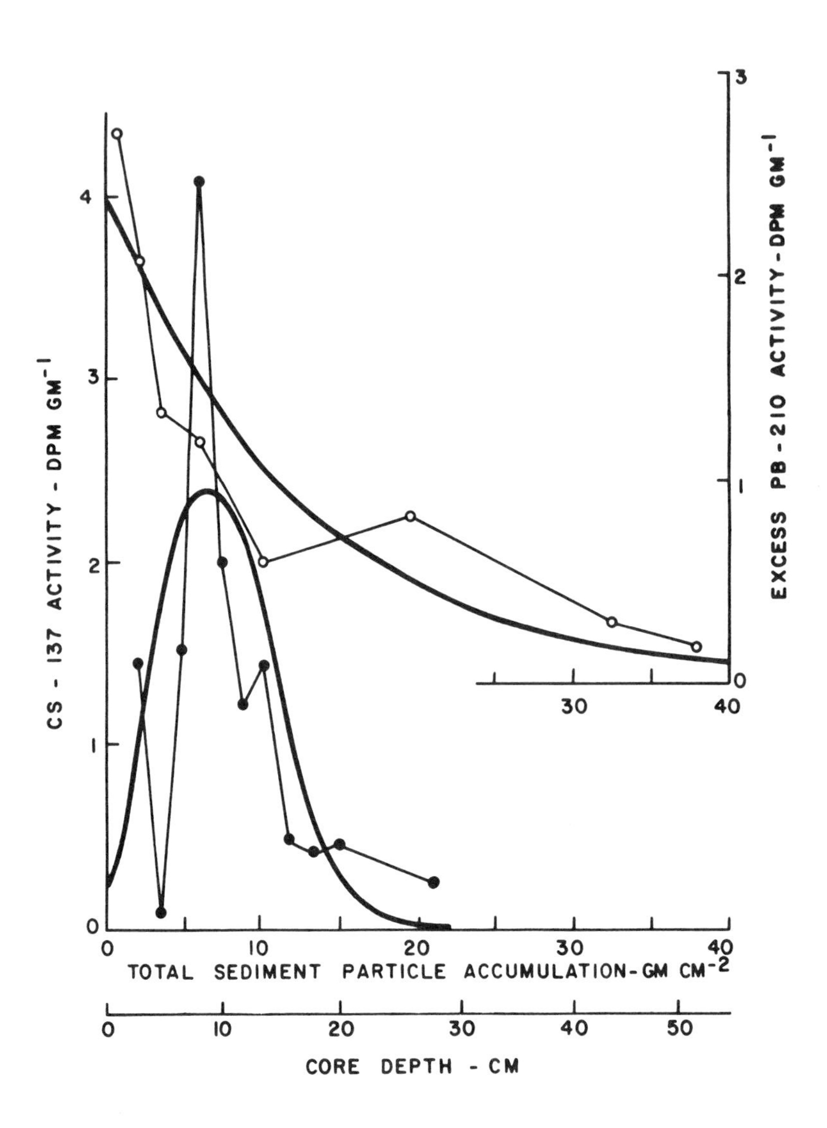

FIGURE 3. Plots of optimum solution curves for Cs-137 and Pb-210 for core 4. Optimum solution parameter values and confidence intervals are given in Table 2. ● — Cs-137 data points; 0 — Pb-210 data points.

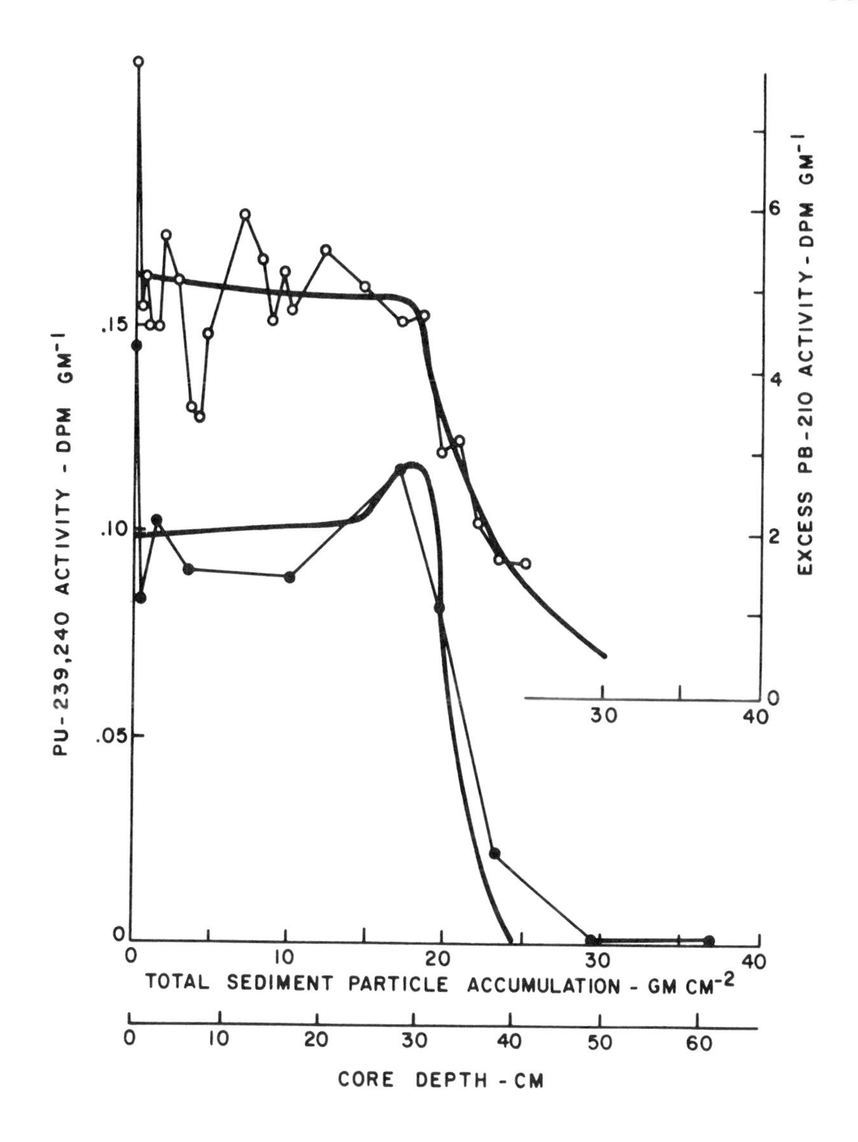

FIGURE 4. Plots of optimum solution curves for Pu-239, 240 and Pb-210 for core 747A. Optimum solution parameter values and confidence intervals are given in Table 2. ● — Pu-239, 240 data points; 0 — Pb-210 data points.

shown in Table 1, this is not the case. There is a general correlation of flux with sedimentation rate, indicating that suspended sediment transport of particles with a given ^{210}Pb signature to the final depositional site is important. For the more limited data for ^{137}Cs and $^{239,240}Pu$ total masses in the cores the same correlation of core mass with sedimentation rate is apparent.

DISCUSSION

Sediment Sources

On the assumption that the minimum of the mass sedimentation trend curve of Fig. 2 represents the approximate extent of both the northern- and southern-derived sediments, the relative magnitudes of each may be determined. From the respective areas under the curve the Susquehanna River and associated northern derived sediments represent 61% of the total bay sedimentation as compared with 39% for the ocean and associated southern-derived sediments. The southern source, which unfortunately is poorly defined by our data, appears to be 0.6 times that of the northern source, a smaller but nevertheless substantial contribution. Considered as a "filter", Chesapeake Bay traps essentially all the river-derived sediments and is also a sink for ocean-derived sediments. Schubel and Carter (1977) arrived at a similar conclusion from their analysis of the suspended sediment distribution and gravitational circulation fluxes in the bay.

From the trend curve of Fig. 2 the average mass sedimentation rate for the entire bay is 0.23 gm $cm^{-2}yr^{-1}$. From the ω values in Table 1 the direct average without weighting is 0.46 gm $cm^{-2}yr^{-1}$.

From these data alone we cannot state whether the bay acts as a source or sink for the tributary estuaries. Schubel and Carter (1977) concluded that during normal conditions the tributary estuaries are sinks for suspended sediment from the bay. Officer and Nichols (1980) reached the same conclusion for the James and Rappahannock estuaries. However, it is possible that episodic events may lead to a net transport from the tributaries to the bay which could dominate the normal trend.

Episodic and Normal Sedimentation

The analysis given here produces sedimentation rates averaged over 40-100 years; it does not, in itself, distinguish between episodic and normal sedimentation. It is possible to identify what appear to be the residue from major episodic events in some of the ^{210}Pb profiles, but we have not attempted such an analysis since much of the core data is not sufficiently detailed and since there are a number of other possible variables which could account for the departures of the data from a reference curve. For example, the ^{210}Pb profile of Fig. 3 could be interpreted to indicate a relatively uniform sedimentation rate to a depth of 10 gm cm^{-2}, a large episodic event from 10 to 20 gm cm^{-2}, followed by relatively uniform sedimentation thereafter. Also, Hirschberg and Schubel (1979) interpret their ^{210}Pb profile as indicating a residue from tropical storm Agnes from the sediment-water interface down to 16 cm and an additional major episodic residue from 32 to 68 cm.

A definitive delineation of the importance of episodic sedimentation events to the bay system has been given by Gross et al. (1978). Between 1966 and 1976 the Susquehanna River discharged approximately 50 million metric tons of suspended sediment to the bay, of which 40 million metric tons were discharged in two floods associated with tropical storm Agnes (24-30 June 1972) and hurricane Eloise (26-30 September 1975). In addition, during normal years, about 50% of the annual sediment discharge was associated with the spring freshet of about two weeks duration. As discussed by the authors, the period of their study was a time of unusually high sediment discharge related to the two major floods. For the thirty years prior there were no major episodic events; there was a major event in 1936. It would seem appropriate, then, to take as the average episodic contribution over the past forty years a value of around 1 million metric tons per year and a normal contribution exclusive of episodic events of the same amount. Thus, as a crude guide we estimate that on average 50% of the sediment discharged to the bay is associated with normal year effects and 25% is associated with discharges exclusive of the spring freshet.

Biggs (1970) made an analysis of sediment effects in the upper and middle portions of Chesapeake Bay from suspended sediment distributions and gravitational circulation effects for calendar year 1966. He obtained average mass sedimentation rates of 0.135 gm $cm^{-2} yr^{-1}$ for the region from the bay head to latitude 39°04'N and 0.040 gm $cm^{-2} yr^{-1}$ for the region from latitude 39°04'N to 38°16'N. From the trend curve of Fig. 2 the corresponding total sedimentation rates over the past 50 years or so are 0.43 and 0.16 gm $cm^{-2} yr^{-1}$. The ratio of Biggs' rates for a normal year to the total sedimentation rates are, respectively, 31 and 25%. Depending on how well Biggs was able to estimate the spring freshet effects, the above percentages are in reasonable agreement with the range of 25-50% anticipated from the Gross et al. (1978) analysis.

Episodic sedimentation is a dominant characteristic of the northern portion of the bay and probably for the southern portion as well although to our knowledge no estimates have been made as to the importance of storm-derived events for the ocean source sedimentation.

Sediment Budget

From Gross et al. (1978) we have estimated that the average sediment discharge to the bay from the Susquehanna River including episodic events is 2.0 million metric tons per year. Biggs (1970) estimates an additional 0.6 million metric tons per year from shore erosion, skeletal material and primary production from the bay head to latitude 38°16'N, the approximate limit of Susquehanna River sedimentation effects discussed previously. From the trend curve of Fig. 2 and the surface areas given in Biggs (1970), the total sediment input to latitude 38°16'N is 3.1 million metric tons per year. This value is in reasonable agreement with the total source budget figure of 2.6 million metric tons per year.

We do not have any corresponding estimate of ocean-derived sediment sources to make a similar comparison for the lower bay where the dominant sediment source appears to be the ocean and perhaps also the tributary estuaries and coastal and shore erosion.

Historical Bathymetric Depth Changes

At the outset it might appear that the use of historical bathymetric depth changes could be a useful method for estimating sedimentation rates. Unfortunately this is not necessarily the case for regions undergoing moderate sedimentation such as Chesapeake Bay. From the values of ω in Table 1 and using the porosity values at depth in the cores as an estimate of ϕ', average sedimentation rates of $v' = 0.76$ cm

yr^{-1} and $v' = 0.35$ cm yr^{-1} are obtained for the Maryland and Virginia portions of the bay, respectively.

There are two concerns. The first regards the data itself and considerations of reliability for intercomparing recent data with that taken 50 to 100 years ago. The second regards the corrections that have to be made to the depth change data to arrive at sedimentation rates. The true sedimentation rate, v', is given in terms of the bathymetric depth change rate, v'', positive for decreasing depths, by the relation

$$v' = v'' + e + s \tag{7}$$

where e is the eustatic sea level rise rate and s is the tectonic subsidence rate. It has only recently been appreciated that epeirogenic plate movements for relatively stable areas can be substantial with values in the range of a fraction of a centimeter per year. Officer and Drake (1982) have shown from variations in the continental shelf break depth off the East coast that over the past 18,000 years the Scotian shelf has been subsiding at a rate of 0.8 cm yr^{-1} with respect to Miami and with the major changes occurring in the region from New York to Cape Hatteras. Of more particular interest here, Holdahl and Morrison (1974) obtain subsidence rates of 0.2-0.3 cm yr^{-1} for Chesapeake Bay from an analysis of tide gauge records over the past 30 years. In the U.S. Geodynamics Committee report (1973) Holdahl included a figure of probable vertical movements for the U.S. which shows subsidence rates of around 0.5 cm yr^{-1} for the Chesapeake Bay region, increasing from the southern to the northern end of the bay. From another analysis of tide gauge records over the past 30-70 years Brown (1978) obtained tectonic subsidence rates of 0.2-0.4 cm yr^{-1}. From an analysis of relevelings in the same region Brown (1978) obtained considerably higher rates than those from the tide gauge records which indicate subsidence rates of around 0.6-1.0 cm yr^{-1} depending on the reference base level chosen and which increase by 0.5 cm yr^{-1} from the mouth of the bay to its head. For use in equation (7), a principal result from these investigations is that the bay region has been undergoing substantial subsidence over the recent historical past with rates in the range of 0.2-1.0 cm yr^{-1}; the problem is that these rates are not sufficiently well known to permit reliable estimates of v' to be made from observations of v". It could be argued that the question might be more appropriately stated to obtain reliable estimates for the least well known variable, s, from geochemical tracer determinations for v' and the bathymetric depth change rates for v".

Kerhin et al. (1982) and Byrne, Hobbs and Carron (1982) made a study of bathymetric depth changes for the Maryland and Virginia portions of Chesapeake Bay, respectively. Kerhin et al. (1982) included a eustatic sea level correction of 0.1 cm yr^{-1} but no tectonic subsidence correction. Byrne et al. (1982) included the same eustatic correction and a subsidence correction of 0.1-0.2 cm yr^{-1}. In particular, Kerhin et al. (1982) obtained results which show approximately equal areas of deposition and erosion with an average deposition rate of 0.76 cm yr^{-1} and an erosion rate of 0.69 cm yr^{-1} for a net deposition rate of 0.07 cm yr^{-1}. This rate is in substantial disagreement with the geochemical tracer rate of 0.76 cm yr^{-1} from a direct average of the data in Table 1 or a rate of about half this amount from the trend curve of Fig. 2. A subsidence rate correction in the range of 0.3-0.7 cm yr^{-1} for the Maryland portion of the bay would bring the deposition-erosion pattern result and the average sediment rate into agreement with the geochemical tracer results.

REFERENCES CITED

Biggs, R.B. 1970. Sources and distribution of suspended sediment in northern Chesapeake Bay. *Mar. Geol. 9*:187-201.

Brown, L.D. 1978. Recent vertical crustal movement along the East coast of the United States. *Tectonophysics 44*:205-231.

Brush, G.S., E.A. Martin, R.S. De Fries and C.A. Rice. 1982. Comparisons of ^{210}Pb and pollen methods for determining rates of estuarine sediment accumulation. *Quaternary Res. 18*:196-217.

Byrne, R.J., C.H. Hobbs and M.J. Carron. 1982. Baseline sediment studies to determine distribution, physical properties, sedimentation budgets and rates in the Virginia portion of Chesapeake Bay. Final draft report from Virginia Institute of Marine Science, Gloucester Point, Virginia to Chesapeake Bay Program, Environmental Protection Agency, Annapolis, MD, 188 pp.

Carpenter, R., M.L. Peterson and J.T. Bennett. 1982. ^{210}Pb derived sediment accumulation and mixing rates for the Washington continental slope. *Mar. Geol. 48*:135-164.

Cutshall, N.H., I.L. Larsen and M.M. Nichols. 1981. Man-made radionuclides confirm rapid burial of kepone in James river sediments. *Science 213*:440-442.

Goldberg, E.D., V. Hodge, M. Koide, J. Griffin, E. Gamble, O.P. Bricker, G. Matisoff, G.R. Holdren and R. Braun. 1978. A pollution history of Chesapeake bay. *Geochim. Cosmochim. Acta 42*:1413-1425.

Gross, M.G., M. Karweit, W.B. Cronin and J.R. Schubel. 1978. Suspended sediment discharge of the Susquehanna River to northern Chesapeake Bay, 1966 to 1976. *Estuaries 1*:106-110.

Helz, G.R., S.A. Sinex, G.H. Setlock and A.Y. Cantillo. 1981. Chesapeake Bay sediment trace elements. Research in Aquatic Geochemistry Report, Department of Chemistry, University of Maryland, College Park, MD., 202 pp.

Hirschberg, D.J. and J.R. Schubel. 1979. Recent geochemical history of flood deposits in the northern Chesapeake Bay. *Estuarine Coastal Mar. Sci. 9*:771-784.

Holdahl, S.R. and N.L. Morrison. 1974. Regional investigations of vertical crustal movements in the U.S. using precise relevelings and mareograph data. *Tectonophysics 23*:373-390.

Kerhin, R.T., J.T. Halka, E.L. Hennessee, P.J. Blakeslee, D.V. Wells, N. Zoltan and R.H. Cuthbertson. 1982. Physical characteristics and sediment budget for bottom sediments in the Maryland portion of Chesapeake Bay. Final draft report from Maryland Geological Survey, Baltimore, Maryland to Chesapeake Bay Program, Environmental Protection Agency, Annapolis, MD., 190 pp.

Knebel, H.J., E.A. Martin, J.L. Glenn and S.W. Needell. 1981. Sedimentary framework of the Potomac River estuary, Maryland. *Geol. Soc. Am. Bull. 92*:578-589.

Ludwig, J.C. 1981. Bottom sediments and depositional rates near Thimble shoals, lower Chesapeake Bay, Virginia. *Geol. Soc. Am. Bull. 92*:496-506.

Lynch, D.R. and C.B. Officer, 1984. Nonlinear parameter estimation for sediment cores. *Chem. Geol.* (in press).

Officer, C.B. 1982. Mixing, sedimentation rates and age dating for sediment cores. *Mar. Geol. 46*:261-278.

Officer, C.B. and C.L. Drake. 1982. Epeirogenic plate movements. *Jour. Geol. 90*:139-153.

Officer, C.B. and D.R. Lynch. 1982. Interpretation procedures for the determination of sediment parameters from time dependent flux inputs. *Earth Planet. Sci. Lett. 61*: 55-62.

Officer, C.B. and D.R. Lynch. 1983. Determination of mixing parameters from tracer distributions in deep sea sediment cores. *Mar. Geol. 52*:59-74.

Officer, C.B. and M.M. Nichols. 1980. Box model application to a study of suspended sediment distributions and fluxes in partially mixed estuaries, pp. 329-340. *In*: V.S. Kennedy (ed.), *Estuarine Perspectives*. Academic Press, New York.

Reinharz, E., K.J. Nilsen, D.F. Boesch, R. Bertelsen, and A.E. O'Connell. 1982. A radiographic examination of physical and biogenic sedimentary structures in the Chesapeake Bay. Report of Investigations No. 36, Department of Natural Resources, Maryland Geological Survey, Baltimore, Maryland, 58 pp.

Robbins, J.A. 1978. Geochemical and geophysical applications of radioactive lead, pp. 285-393. *In*: J.O. Nriagu (ed.), *The Biogeochemistry of Lead in the Environment*. Elsevier Press, Amsterdam.

Robbins, J.A. and D.N. Edgington. 1975. Determination of recent sedimentation rates in Lake Michigan using Pb-210 and Cs-137. *Geochim. Cosmochim. Acta 39*:285-304.

Schubel, J.R. and H.H. Carter. 1977. Suspended sediment budget for Chesapeake Bay, pp. 48-62. *In*: M.L. Wiley (ed.), *Estuarine Processes, Vol. II*. Academic Press, New York.

Schubel, J.R. and D.J. Hirschberg. 1977. Pb^{210} determined sedimentation rate and accumulation of metals in sediments at a station in Chesapeake Bay. *Chesapeake Sci. 18*:379-382.

Shideler, G.L. 1975. Physical parameter distribution patterns in bottom sediments of the lower Chesapeake Bay estuary, Virginia. *Jour. Sed. Pet. 45*:728-737.

U.S. Geodynamics Committee. 1973. U.S. program for the Geodynamics project. National Academy of Sciences, Washington, 235 pp.

THE ROLE OF FLOCCULATION IN THE FILTERING OF PARTICULATE MATTER IN ESTUARIES

Kate Kranck

Atlantic Oceanographic Laboratory
Bedford Institute of Oceanography
Dartmouth, Nova Scotia

Abstract: The structuring of suspended particulate matter in estuaries into flocculated settling entities is described and discussed. A hierarchy of three particle distribution types is described: *In situ* distributions contain abundant large low density fragile aggregates (macro-flocs) with high settling rates, the formation of which is promoted by low turbulence and high particulate concentrations. During periods of high currents in nature and during shipboard and laboratory sample handling, these macro-flocs break up into more stable distributions of smaller flocs which are the basic building blocs of the larger units. Laboratory oxidation of organic matter and disaggregation of aggregates allow the size analysis of individual single mineral grain distributions for purposes of direct comparison between bottom and suspended sediment spectra.

The constituent grain-size distribution of the inorganic component of a floc replicates the grain size of the suspension as a whole. The diverse grain size and chemical nature of the organic matter makes its flocculation kinetics more complicated. Past laboratory experiments have indicated the existence of organic-inorganic proportions optimum for flocculation. This is substantiated by ash-loss data from three estuaries with different relative inputs of organic and inorganic matter. The particulate matter in each was dominated by similar organic-inorganic proportions (65-75% organic matter by volume) indicating that the component in excess of this value had been preferentially exported from the estuary. Settling of suspended sediment as macro-flocs and

ISBN 0-12-405070-0

near bottom break-up and resuspension of floc fragments control the very dynamic equilibrium between particle trapping and particle flushing in estuaries.

INTRODUCTION

Estuaries act as catchment basins for a large portion of the sediment load of rivers flowing into the sea. It has been estimated that 90% of the 15 x 10^{15} $g.y^{-1}$ terrestrial sediment transported by world rivers is deposited close to the continents, mostly in estuaries (Judson 1968; Milliman and Meade 1983).

Particles of terrestrial origin are flushed out to sea if their settling rates are sufficiently low to allow them to remain in the net seaward-moving surface layers. Particles of marine origin are expelled from the estuary if they are mixed upward into the surface layers. Rapidly settling particles which reach the bottom of the net landward-moving bottom layer are retained as bottom sediment or become part of the turbidity maximum. Whether or not a given particle becomes trapped in an estuary or is flushed out to sea is primarily dependent on its settling rate relative to the patterns of water motion (Postma 1967; Meade 1968, 1972; Dyer 1971; Festa and Hansen 1978).

Settling rates in still water depend primarily on particle size and density. Naturally occurring suspended particles range in diameter from less than one micrometer to many centimeters and in specific gravity from barely 1 to over 2.5. This results in settling rates from less than 1 x 10^{-4} $cm.sec^{-1}$ to over 100 $cm.sec^{-1}$. From this six orders of magnitude variation in the settling rates, large differences in the fate of particle associated substances may be expected and settling behaviour must be a prime variable in any models of estuarine transport of particulate matter. The settling properties of the particle entities with which substances occur are thus a basic factor in predicting the relative proportions of material filtered or flushed out of estuaries. This paper discusses the physical nature and settling behaviour of suspended particulate matter, with special emphasis on particle size and flocculation properties.

PARTICLE SIZE

A basic concept in studies of particle dynamics is the particle size distribution, or size spectra, which is usually presented as the frequency of occurrence of different grain sizes in a sample. Many researchers have measured particle size distributions of suspended sediment using a variety of methods and data presentation formats (eg. Sheldon 1968; Schubel 1971; Eisma and Gieskes 1977; Kranck and Milligan 1979; Nelsen 1981). In intercomparison and discussion of these results, problems arise from the fact that different results are obtained depending on sample handling and particle treatment prior to analysis. To clarify concepts of size distribution, three types of size analysis are defined below, each of which has different significance in terms of particle dynamics and estuarine transport processes.

In Situ Macro-floc Distributions

Recent studies have shown that particulate matter suspended in the natural environment is structured into relatively large very fragile aggregates. In the open ocean, divers and observers in submersibles have described and sampled visible clumps and stringers of mainly organic detritus with settling rates many orders of magnitude greater than that of the individual organic components forming these macro-flocs (Suzuki and Kato 1953; Nishizwa, Fukuda and Inoue 1954; Inoue, Nichizawa and Fukuda 1955; Jannasch 1973; Aldredge 1976, 1979; Bishop et al. 1977; Trent, Shanks and Silver 1978; Shanks and Trent 1979; Syvitski et al. 1983). Similar material has been observed by divers in near shore environments, although high turbidity makes it more difficult to distinguish (Kranck, pers. obs.).Recently a Benthos-Edgerton Model 373 Plankton Camera has been used to photograph the large *in situ* aggregated flocs formed by estuarine particulate matter (Fig. 1 and 2). Eisma et al. (1984) counted over 75 particles > 1000 μm in 80 ml of water. Experimental production of macro-flocs demonstrated their origin from physical flocculation of fine grained particulate matter in a turbulent environment. Size, shape and density varied greatly with flow rates, particle concentration and particle composition. The most common shapes were equi-dimensional blobs and long strings or chains. Sizes ranged up to several centimeters (Kranck and Milligan 1980).

The fragile nature of these particles prevents their intact sampling. They can withstand high rates of shear but disrupt when the dynamic environment is altered. This ephemeral nature demands that size measurements be carried out without physical sampling of the material. To date no complete in situ spectra are available for estuarine particulate matter.

Stable Floc Distributions

Many studies contain analyses of the particle size of suspended matter in untreated water which has been placed in bottles and subjected to some handling. Some aggregate structures are retained by this method as shown by microscopic studies and differences between floc spectra and single grain spectra. The size distributions of these samples are relatively stable and duplicate analysis can be obtained even after repeated shaking of a sample (Kranck 1975; McCave 1979;

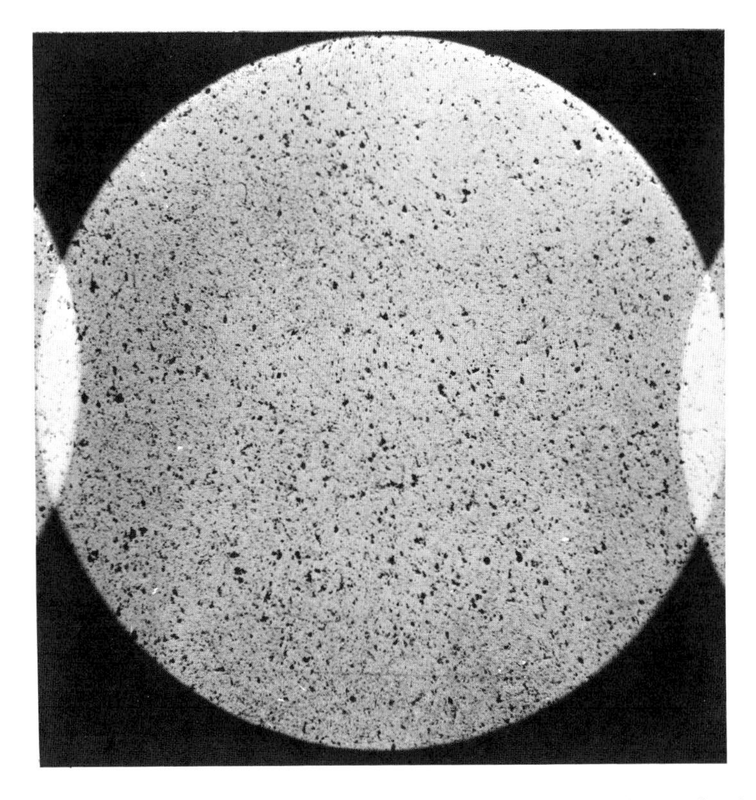

FIGURE 1. In situ photograph from Benthos Plankton Camera. Field of view represents a slice of water 10 cm in diameter and 4 cm thick.

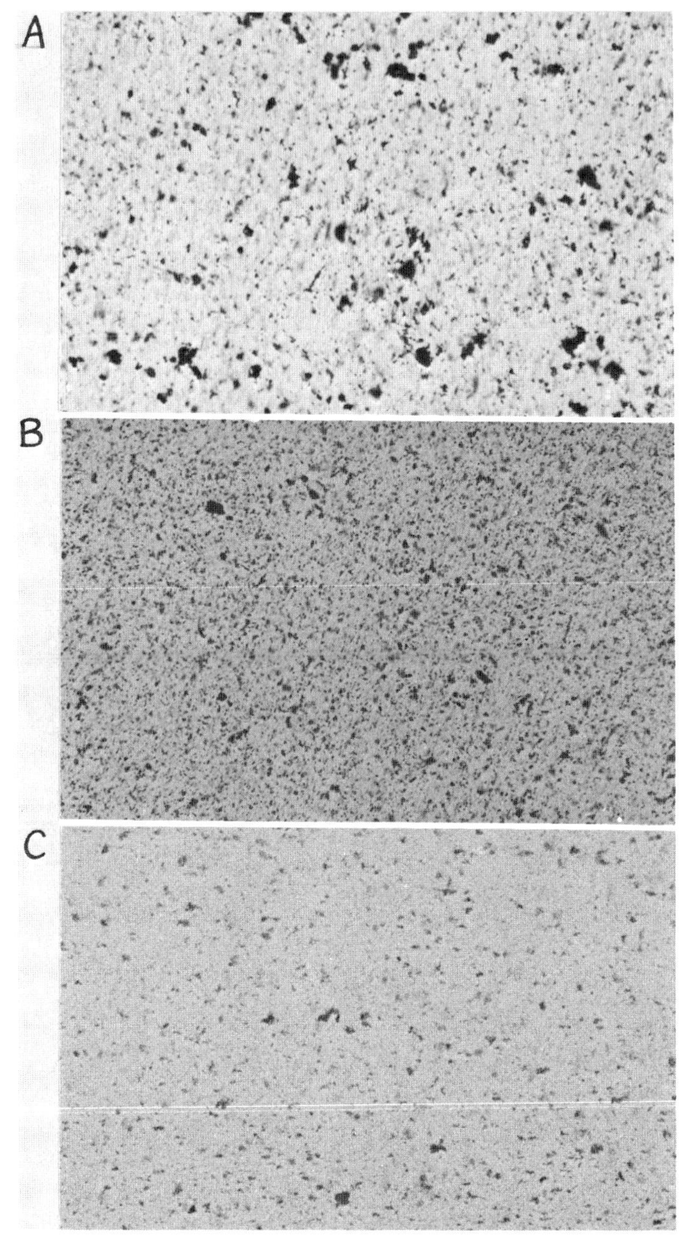

FIGURE 2. Estuarine suspended sediment showing presence of macro-flocs. A. Ems Estuary, detail of Figure 1. Plankton Camera photograph 3.5X. B. Puget Sound, Plankton Camera photograph 1.5X. C. Laboratory produced distribution, Conventional photograph 2.6X.

Eisma et al. 1984). Hard stirring and sonification or lengthy storage however will change floc distributions (Hannah, Cohen and Robech 1967; Kranck and Milligan 1979; Gibbs 1982).

Most estuarine materials have stable distributions which are broad, unimodal, and nearly symmetrical with modal sizes ranging from 5 μm to 50 μm. Very little material > 100 μm is found even in samples from water containing abundant macro-flocs, an order of magnitude greater than this (Eisma et al. 1984). Floc distributions will frequently reflect the presence of particles which occur unflocculated, such as newly eroded sediment or plankton (Fig. 3). In very high energy environments the *in situ* distribution may be the same as the floc distribution. More often however floc distributions represents fragments formed from larger *in situ* particle structures. It may be concluded that the relatively stable flocs are the basic building blocks of suspended particulate matter. The majority of suspended sediment size analyses performed by workers to date probably represent floc distributions.

Single Grain Distributions

Geologists studying the grain size of recent and ancient bottom sediments generally work with the single mineral constituent size distribution, i.e., the size after any flocs are broken up and absorbed or aggregated organic matter removed. An essential component of their size analysis is optical examination to ensure that all flocs and aggregates are broken up and only discrete grains remain. Analysis of grain size of suspended sediment allows direct comparison between bottom sediment and suspended sediment size characteristics. Studies of well flocculated sediment have shown a direct relationship between the grain size and floc size of suspended sediment (Kranck 1975).

SETTLING BEHAVIOUR

Since floc or grain distributions do not usually represent the true *in situ* particle distribution, neither can be used to calculate settling rates even if particle density is known. Most available information on settling behaviour of fine grained sediment is based on laboratory observations of artificial or natural suspensions (Krone 1962; Migniot 1968; Owen 1971; Kranck 1980). All these studies have recorded higher settling rates than predicted from the mineral grain size of the experimental suspension. Another significant

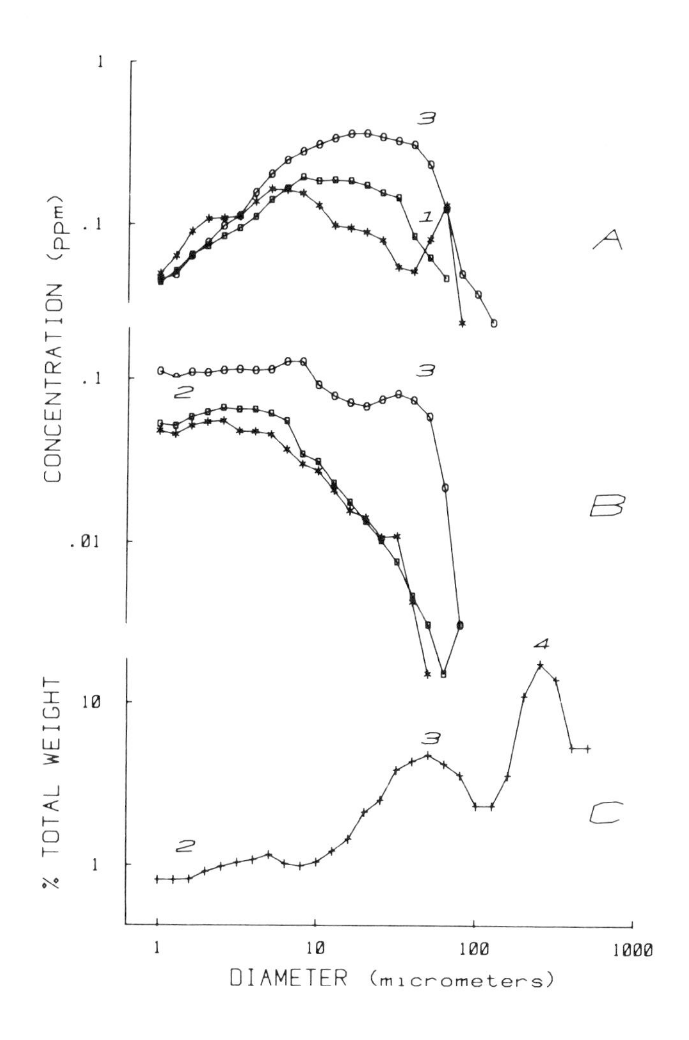

*FIGURE 3. Bay of Fundy suspended and bottom size spectra from Coulter Counter analysis. A. Floc size spectra of suspended sediment. B. Grain size spectra of suspended sediment. C. Grain size spectra of bottom sediment. * = surface; [] = 22 M depth; 0 = 42 M depth. 1. Plankton peak present in surface floc spectra but ashed away in grain spectra. 2. Similar shaped tails composed of grains deposited as flocs. 3. Modal hump of coarse grains deposited as unflocculated single grains. 4. Modal peak from bedload material not present in suspension.*

property of cohesive sediment settling noted in most of these studies is the strong dependency of settling flux on total concentration. Kranck (1980) compared decrease in concentration in three suspensions with different initial concentrations (Fig. 4). After an initial period of slow settling, flocculation caused a rapid settling rate during which the decrease in concentration with time followed a power law with a -4/3 slope. The time of start of rapid floc settling was also inversely related to concentration by the same -4/3 power law so that the rapid settling phase in any one suspension did

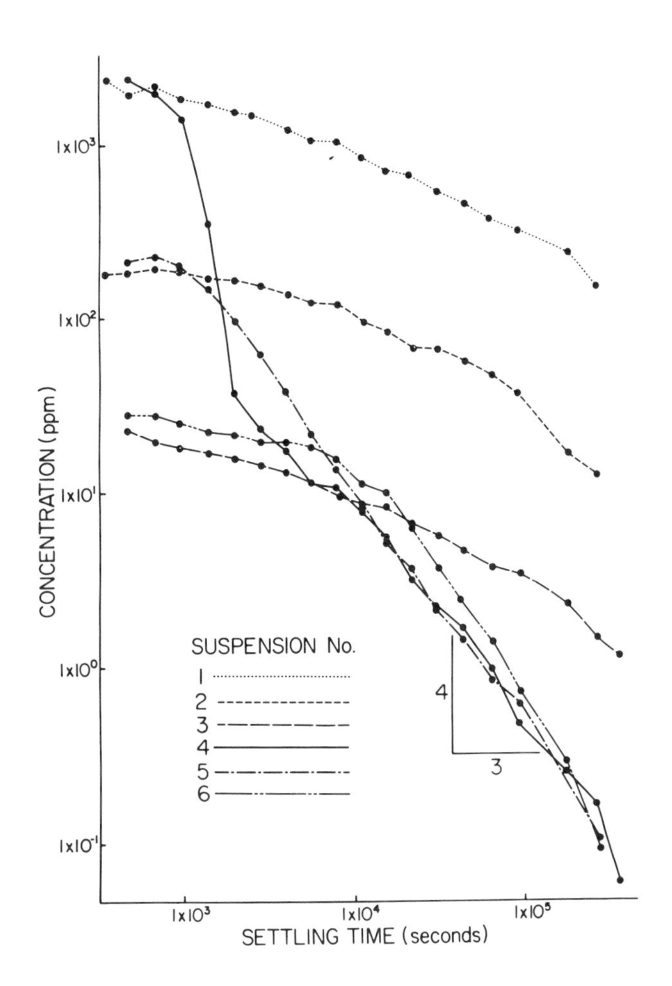

FIGURE 4. Change in total concentration with time in six different laboratory seidment suspensions. Suspensions 1-3 settled in fresh water and Suspensions 4-6 settled in salt water (from Kranck 1980).

not start until the suspensions with higher initial concentrations had decreased to the level of the less concentrated suspensions. The exact mechanism causing this concentration equalization is not known but it follows that localized or temporary elevations in suspended sediment concentrations in nature tend to return to relatively constant background levels.

The role of organic matter in flocculation and settling dynamics is poorly known. Two types of processes can reconstitute suspended particulate carbon into larger faster setting units. [1.] Organisms themselves repackage particulate matter into regularly shaped, relatively compact fecal pellets containing varying proportions of organic and inorganic matter depending on environment and species. This process, which cannot be strictly classified as flocculation, is responsible for up to 40% of particle transfer to the bottom in some coastal waters. [2.] Flocculation more or less permanently aggregates particles brought together by chance collision while suspended in a turbulent media. A variety of physio-chemical forces have been held responsible for interparticle adhesion but the role of organic coatings, bacteria and extracellular strands and monofilaments in aggregating particles or reinforcing other binding mechanisms is also well documented (eg. Neihof and Loeb 1974; Zabawa 1978).

The function of organic matter in promoting settling was demonstrated during experimental production of macro-flocs (Kranck and Milligan 1980). At a given turbulence level in a recirculating elutriator, a suspension of half organic matter and half inorganic sediment was found to settle faster than suspensions of all organic or all inorganic matter of the same initial concentration (Fig. 5). Of the three suspensions, the inorganic matter settled slowest despite its higher density. It appears that greater "stickability" of organic particles holds together inorganic particles while the inorganic grains function as sinkers for the organic fraction. These results suggest the existence of an organic-inorganic ratio at which floc formation and settling is most efficient. In an estuary, material with this composition is most likely to form macro-flocs and sediment out or become part of the turbidity maximum. The component, whether organic or inorganic, in excess of this ratio is most likely to be flushed out of the estuary. Figure 6 presents the percent organic matter (% of ash loss by volume) in samples of suspended sediment collected along the length of three estuaries in eastern Canada. All three are dominated by particulate matter with 65% organic matter, pointing to this value as an optimum composition for flocs formation.

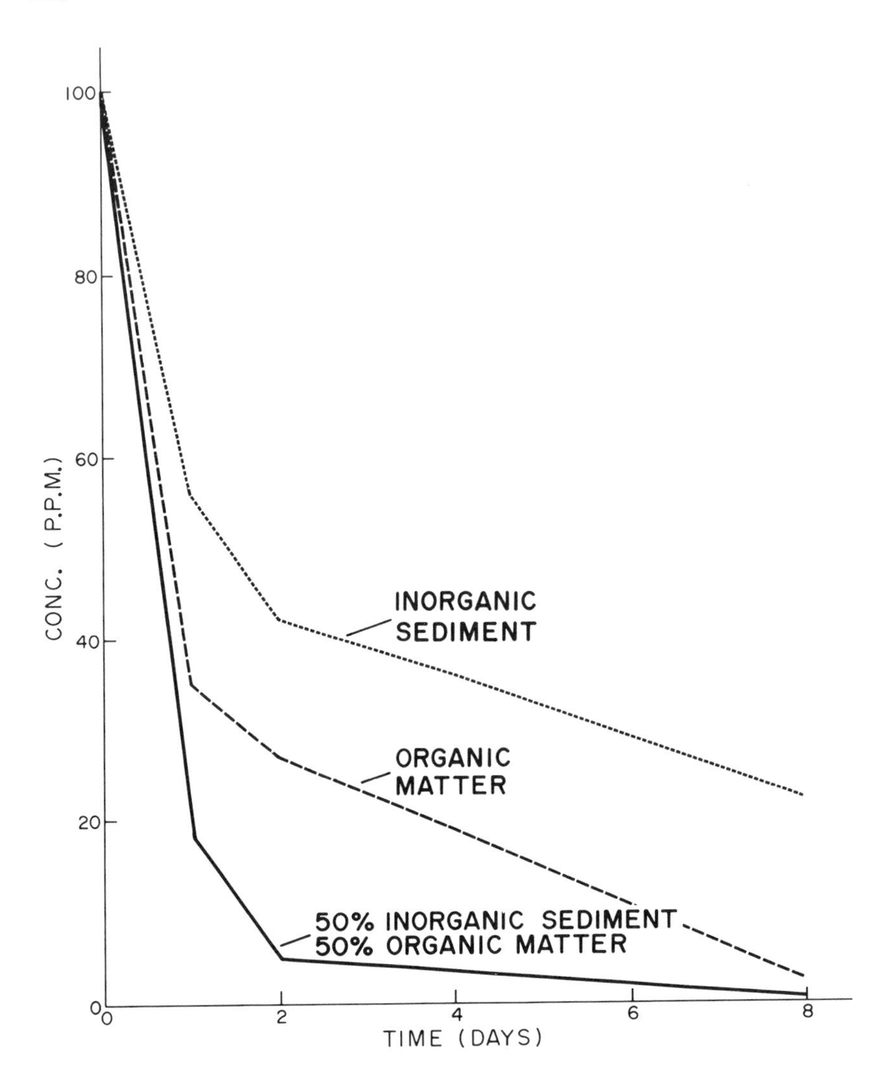

FIGURE 5. Decrease in concentration with time due to flocculation and settling in suspensions with different organic-inorganic compositions.

DISCUSSION AND CONCLUSIONS

The filtering efficiency of estuaries in controlling the transfer of substances between rivers and the sea is intimately tied up with flocculation processes within the

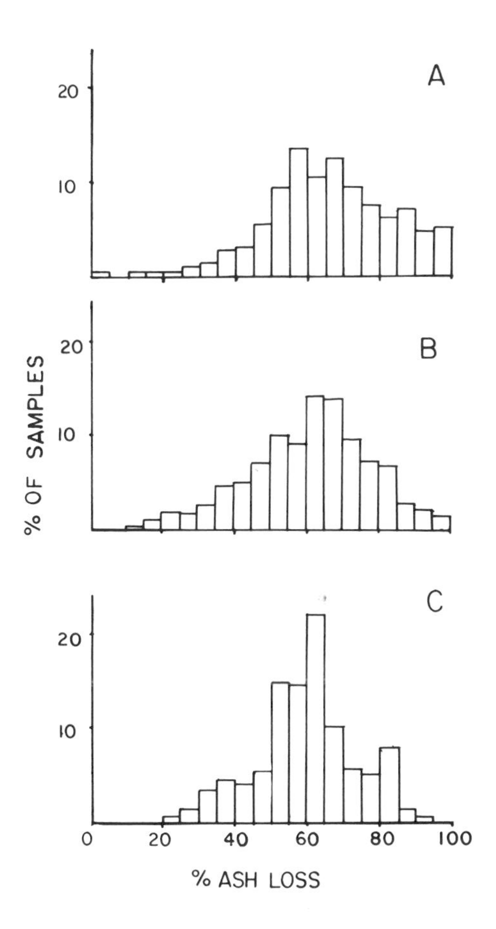

FIGURE 6. Organic-inorganic composition (volume ash loss) of samples collected along the length of three estuaries. A. Bay of Fundy, B. St. Lawrence Estuary, C. Miramichi Bay. A, B, and C represent 331, 230, and 302 samples, respectively.

turbidity maximum. In theory, a turbidity maximum may form solely by the trapping action of a two layer estuarine flow (Festa and Hansen 1978). This should concentrate a narrow range of particle sizes in the turbidity maximum (Postma 1967) as confirmed by Schubel (1969) in measurements of stable distributions of suspended sediment in Chesapeake Bay. When analyzed as mineral grains, however, estuarine suspended particulate matter is found to be composed of a wide range of grain sizes including the finest submicron size fractions showing that flocculation is important in turbidity maximum formation. This fine labile sediment contains the highest relative concentrations of colloidal substances and is of prime importance in environmental considerations.

The hypothetical path a fine mineral grain might follow within an estuary illustrates the effect of flocculation in particulate trapping. When grains enter the estuary, the increase in salinity, total suspended sediment concentration, and relative proportion of organic matter, as well as the decrease in current turbulence all promote flocculation. If conditions are correct, the grain will become part of a macro-floc with sufficiently high settling rate to overcome upward and seaward advection and settle into the more saline bottom flow. At the bottom the fragile low-density macro-flos may be torn apart by high shear forces. The floc fragments are resuspended to become part of the turbidity maximum and contribute to trapping of other grains. The regular tidal variations in current speeds and water depths promote alternating settling and upward advection of sediment. At times both upward advection of fine particles and downward settling of macro-flocs may occur in the same flow. If during periods of slack water or low current, a sufficiently thick and dense bottom layer is formed to withstand the next high current stage, a permanent deposit results (Creutzberg and Postma 1979). Stationary deposition may also occur on a neap-spring cycle (Kirby and Parker 1983). Mud silt interlayering and fine laminae characterize tidal and estuarine sediment (eg. VanStraaten 1963) and probably result from a process of incomplete resuspension of flocculated material similar to that proposed by Stow and Bowen (1978, 1980) for the origin of turbidite layering. By this mechanism, the coarsest constituent of macro-flocs as well as any grains which have settled singly, sediment out continuously. The fines, however, deposit as an intermittent layer whenever the near bottom concentrations have built up sufficiently to dampen turbulence or create more resistant flocs.

The balance between sediment supply, trapping, and loss to the sea is a dynamic equilibrium which shifts with tidal currents and seasonal effects, as well as with irregular variables such as storm events and man-made changes including dredging activities. The controlling factors are not sufficiently well-known to allow modelling of the sort possible for other oceanographic variables. The effect of flocculation in general is to decrease the time over which elevated sediment concentrations persist. Measurements over tidal cycles show that high suspended sediment concentrations formed during high current phases return to a relatively constant base level concentration characteristic for all depths during slack conditions and for near surface depths during most of the tidal cycle (Schubel 1971; Krone 1972; Kranck 1979, 1981).

Nichols (1977) found that a greater proportion of storm sediment than of normal runoff sediment becomes trapped in an estuary. The constant dampening of concentration levels by concentration equalization greatly limits particulate flushing. Labile sediment once introduced to an estuary whether by natural or artificial means tends to remain there. Many environmental problems such as fluid mud formation and high concentrations of toxic chemicals are the result.

ACKNOWLEDGMENTS

I am grateful to D. Eisma for discussions on floc dynamics and for the use of Figures 1 and 2A. T.G. Milligan performed all the size analysis reported in the paper. Figure 4 is reprinted from Kranck (1980) with permission of the National Research Council of Canada.

REFERENCES CITED

Alldredge, A. L. 1976. Discarded appendicularian houses as source of food, surface habitats and particulate organic matter in planktonic environments. *Limnol. Oceanogr. 21:* 14-23.

Alldredge, A. L. 1979. The chemical composition of macroscopic aggregates in two neritic seas. *Limnol. Oceanogr. 24:* 855-866.

Bishop, J. K. B., J. M. Edmonds, D. R. Kellon, M. P. Bacon and W. B. Silker. 1977. The chemistry, biology and vertical flux of particulate matter from the upper 400 m of the equatorial Atlantic Ocean. *Deep-Sea Res. 24:* 511-548.

Creutzberg, F. and H. Postma. 1979. An experimental approach to the distribution of mud in the Southern North Sea. *Netherlands Jour. Sea Res. 13:* 99-116.

Dyer, K. R. 1971. Sedimentation in estuaries, pp. 10-32. *In:* R.S.K. Barnes and J. Green (eds.), *The Estuarine Environment.* Applied Science Publishers, London.

Eisma, D., J. Boon, R. Groenewegen, V. Ittekkot, J. Kalf and W. G. Mook. 1984. Observations on macro-aggregates, particle size and organic composition of suspended matter in the Ems Estuary. *Mitteilungen aus dem Geologisch-Paläontologischen Institut der Universität Hamburg,* Heft 55, in press.

Eisma, D. and W. W. C. Gieskes. 1977. Particle size spectra of non-living suspended matter in the southern North Sea. *Interne Verslagen, Nederlands Instituut voor Ondenzoek der Zee, Texel,* 22 p.

Festa, F. and D. V. Hansen. 1978. Turbidity maxima in a partially mixed estuary: a two dimensional numerical model. *Estuar. Coastal Mar. Sci. 7:* 347-359.

Gibbs, R. J. 1982. Floc breakage during HIAC Light-Blocking Analysis. *Environ. Sci. Technol. 16:* 298-299.

Hannah, S. A., G. G. Cohen and G. G. Robech. 1967. Measurement of floc strength by particle counting. *Jour. Amer. Water Works. Assoc. 59:* 843-858.

Inoue, N., S. Nishizawa and M. Fukuda. 1955. The perfection of a turbidity meter and the photographic study of suspended matter and plankton in the sea using an undersea observation chamber, pp. 53-58. *In: Proceedings UNESCO symposium on physical oceanography,* Tokyo, UNESCO, New York.

Jannasch, H. W. 1973. Bacterial content of particulate matter in offshore surface waters. *Limnol. Oceanogr. 18:* 340-342.

Judson, S. 1968. Erosion of the land or What's happening to our continents? *Amer. Sci. 56:* 356-379.

Kirby, R. and W. R. Parker. 1983. Distribution and behaviour of fine sediment in the Severn Estuary and inner Bristol Channel, U.K. *In:* Gordon, D. C. Jr. and A. S. Houston (eds.), Proceedings of the Symposium on the Dynamics of Turbid Coastal Environments. *Can. Jour. Fish. Aquatic Sci. 40* Suppl. 1: 83-95.

Kranck, K. 1975. Sediment deposition from flocculated suspension. *Sedimentology 22:* 111-123.

Kranck, K. 1979. Dynamics and distribution of suspended particulate matter in the St. Lawrence Estuary. *Nat. Can. 106:* 163-179.

Kranck, K. 1980. Experiments on the significance of flocculation of fine-grained sediment in still water. *Can. Jour. Earth Sci. 17:* 1515-1526.

Kranck, K. 1981. Particulate matter grain-size characteristics in a partially mixed estuary. *Sedimentology 28:* 107-114.

Kranck, K. and T.G. Milligan. 1979. Methods of particle size analysis using a Coulter counter. Bedford Institute of Oceanography Report Series BI-R-79-7, 48 pp.

Kranck, K. and T. G. Milligan. 1980. Macro-flocs; the production of marine snow in the laboratory. *Mar. Ecol. Progr. Ser. 3:* 19-24.

Krone, R.B. 1962. Flume studies of the transport of sediment in estuarial shoaling processes. Final report to U.S. Army Corps. of Engineers, Contract No. Da-04-203: 110 pp.

Krone, R. B. 1972. A field study of flocculation as a factor in estuarial shoaling processes. Technical Bulletin 19. Committee on Tidal Hydraulics, Corps of Engineers, U.S. Army. Waterways Experimental Station, Vickburg, Mississippi.

McCave, I. N. 1979. Suspended sediment, pp. 131-164, In: K. R. Dyer (ed.), *Estuarine Hydrography and Sedimentation.* Cambridge Univ. Press, Cambridge.

Meade. R. H. 1968. Relation between suspended matter and salinity in estuaries of the Atlantic seaboard. U.S.A. Int. Ass. Sci. Hydrol., Gen. Ass. Bern 1967, vol. 4 (I.A.S.H. Publ. 78).

Meade, R. H. 1972. Transport and deposition of sediments in estuaries. *Mem. Geol. Soc. Am. 133:* 91-119.

Mignoit, C. 1968. Etude de propriétés physiques de différents sédiments très fins et de leur comportement sous des actions hydrodynamiques. La Houille Blanche, 7, 591-620.

Milliman, J.D. and R.H. Meade. 1983. World-wide delivery of river sediment to the oceans. *Jour. of Geology 1:* 1-21.

Neihof, R. A. and G. L. Loeb. 1974. Dissolved organic matter in seawater and the electric charge of immersed surfaces. *J. Mar. Res. 32:* 5-12.

Nelsen, T. A. 1981. The application of Q-mode factor analysis to suspended particulate matter studies: Examples from the New York Bight Apex. *Mar. Geol. 39:* 15-31.

Nichols, M. M. 1977. Response and recovery of an estuary following a river flood. *J. Sedimentary Petrology 47:* 1171-1186.

Nishizawa, S., M. Fukuda and N. Inoue. 1954. Photographic study of suspended water and plankton in the sea. *Bull. Fac. Fish Hokkaido Univ. 5:* 36-40.

Owen, M.W. 1971. The offset of turbulence on the settling velocities of silt flocs: *Int. Assoc. Hydraul. Res., Congr., Proc.,* No. 14, Vol. 4: 27-32.

Postma, H. 1967. Sediment transport and sedimentation in the estuarine environment, pp. 158-179. *In:* G. H. Lauff (ed.), *Estuaries.* Am. Ass. Adv. Sci. Publ. 83, Washington, D.C.

Schubel, J. R. 1969. Size distributions of the suspended particles of the Chesapeake Bay turbidity maximum. *Netherlands J. Sea Res. 4:* 283-309.

Schubel, J. R. 1971. Tidal variation of the size distribu tion of suspended sediment at a station in the Chesapeake Bay turbidity maximum. *Netherlands J. Sea Res. 5:* 252-266.

Shanks, A. L. and J. D. Trent. 1979. Marine show: Microscale nutrient patches. *Limnol. Oceanogr. 24:* 850-854.

Sheldon, R. W. 1968. Sedimentation in the Estuary of the River Crouch, Essex, England. *Limnol. Oceanogr. 13:* 72-83.

Stow, D. A. V. and A. J. Bowen. 1978. Origin of lamination in deep-sea, fine grained sediments. *Nature, 274:* 324-328.

Stow, D. A. V. and A. J. Bowen. 1980. A physical model for the transport and sorting of fine-grained sediment by turbidity currents. *Sedimentology 27:* 31-46.

Suzuki, N. and K. Kato. 1953. Studies on suspended materials marine snow in sea. Part I. Sources of marine snow. *Hakkaido Univ. Fac. Fish.* Bull. 4: 132-135.

Syvitski, J. P. M., G. B. Fader, H. W. Josenhans, B. MacLean and D.J.W. Piper. 1983. Seabed Investigations of the Canadian East Coast and Arctic using *Pisces IV*. *Geoscience Canada 10:* 59-67.

Trent, J. D., A. L. Shanks and M. W. Silver. 1978. *In situ* and laboratory measurements on macroscopic aggregates in Monterey Bay, California. *Limnol. Oceanogr. 23:* 626-635.

Van Straaten, L. M. J. U. 1963. Aspects of Holocene sedimentation in the Netherlands. *Ver. K. Ned. geol.-mijnb. Genoot. Geol. Ser. 21:* 149-172.

Zabawa, C. F. 1978. Microstructure of agglomerated suspended sediments in Northern Chesapeake Bay Estuary. Science *20:* 49-51.

CHEMICAL–GEOCHEMICAL PROCESSES

THE FATES OF URANIUM AND THORIUM DECAY SERIES NUCLIDES IN THE ESTUARINE ENVIRONMENT.

J. Kirk Cochran

Marine Sciences Research Center
State University of New York
Stony Brook, New York

Abstract: Naturally occurring radionuclides of the uranium and thorium decay series are useful in studying estuarine geochemical processes ranging from the removal of reactive nuclides from the water column to the mobilization and transport of chemical species in the sediment column. Although uranium isotopes generally show conservative behavior during estuarine mixing, removal from solution can occur at low salinities or in estuaries high in dissolved organic matter. Uranium in estuarine sediments also displays a diagenetic chemistry linked to oxidation-reduction. In contrast to uranium, radium isotopes are released during estuarine mixing. Mechanisms of release include desorption from suspended sediments and the mobilization of radium from bottom sediments. The latter is demonstrated by measurements of ^{226}Ra and ^{228}Ra in sediment pore waters. The thorium isotopes ^{234}Th and ^{228}Th, produced by decay of soluble parents ^{238}U and ^{228}Ra, are good analogs for chemical species which interact strongly with particles in the estuary. Both have been used to estimate removal times of thorium from solution

ISBN 0-12-405070-0

to particles. The distribution of ^{234}Th and ^{228}Th in sediments is governed primarily by particle mixing by the benthic fauna, and both isotopes serve as chronometers to determine mixing rates. ^{210}Pb and its daughter, ^{210}Po, also serve as tracers of reactive nuclides in the estuarine environment. Both are removed rapidly to bottom sediments. Although ^{210}Pb is a longer lived tracer than ^{234}Th or ^{228}Th, its depth distribution in estuarine sediments is also affected by bioturbation. Successful reconstruction of sediment chronology in the estuary thus requires the application of several tracers with different half-lives and independent information on the sediment budget of the area.

INTRODUCTION

Successful understanding of the estuary as a geochemical filter requires knowledge not only of the distributions of dissolved and particulate chemical species but also of the rates of their input, release, and accumulation. While synoptic observations may serve to document the former, the latter is less readily determined. The naturally occurring radionuclides of the uranium and thorium decay series form a unique class of tracers which can be used to document both chemical behavior and the rates of the processes which affect it. The decay series contain isotopes of elements whose properties range from soluble (U, Ra) to particle-reactive (Th, Po, Pb) and which display both conservative and non-conservative behavior in estuaries.

Useful information about the rates of estuarine processes is gained because the decay series contain several instances of parent-daughter radionuclide pairs in which the parent possesses a strikingly different chemical behavior from the daughter. In a closed system, the daughter/parent activity ratio will approach a value governed by radioactive equilibrium (usually the secular equilibrium value of 1.0). Perturbation of the equilibrium state by estuarine geochemical processes sets up a condition of radioactive disequilibrium which allows the rates of processes causing it to be determined. The parent/daughter couples which have been useful in deciphering estuarine geochemistry are summarized in Table 1.

There is a considerable body of literature detailing uranium and thorium series nuclide distributions in rivers, estuaries and coastal waters of the continental shelf and slope. This paper focuses on examples of efforts to use the decay series nuclides in understanding the geochemistry of estuarine waters and sediments.

Methods of analysis usually involve chemical separations of the isotopes by element followed by determination of the radioactivity (or, more simply, activity) using ionization detectors. Methods for the radiochemical separations are summarized in Santschi, Li and Ball (1979) for water samples, Krishnaswami and Sarin (1976) and Koide and Bruland (1975) for sediments, and Cochran (1979), Cochran and Krishnaswami (1980) and Sholkovitz, Cochran and Carey (1983) for sediment pore water. For discussions of methods of counting low activity samples, the reader is referred to the review by Ivanovich (1982).

TABLE 1: Parent/Daughter pairs of the uranium and thorium decay series useful in studying estuarine processes (Daughter half-lives are in parentheses).

Parent	Daughter	Application of Disequilibrium
^{238}U	^{234}Th (24 days)	Rate of removal of Th from solution, particle mixing rates in estuarine sediments.
^{228}Ra	^{228}Th (1.9 years)	Rate of removal of Th from solution, particle mixing and short term sediment accumulation rates.
^{226}Ra (^{222}Rn)	^{210}Pb (22.3 years)	Rate of removal of Pb from solution, particle mixing and sediment accumulation rates.
^{210}Pb	^{210}Po (138 days)	Rate of removal of Po from solution.
^{230}Th	^{226}Ra (1622 years)	Desorption of Ra from suspended particles, fluxes across sediment-water interface.
^{232}Th	^{228}Ra (5.75 years)	Desorption of Ra from particles, fluxes out of sediments, mixing rates in coastal waters.
^{228}Th	^{224}Ra (3.6 days)	Fluxes from estuarine sediments, mixing rates in water column.
^{226}Ra	^{222}Rn (3.8 days)	Fluxes from sediments (irrigation rates by infauna), air-water gas exchange, mixing rates in water column.

URANIUM

The uranium isotopes ^{238}U, ^{235}U and ^{234}U, are long-lived and enter the ocean principally by rivers. As primordial isotopes, ^{238}U and ^{235}U are chemically weathered in their abundance ratio. ^{234}U is a decay product in the ^{238}U decay series and is preferentially brought into solution as an indirect result of alpha recoil or through enhanced weathering of mineral lattice uranium sites damaged by radioactive decay. As a result, ^{234}U is present in rivers in excess of the activity expected from equilibrium with ^{238}U. $^{234}U/^{238}U$ activity ratios in river water often fall in the range of 1.20 - 1.30 while the open ocean has a ratio of 1.14 (Scott 1982; Osmond and Cowart 1976). In general, there is a good positive correlation between dissolved uranium and HCO_3^- as well as total dissolved solids in world rivers (Turekian and Cochran 1978; Mangini et al. 1979; Borole, Krishnaswami and Somayajulu 1982; Figuères, Martin and Thomas 1982) suggesting that the extent of chemical weathering is controlling the variation in uranium concentration from river to river.

In the river-estuarine system, the flux of uranium may be modified by uptake and release by mechanisms which are poorly understood. Lewis (1976) demonstrated that uranium removal occurs in rivers by showing that uranium vs. calcium concentrations followed a non-linear trend in the Susquehanna River, USA. Data on the geochemical behavior of uranium in different estuaries appear to show both cases of conservative mixing with sea water and of removal from solution. For example, Borole and co-workers (1977, 1982) have shown that, above chlorosities of 0.14 g/ℓ, uranium is mixed conservatively in three estuaries on the west coast of India. Martin, Nijampurkar and Salvadori (1978) and Figuères et al.(1982)

obtained similar results in studies of the Gironde (France) and Zaire (Africa) estuaries. In the Charente (France) estuary, Martin et al. (1978) noted that river concentrations of uranium were variable due to the effluents from phosphate processing plants, and they used the highest observed river concentration to conclude that removal was occurring. However, the variability in the riverine end member may obscure a conservative mixing trend. Indeed, Scott (1982) demonstrated temporal variability (a factor of $\sim$ 2) in dissolved uranium in the Mississippi River (USA) and suggested that release or uptake of uranium onto particles was slight, if it occurred at all.

While the general behavior of uranium was conservative in the Narbada (India) estuary studied by Borole et al. (1982) and in the Zaire (Africa) (Figuères et al. 1982), in each case removal of uranium at low chlorosities appeared to occur. Uranium concentrations decrease with increasing chlorosity to 0.14 mg/ℓ in the Narbada estuary (Borole et al. 1982) and to $\sim$ .03 mg/ℓ in the Zaire (Figuères et al. 1982). Borole et al. (1982) point out that this pattern could be explained by incomplete mixing of the possible sources of the fresh-water end member. Figuères et al. (1982) attribute the decrease to uranium removal in near-anoxic waters near the head of the Zaire Canyon. More dramatic examples of uranium removal at higher salinities have been presented by Osmond and Cowart (1976) and by Maeda and Windom (1982). In the former case, uranium appears to behave nonconservatively over salinities of 21-33% in Tampa Bay, Florida. However, the situation is complicated by the fact that uranium is released to the bay by a phosphate processing plant, forming a source of uranium in addition to rivers. Maeda and Windom (1982) have shown uranium removal occurring in the Ogeechee (Georgia) estuary during periods of low river discharge. Removal is not apparent during times of high discharge. Maeda and Windom (1982) suggest that uranium removal in these organic-rich estuaries may be linked to the flocculation of organic matter or to the precipitation of iron and manganese. Thus, most of the uranium data from estuaries indicate conservative mixing. Removal of U, if it occurs, takes place in the river, at low salinities, or in estuaries high in dissolved organic matter.

Although much of the estuarine data indicate that uranium passes through the estuary without appreciable removal, the redox processes operating in muddy coastal sediments can serve as a sink for this element. Evidence for uranium uptake in estuarine sediments comes from Long Island Sound and Buzzards Bay, USA (Fig. 1). In Long Island Sound, Thomson, Turekian and McCaffrey (1975) and Aller and Cochran (1976) noticed increases of uranium with depth in the sediment column which they attributed to possible uptake. Aller and Cochran (1976) also observed an increase in the $^{234}U/^{238}U$ activity ratio with increasing concentrations of uranium in the sediments. Such patterns imply removal of uranium (with a sea water $^{234}U/^{238}U$ activity ratio of 1.14) from sediment pore water. Based on their observations of elevated uranium concentrations and $^{234}U/^{238}U$ activity ratios in the upper 5 cm of a Delaware Bay (USA) salt marsh core, Church, Lord and Somayajulu (1981) conclude that a similar removal of uranium may be occurring in salt marsh sediments. As a test of the possibility of uranium removal from pore water in reducing nearshore sediments, Fig. 2 shows pore water uranium profiles from sediment cores taken in Buzzards Bay, USA (Location in Fig. 1). Relative to concentrations in the overlying water, uranium is removed from solution, reaching a minimum which coincides with the depth of the pore water iron maximum within the resolution of the data. Pore water uranium concentrations then increase with depth to $\sim$ 30 cm and again

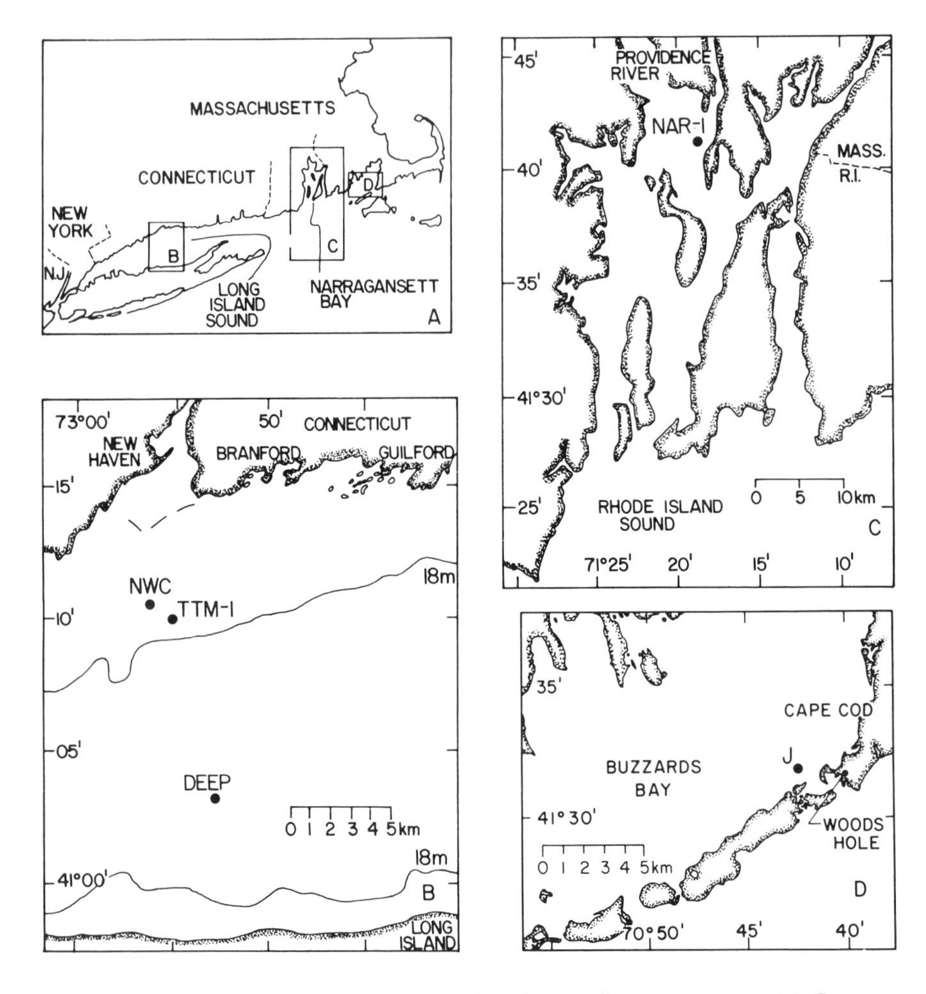

FIGURE 1. Map of Long Island Sound, Narragansett Bay and Buzzards Bay showing locations of coring sites.

FIGURE 2 (opposite). ^{238}U activity (dpm/kg) and Fe (open squares) and Mn (solid triangles) concentrations (μmole/liter) in the pore water of sediment cores collected at Station J in Buzzards Bay. The cores were collected in 12 m of water in muddy sediment. Values for the overlying water are plotted above the dashed lines. The high manganese value in the overlying water for the core taken 1 September 1982 may reflect disturbance of the top ~1 cm of the core, (Data from Cochran, Sholkovitz and Carey, unpublished).

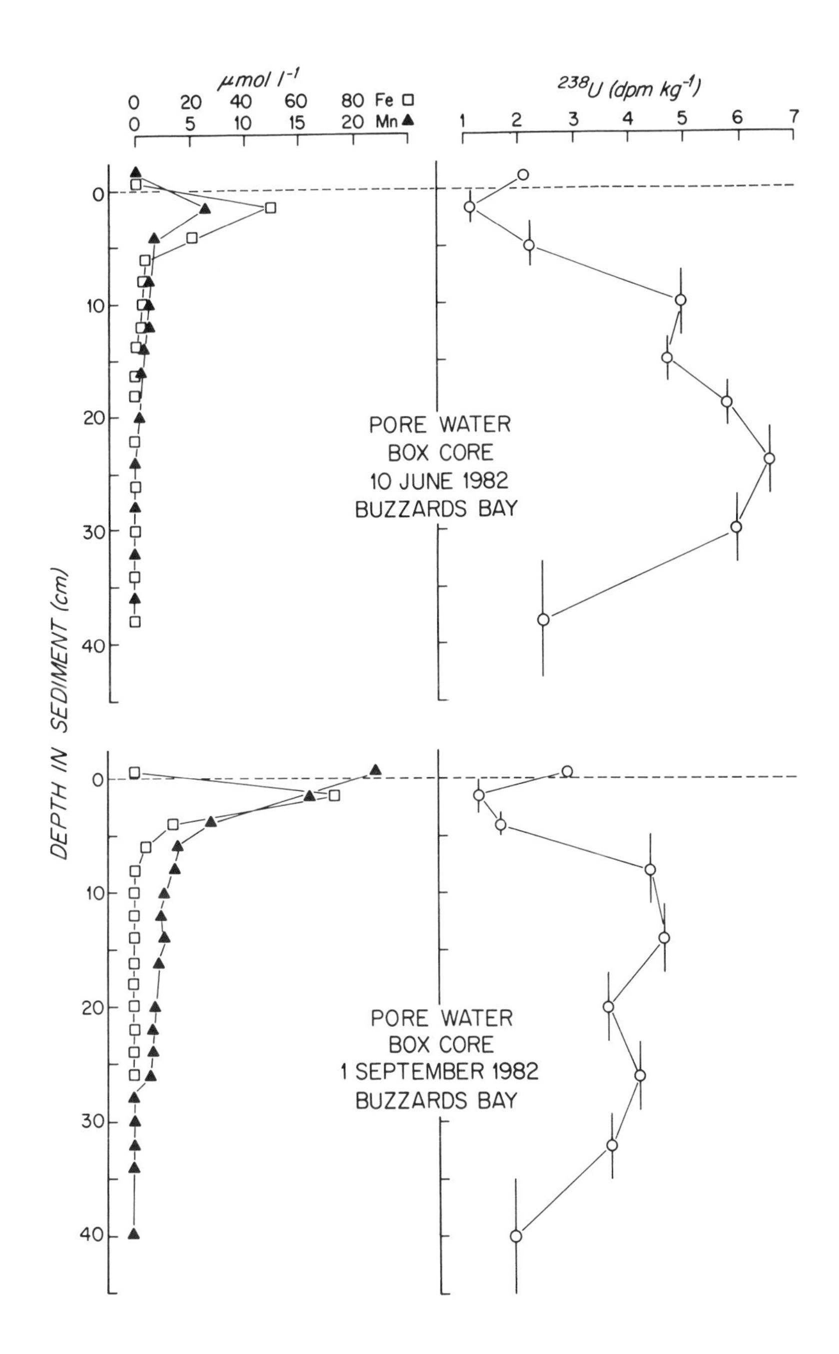
μmol l⁻¹
0 20 40 60 80 Fe □
0 5 10 15 20 Mn ▲
238U (dpm kg⁻¹)
1 2 3 4 5 6 7
DEPTH IN SEDIMENT (cm)
0
10
20
30
40
PORE WATER
BOX CORE
10 JUNE 1982
BUZZARDS BAY
PORE WATER
BOX CORE
1 SEPTEMBER 1982
BUZZARDS BAY

decrease. The coincidence of the uranium minimum and iron maximum suggests that reduction of U^{6+} (present in oxygenated sea water as $UO_2(CO_3)_3^{4-}$) to the relatively insoluble U^{4+} may be occurring. Both alkalinity and dissolved organic carbon increase with depth in the core, and the increase in pore water uranium may be linked to the stabilization of a soluble uranium carbonate complex or an organic-uranium complex in the pore water. Uranium depletion at depths greater than ∿ 30 cm coincides with high concentrations of dissolved sulfide in the pore water. Kolodny and Kaplan (1973) also observed increases in pore water uranium concentrations with depth in Saanich Inlet sediments. They attributed these changes to a combination of the effects of Eh variation and change in ΣCO_2 down the core.

A uranium flux from the overlying water into Buzzards Bay sediments may be calculated by assuming a one-dimensional transport and using the uranium gradient across the sediment-water interface in Fick's first law (Berner 1980). Using values for porosity of 0.8 and a diffusion coefficient of U in pore water of $2 \times 10^{-6} cm^2/s$ (corrected for tortuousity), the calculated flux is ∿ 4×10^{-2} dpm/cm^2y. For an average water depth of 11 m and uranium concentration of 2 dpm/ℓ, the residence time of uranium in Buzzards Bay with respect to its removal into the sediments is ∿ 50 years. It is unlikely that uranium depletions would be observed in the overlying water unless the mean residence time of the water in the Bay is appreciably longer than 50 years. Thus, removal of uranium into the sediment of Buzzards Bay is occurring yet would not be apparent in the estuarine water column.

RADIUM AND RADON

In contrast to uranium, radium shows strongly non-conservative behavior in estuaries. The most widely studied isotopes of radium are ^{226}Ra (half-life - 1622y) and ^{228}Ra (half-life - 5.75y). Both isotopes are brought into groundwater through chemical weathering, desorption, and alpha recoil mechanisms.

Radium is supplied to estuarine waters by inflow from both rivers and the ocean. Two important additional sources have been identified: desorption from river-borne suspended sediments and mobilization and diffusion from estuarine sediments. Release of ^{226}Ra and ^{228}Ra from coastal sediments in general (primarily through diffusion) has been proposed by Koczy et al. (1957), Blanchard and Oakes (1965), Moore (1969) and Feely et al.(1980) to account for high concentrations of these isotopes in coastal water relative to river or open ocean sea water. Li et al. (1977) were the first to demonstrate release of ^{226}Ra during estuarine mixing. A plot of ^{226}Ra concentrations as a function of salinity in the Hudson estuary (USA) showed that the data fell above the conservative mixing line, indicating release of ^{226}Ra in the estuary. Assuming that the salinity distribution of the Hudson estuary was in semi-steady state during the period of observation, Li et al. (1977) calculated the total amount of ^{226}Ra released to be 2.1×10^{12} dpm ^{226}Ra/y (or $\sim$ 2 dpm/cm^{2}y for the area of the Hudson they studied). They compared this value with estimates of the amount of ^{226}Ra released from suspended sediments (obtained from simple desorption experiments) to conclude that about 70% of the release of ^{226}Ra could be accommodated through desorption. The observations of Li et al. (1977) were extended by Li and Chan (1979) to demonstrate that the related element Ba displays non-conservative behavior similar to ^{226}Ra in the Hudson estuary. Refinement of the sorption experiments and balance calculations discussed above again indicated that desorption of ^{226}Ra from suspended sediments was able to account for most of the ^{226}Ra release in the estuary. On a global scale (using the worldwide total suspended flux from rivers estimated by Garrels and Mackenzie, 1971), Li and Chan (1979) estimated that 17 to 43% of the total ^{226}Ra flux from coastal regions could be accounted for by desorption during estuarine mixing.

Non-conservative behavior of ^{226}Ra during estuarine mixing was confirmed by Elsinger and Moore (1980) in the Pee Dee River-Winyah Bay estuary of South Carolina. Their observations on dissolved ^{226}Ra showed departures from the conservative mixing relationship which were matched by decreasing concentrations of ^{226}Ra in the suspended particles. Based on the suspended sediment ^{226}Ra data under average discharge conditions, Elsinger and Moore (1980) concluded that about 75% of the non-conservative fraction of ^{226}Ra was released in low salinity water (< 6 $^{o}/oo$) with the remainder added at salinities up to 12 $^{o}/oo$. Elsinger and Moore (1980) also noted that the total amount of ^{226}Ra supplied by the fresh water accounted for the total ^{226}Ra in the estuary; the primary change was in the partitioning between solution and suspended sediment. This conclusion is essentially the same as that reached by Li et al. (1977) for the Hudson estuary.

Another important source of radium to the estuary is its mobilization in deposited sediments and migration into the overlying water. Cochran (1979) determined pore water and solid phase radium concentrations in sediment cores taken in Long Island Sound and Narragansett Bay (USA). These results are presented in Figs. 3 through 6. Station NWC lies in 15 m of water in central Long Island Sound (Fig. 1) and is characterized by dense populations of the deposit feeding bivalves *Yoldia limatula* and *Nucula annulata* and the mobile, burrowing polychaete *Nephtys incisa*. Aller and Cochran (1976) used excess ^{234}Th distributions to show that the sediments at this station were mixed rapidly to ∿ 4 cm, with the rate of mixing varying seasonally. Benninger et al. (1979) showed that excess ^{210}Pb was constant to ∿ 4 cm depth below which it decreased regularly to 15 cm. The ^{210}Pb profile is consistent with rapid mixing in the upper 4 cm and less rapid mixing to 15 cm. Excess ^{210}Pb is observed associated with burrows up to depths of 1 meter. Such burrows are probably created by the stomatopod shrimp *Squilla* sp.

The radium isotopic data on the solid phase of the sediments in core NWC-R-6 studied by Benninger et al. (1979) are presented in Figure 3. To correct for changes in sediment composition or grain size, ^{228}Ra and ^{226}Ra activities are normalized to their thorium parents, ^{232}Th and ^{230}Th. Both $^{228}Ra/^{232}Th$ and $^{226}Ra/^{230}Th$ activity ratios are less than the closed system secular equilibrium value of 1.0 at the core top. ^{228}Ra attains equilibrium with ^{232}Th by about 30 cm. $^{226}Ra/^{230}Th$ activity ratios increase with depth but do not reach 1.0. The profiles are governed by the rate of particle mixing, the rate of production of radium by decay, and the extent of migration and loss of radium from the core. That the latter is occurring is demonstrated by pore water radium profiles (Fig. 4) obtained from cores taken at Station NWC about two months prior to core NWC-R-6. Substantial enrichments of radium are observed in pore water compared to bottom water and the gradients are such that a flux to the overlying water is produced. Radium is mobilized in estuarine sediments by several mechanisms. A radium atom may be recoiled from the sediment grain to the pore fluid during its production

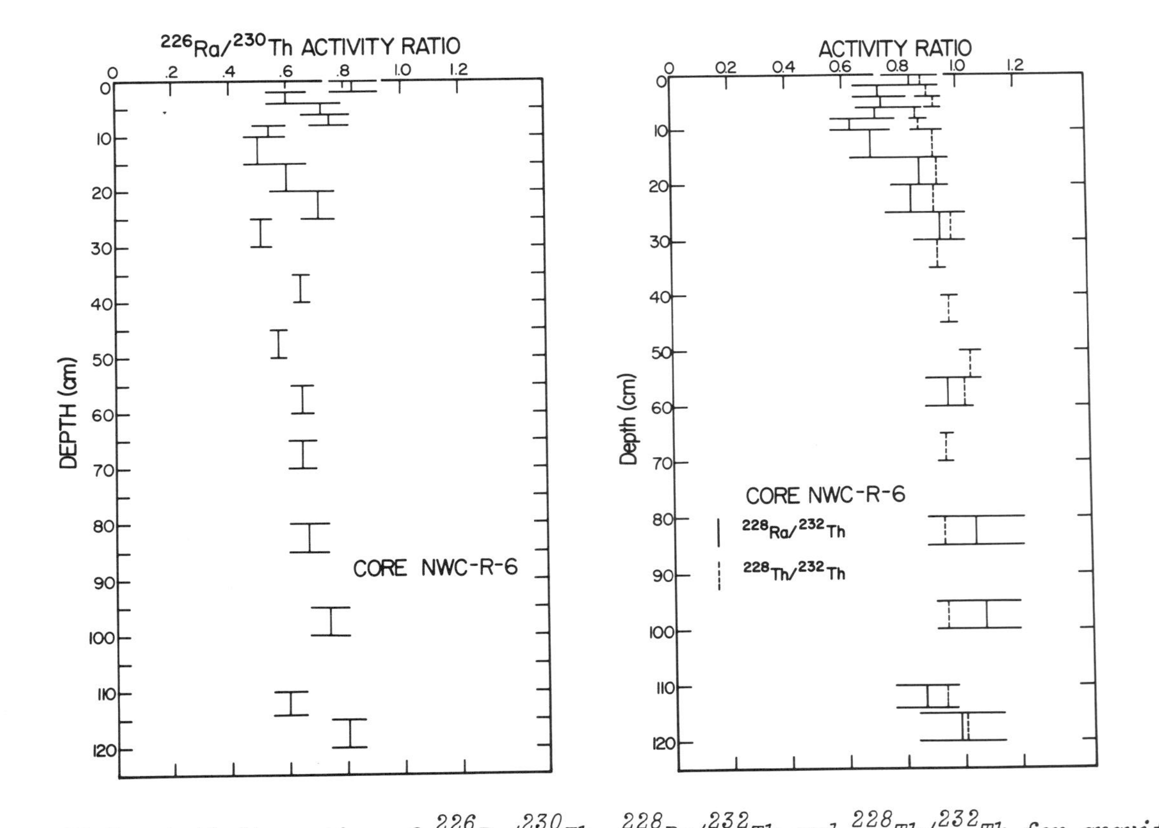

FIGURE 3. Activity ratios of $^{226}Ra/^{230}Th$, $^{228}Ra/^{232}Th$ and $^{228}Th/^{232}Th$ for gravity core (Oct. 29, 1975) at Station NWC, Long Island Sound. Values represent analyses of sediment solid phase. Vertical bars—sampling interval, horizontal bars—1σ counting errors.

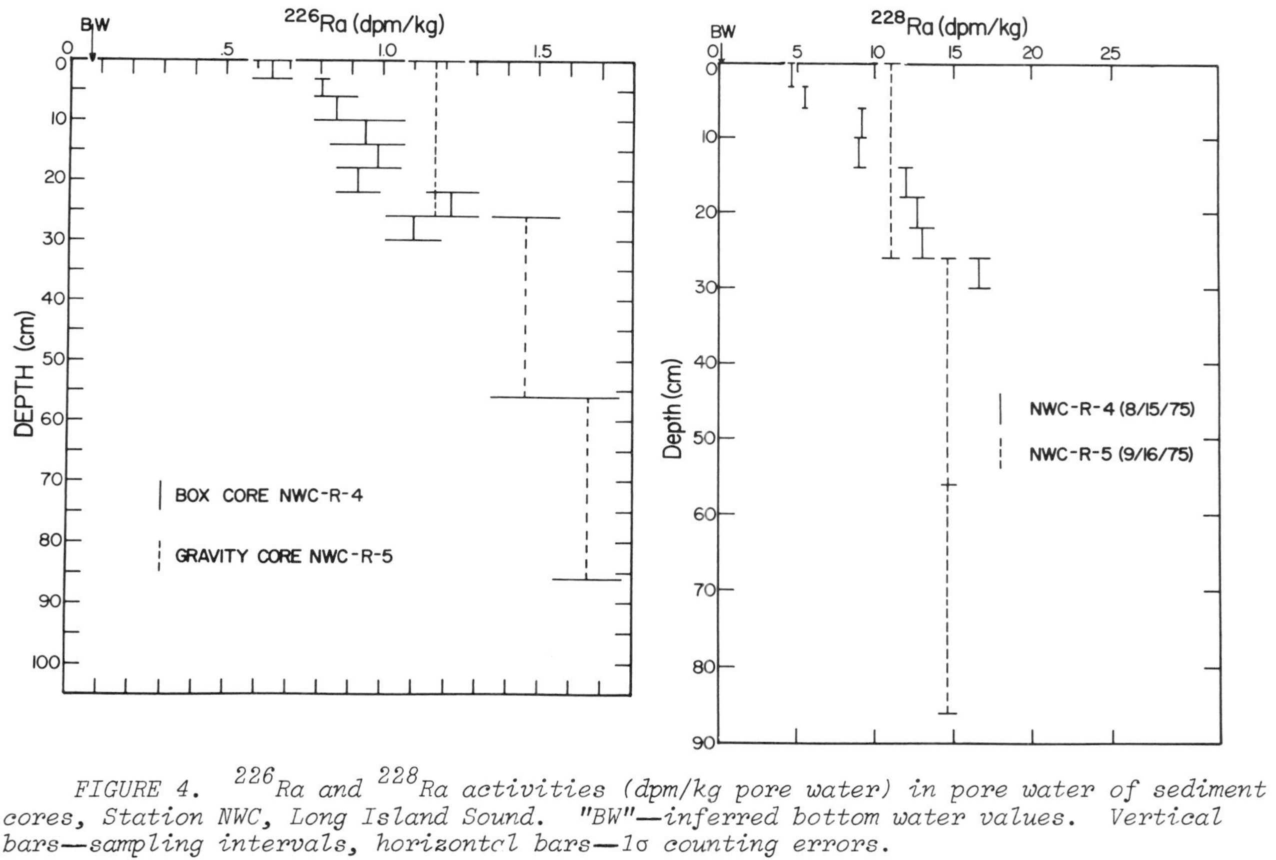

FIGURE 4. ^{226}Ra and ^{228}Ra activities (dpm/kg pore water) in pore water of sediment cores, Station NWC, Long Island Sound. "BW"—inferred bottom water values. Vertical bars—sampling intervals, horizontal bars—1σ counting errors.

from Th decay. Radium also may be removed from the overlying water column, for example in association with Mn oxides, and the decomposition of the carrier phase once deposited in the sediments (such as by reduction of Mn^{4+}) may effect release of radium to the pore water. The different concentrations of radium isotopes in the pore waters are set by their relative release rates to solution. Because of its shorter half-life, ^{228}Ra more quickly achieves a steady state concentration, at least in sediments like those at Station NWC which show little change in solid phase ^{232}Th with depth.

The core from Narragansett Bay was taken in 7 m of water near the mouth of the Providence River (Fig. 1). The bottom fauna in this area, based on the observations of Myers and Phelps (1978), is similar to that at Station NWC. Benthic camera studies by D. Rhoads (presented in Myers and Phelps 1978) suggest a high degree of particle reworking and biogenic exchange of pore water at this site. The pore water radium profiles in Fig. 5 show more gradual increases with depth relative to Station NWC in Long Island Sound. Values of ^{228}Ra at depth in the core are similar to those at Station NWC, however. Likely explanations for the difference include the possibilities that the Narragansett Bay site is more frequently subject to major physical and biological mixing of the sediment to depths of ∿ 25 cm or that the "ventilation" of the sediment through irrigation of faunal burrows is more intense. In support of the former possibility, X-radiographic and ^{210}Pb data from a nearby core taken at the same time (Goldberg et al. 1977) show discrete shell layers and

FIGURE 5. ^{226}Ra and ^{228}Ra activities (dpm/kg pore water) in the pore water of box core NAR-1 taken in Narragansett Bay. See Fig. 1 for sampling location.

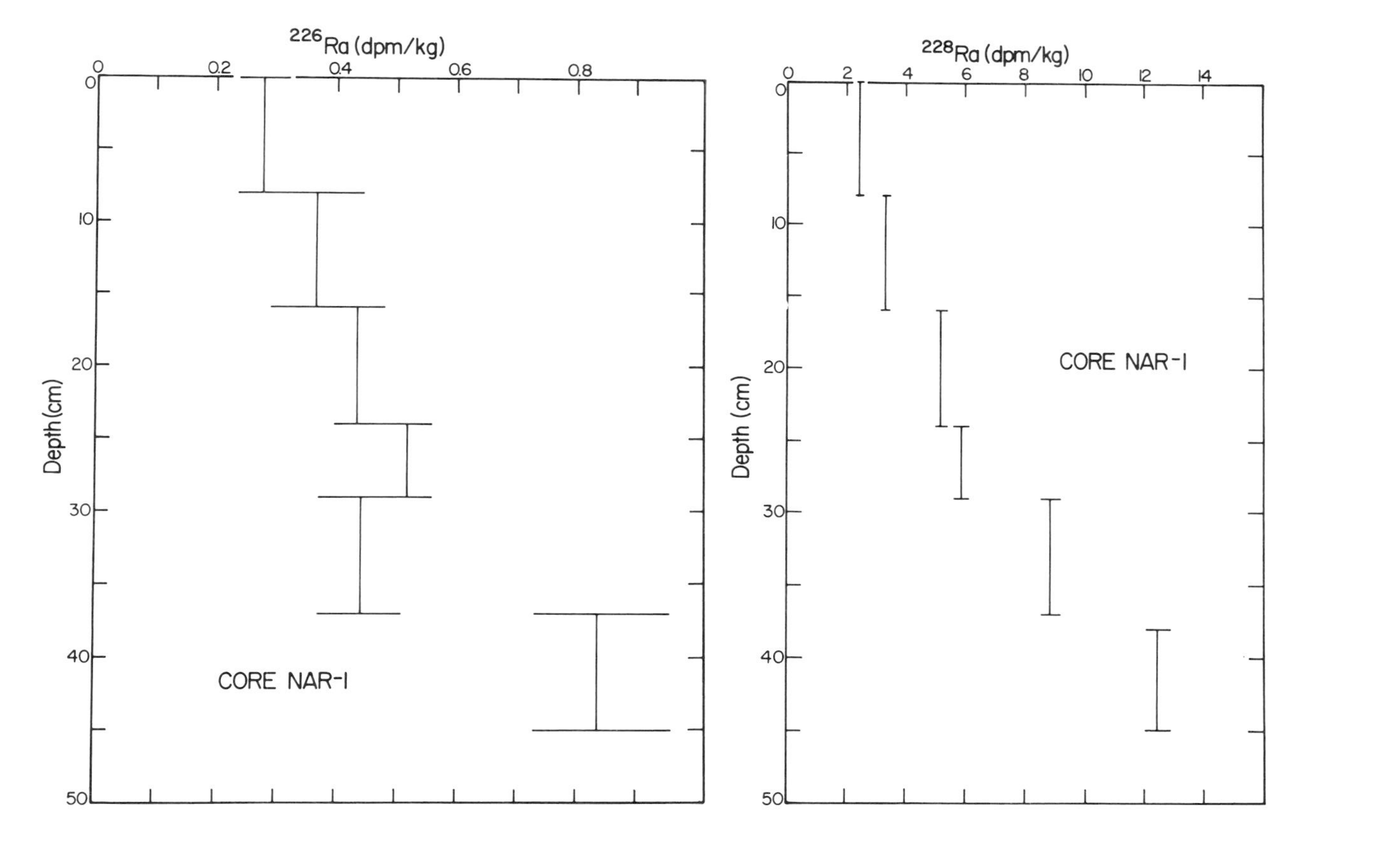

226Ra (dpm/kg)
0
0.2
0.4
0.6
0.8
Depth (cm)
0
10
20
30
40
50
CORE NAR-1
228Ra (dpm/kg)
0
2
4
6
8
10
12
14
Depth (cm)
0
10
20
30
40
50
CORE NAR-1

constant excess ^{210}Pb activity from 15-30 cm depth.

The calculation of fluxes from estuarine sediments is difficult based on pore water data alone. Aller (1977) pointed out that fluxes of ammonia, phosphate, iron, and manganese calculated from vertical pore water gradients at Station NWC required apparent diffusion coefficients of 1×10^{-5} cm^2/s to match the measured values. Fluxes for ^{228}Ra can be calculated from solid phase data, however. Based on data such as Fig. 3, Cochran (1979) calculated values of 1.3 dpm/cm^2y for Station NWC in Long Island Sound and 0.8 dpm/cm^2y for the Narragansett Bay site. The latter compares with the value of 1.2 dpm/cm^2y obtained by Santschi et al. (1979) from water column radium data in Narragansett Bay. Fluxes of ^{226}Ra are more difficult to determine from solid phase data, but based on the pore water profiles of Figs. 4 and 5, appear to be lower than those of ^{228}Ra. Moore (1981) made a similar observation on radium fluxes in Chesapeake Bay (USA). He attributed the release of radium isotopes to the Bay waters to both diffusion from bottom sediments and desorption from suspended particles. Because of the slower production rate of ^{226}Ra from ^{230}Th decay relative to that of ^{228}Ra from ^{232}Th decay, Moore (1981) suggested that desorption from suspended sediments was more important for ^{226}Ra while diffusion from bottom sediments was more significant for ^{228}Ra. The difference in the fluxes of ^{226}Ra and ^{228}Ra to the water column is apparent also in the New York Bight (Li, Feely and Santschi 1979) and Narragansett Bay (Santschi et al. 1979). Elsinger and Moore (1983) have extended observations on the magnitude of fluxes of ^{226}Ra and ^{228}Ra in estuarine systems by also analyzing estuarine water samples for ^{224}Ra (half-life = 3.6 days). All three radium isotopes showed non-conservative behavior in Winyah Bay and Delaware Bay (USA). Elsinger and Moore (1983) calculated that desorption from suspended particles could account for all of the ^{226}Ra, ∿ 80% of ^{228}Ra and ∿ 50% of the ^{224}Ra released during estuarine mixing in Winyah Bay. The balance comes from bottom sediments and, for ^{224}Ra, possibly from ^{228}Th decay on suspended particles. Elsinger and Moore (1983) also identified input from salt marshes as being of potential importance for the input of radium isotopes to Delaware Bay. Quantitative comparison of the magnitude of radium fluxes from bottom sediments, suspended particles, and salt marshes requires data on all three systems in the same estuary.

The radionuclide ^{222}Rn also is produced primarily in sediments from ^{226}Ra decay. It is mobilized to the sediment pore waters in a fashion similar to radium, but, unlike radium, it does not interact with the sediment during diffusion. Its distribution in the estuarine sediment and water columns makes it useful as a tracer for determining transport rates across the sediment-water and air-water interfaces and for determining mixing rates in the estuarine water column. Hammond, Simpson and Mathieu (1975, 1977) determined that virtually all of the ^{222}Rn was input to the Hudson River estuary from the sediments and was lost by decay (∿ 50%) and evasion to the atmosphere (∿ 50%). They also noted that only about 40% of the loss from the sediments could be explained by molecular diffusion. The sediment flux is enhanced by stirring of the upper few centimeters by currents or bioturbation (specifically irrigation of burrows). The importance of irrigation to the fluxes of dissolved chemical species across the sediment-water interface has been stressed by Aller (1980), and ^{222}Rn is a useful tracer of the rate and depth of irrigation in marine sediments. For example, depth profiles of ^{222}Rn in coastal sediments (e.g., Smethie, Nittrouer and Self 1981) show depletions to depths of up to 25 cm. much greater than would be expected from one dimensional molecular diffusion. In sediments which are not actively bioturbated, such as the laminated sediments of Santa Barbara Basin, the radon flux frequently approaches the value expected from one-dimensional diffusive transport (Berelson, Hammond and Fuller 1982). However, enhanced radon flux can occur from sediments which are not bioturbated through the formation of bubble tubes resulting from CH_4 production in the sediments (Martens, Kipphut and Klump 1980). In such instances the radon flux discrepancy gives information on physical processes related to the sediment geochemistry. The radon which emanates from sediments may be used as a tracer for mixing rates in the estuarine water column through observations of its depth profile above the bottom, and it has been applied in this manner in fjords (Smethie 1981).

THORIUM ISOTOPES

Thorium has four isotopes whose chemistry has been studied in the estuarine environment. The long-lived isotopes ^{232}Th (half-life = 1.4 x 10^{10} years) and ^{230}Th (half-life = 75,200 years) are added to the estuary with particles and their transport through the estuary is linked to that of the particles. The shorter-lived isotopes ^{234}Th (half=life = 24 days) and ^{228}Th (half-life = 1.9 years) are also associated with particles but, more importantly, are produced from radioactive decay of ^{238}U and ^{228}Ra, respectively. As discussed above, both uranium and radium are soluble in sea water and the production of ^{234}Th and ^{228}Th permits a unique chronometer to measure the rate of removal of thorium (and presumably other similarly reactive elements, Santschi et al. 1980) from solution. Deposition of ^{234}Th and ^{228}Th in excess of their parents on the estuarine sea floor provides chronometers for the rate of mixing of the surface sediments and, in some instances, the rate of sediment accumulation.

The residence times for ^{234}Th and ^{228}Th with respect to removal onto particles are short in nearshore waters. Li et al. (1979, 1981) determined that the scavenging residence time of ^{228}Th was about 20 to 70 days in shelf and slope water of the New York Bight (USA). A trend of decreasing mean residence times for Th removal toward shore had been observed by Broecker, Kaufman and Trier (1973), Li et al. (1979, 1981), Kaufman, Li and Turekian (1981), and McKee, DeMaster and Nitrouer (1983). This trend continues into the estuary. Aller and Cochran (1976) determined that ^{234}Th had a mean residence time of $\sim$ 1 day based on a sample taken in the fall in central Long Island Sound. Santschi et al. (1979) studied the seasonality of the removal of both ^{234}Th and ^{228}Th in Narragansett Bay. They observed half-removal times of less than about 20 days for the removal from solution to particles. The scavenging rate varied seasonally and was inversely correlated with the rate of sediment resuspension. Sediment resuspension may be the ultimate control on Th removal or removal may be linked to another related process such as the flux of reduced iron or manganese from estuarine sediments (see Turekian 1977). Indeed, Santschi et al. (1980) have shown that removal rates of Th and Fe are strikingly similar in the coastal environment.

Once removed from the water column, ^{228}Th and ^{234}Th are distributed in estuarine sediments by the processes of sediment mixing by infauna, physical mixing and resuspension of the bottom sediment, and, in some instances, sediment accumulation. The decrease in excess ^{234}Th activity (= measured ^{234}Th activity - measured ^{238}U activity) with depth in the sediment column has provided estimates of rates of particle mixing by the benthic fauna (Aller and Cochran 1976; Cochran and Aller 1979; Aller, Benninger and Cochran 1980; Santschi et al. 1980; Nittrouer et al. 1984). Excess ^{228}Th can be used as an indicator of mixing in a fashion similar to ^{234}Th but is less commonly applied because of the necessity to make ^{228}Ra measurements to correct for the supported ^{228}Th activity. Figure 6 shows the relationships

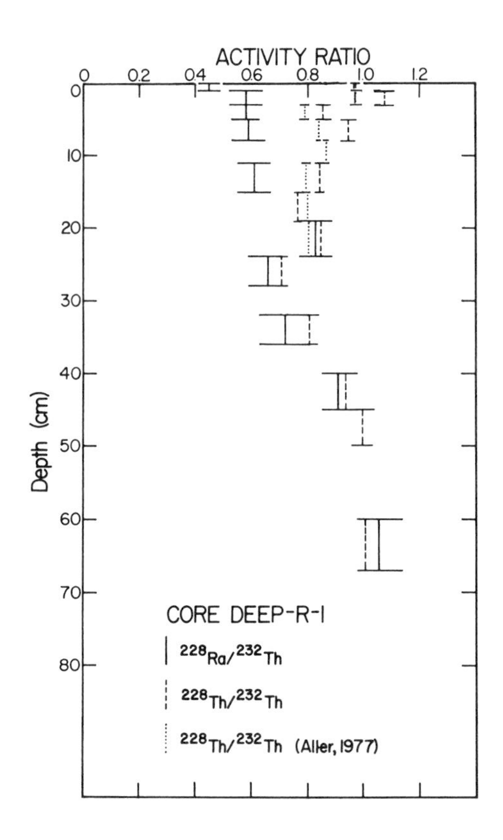

FIGURE 6. Disequilibrium relations between ^{232}Th and its daughters ^{228}Ra and ^{228}Th in the solid phase of a sediment core from station DEEP in Long Island Sound. An activity ratio of 1.0 represents the expected equilibrium value. The core is a large-area gravity core taken October 23, 1975.

between ^{228}Th, ^{228}Ra and ^{232}Th in a core from Long Island Sound (data from Cochran 1979). The $^{228}Ra/^{232}Th$ activity ratios are explainable due to ingrowth and loss of ^{228}Ra from the core (see earlier section on Radium). The $^{228}Th/^{232}Th$ activity ratio profile is similar to one determined by Thomson et al. (1975) elsewhere in the Sound and is controlled by addition of excess ^{228}Th from the overlying water and particle mixing in the surface sediments. ^{210}Pb results discussed below demonstrate that sediment accumulation alone cannot produce the observed profile of excess ^{228}Th. The distribution of excess ^{228}Th in this core is plotted in Fig. 7. Assuming that particle mixing takes place in a manner analogous to eddy diffusion and fitting the model of Benninger et al. (1979) to the data yields a mixing coefficient of $\sim 1 \times 10^{-6}$ cm^2 s^{-1}. In cases where particle mixing is absent,

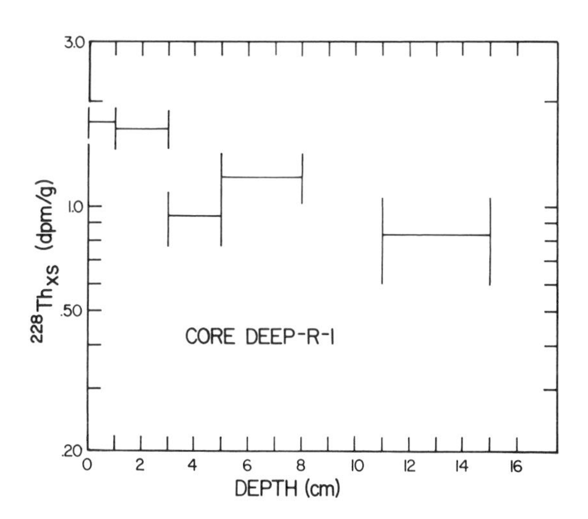

FIGURE 7. Excess ^{228}Th (= measured ^{228}Th - measured ^{228}Ra) as a function of depth in core DEEP-R-1 from Long Island Sound. The activity ratio data for this core are shown in Fig. 6.

as in sediments accumulating below an anoxic water column, excess ^{228}Th can be used to derive a sediment accumulation rate over the last 10 years. Koide, Bruland and Goldberg (1973) have applied it in this manner in California coastal sediments.

One interesting application of excess ^{234}Th and ^{228}Th distributions in estuarine sediments is in evaluating rates of diagenetic reactions. The upper few centimeters of the sediment column are mixed on time scales of months based on ^{234}Th data, yet sharp gradients in solid-phase metals like Fe and Mn which undergo oxidation-reduction reactions also occur in this zone (Aller 1977). These gradients must be maintained by redox reactions and diffusive transport whose rates are rapid relative to sediment mixing.

In addition to determinations of particle mixing rate, the short-lived thorium isotopes may be used to delineate the repositories of particle-reactive chemical species in the estuary. Aller et al. (1980) conducted a survey of ^{234}Th distributions in the sediments at 12 stations in Long Island Sound. This study confirmed the earlier observations of rapid removal of ^{234}Th from the water column to the sediments but demonstrated that removal was not acting in a simple vertical sense. Instead, the integrated activity of excess ^{234}Th (= inventory) in the sediments was linked to the grain size. This is shown in Fig. 8a in which the excess ^{234}Th inventory is correlated with the ^{232}Th activity. The latter is introduced to the estuary on particles and its concentration is linked to the proportion of fine grained sediment. Thus, Fig. 8a shows that more excess ^{234}Th is present in areas of fine grained sediment. The second important point noted by Aller et al. (1980) is that ^{234}Th inventory is related to the intensity of particle mixing by the benthic fauna (Fig. 8b). This implies that fine particles can

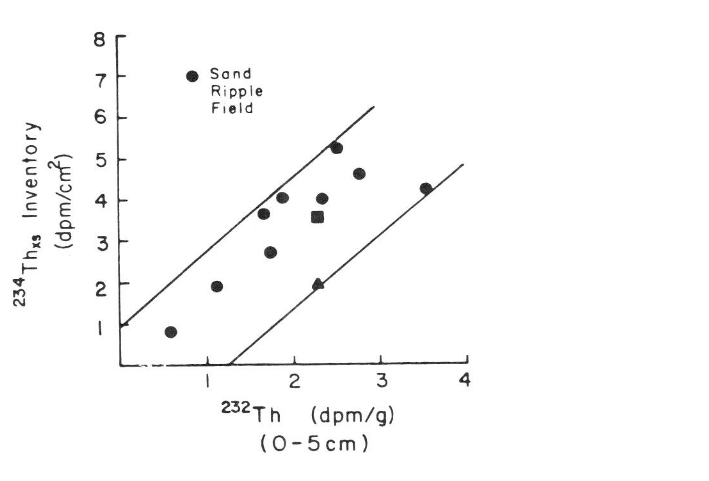

FIGURE 8a. *Integrated ^{234}Th activity in the sediment column (inventory) versus ^{232}Th activity (dpm/g) for twelve stations in Long Island Sound. Square represents station NWC, triangle represents station DEEP. The cores were collected July-August, 1977 (from Aller et al. 1980).*

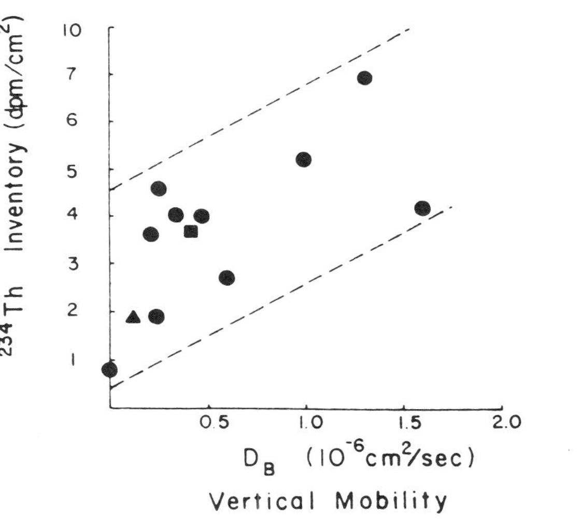

FIGURE 8b. *^{234}Th inventory vs. the particle mixing coefficient, D_B, for sediment cores collected in Long Island Sound. The value of D_B is a measure of the rate of bioturbation in the surface sediment (from Aller et al. 1980).*

scavenge reactive nuclides and that these particles are mixed into the sediment column in areas of rapid mixing. Resuspension of surface sediment continues the cycle of scavenging, with inventories of reactive species enhanced in areas of high mixing at the expense of adjacent areas where mixing is slower. This work implies that particle-reactive chemical species will be concentrated in areas of fine grained sediment in which particle mixing is rapid (and deep as well). The relationship holds for ^{210}Pb in Long Island Sound (Turekian et al. 1980; see section on ^{210}Pb and ^{210}Po) and on a larger scale has been documented for plutonium (an artificial element produced during atomic weapons testing and added to the oceans principally from the atmosphere) in both nearshore and offshore sediments (Santschi et al. 1980).

Just as ^{234}Th may be used as an analog for particle-reactive species in the water column, a similar use is possible in the interstitial water of estuarine sediments. Figure 9 shows a pore water profile of the ^{234}Th/^{238}U activity ratio in Buzzards Bay sediments (location in Fig. 1). The pore water ^{234}Th/^{238}U activity ratio is greater in the top 5 cm of the core than at depth. This depth region is one in which ^{234}Th removed from the overlying water column is present in the solid phase and in which Fe and Mn are released to solution (primarily upon reduction of their oxyhydroxides). Thus the ^{234}Th in solution could be released from the solid phase either as Fe and Mn are reduced or by recoil from ^{238}U decay on particle surfaces. Alternatively the ^{234}Th could be produced from ^{238}U in solution. In the latter case, the greater ^{234}Th/^{238}U activity ratios in the top 5 cm could be due to slower uptake of ^{234}Th on particle surfaces over depths where the surfaces are being continually solubilized (again by reduction of oxides). In fact a combination of these possibilities may exist.

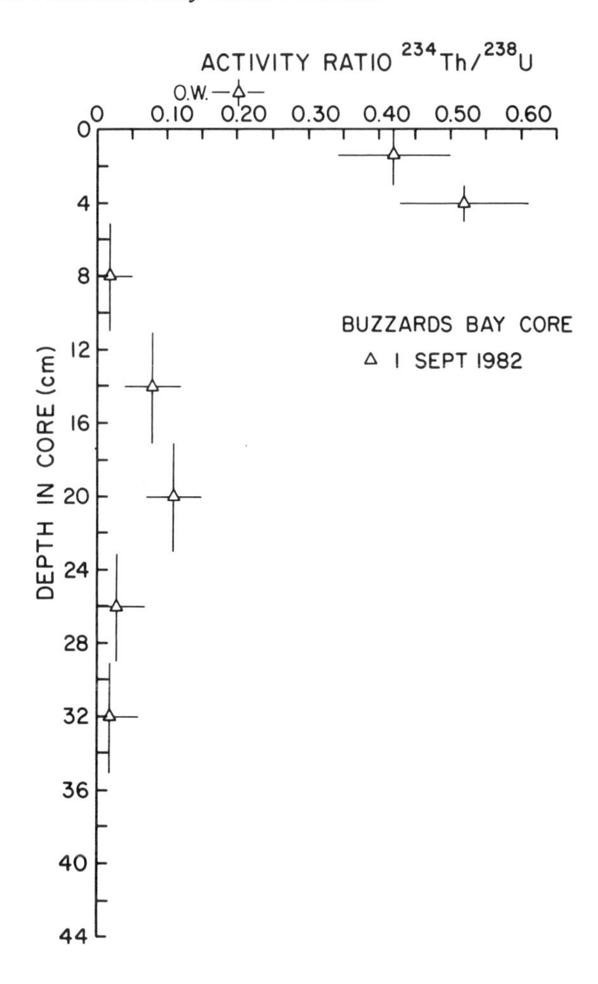

FIGURE 9. $^{234}Th/^{238}U$ *activity ratio in the pore water of a box core collected at station J in Buzzards Bay. Vertical line represents sampling interval, horizontal line is 1σ counting uncertainty.* ^{234}Th *values have been corrected to the date of core collection. Value above the axis is the measured overlying water value.*

LEAD-210 AND POLONIUM-210

The principal pathways by which ^{210}Pb enters the estuary are supply from the atmosphere and by streams. The atmospheric pathway arises because ^{222}Rn emanating from rocks and soils decays to produce ^{210}Pb which is removed from the atmosphere by precipitation. Although some dissolved ^{210}Pb is supplied by rivers, studies of the behavior of ^{210}Pb in the Colorado River, USA (Rama, Koide and Goldberg 1961) and in the Susquehanna River (Lewis 1977) have shown that ^{210}Pb is rapidly removed from solution. As a consequence, the ^{210}Pb supplied to the estuary by streams and rivers is dominantly in the particulate phase. ^{210}Pb also may be supplied to coastal areas by advection from offshore (where it is less rapidly removed). Such a process has been demonstrated by Carpenter, Bennett and Peterson (1981) to be occurring on the Washington (USA) continental shelf and slope and helps account for the large integrated amounts of ^{210}Pb seen in the sediments there. ^{210}Po is also supplied to the surface oceans from the atmosphere, but because the mean residence time of aerosols (and associated ^{210}Pb) in the atmosphere is relatively short, $^{210}Po/^{210}Pb$ activity ratios in precipitation are low (< 0.10, see for example, Turekian and Cochran 1981). ^{210}Po is thus introduced *in situ* into the estuarine environment by ^{210}Pb decay.

The uses of ^{210}Pb in the estuarine regime are twofold: 1) as a tracer of heavy metals and 2) as a means to determine the chronology of the estuarine sediment column. The first use permits determination of the rates of removal of ^{210}Pb and similar metals from the water column and also an evaluation of the repositories of the scavenged metals (Benninger, Lewis and Turekian 1975; Turekian et al. 1980). The second use has included efforts to use ^{210}Pb to measure the accumulation rate of estuarine sediments and reconstruct the pollution history of the area (Goldberg et al. 1977, 1978). However, studies on the ^{210}Pb distributions in Long Island Sound sediments (Thomson et al. 1975; Benninger et al. 1979) showed that bioturbation of the sediment column can significantly alter ^{210}Pb gradients in the sediments and complicate the determination of sediment accumulation rate and pollution history.

Removal of ^{210}Pb from the estuarine water column has been studied by Benninger (1978) in Long Island Sound and by Santschi et al. (1979) in Narragansett Bay. The latter work showed half-removal times of 2 to 29 days for ^{210}Pb onto particles and $<$ 1 to 60 days for ^{210}Po. Santschi et al. (1979) suggested that the seasonal variation in ^{210}Pb and ^{210}Po activities which they observed in the water column was due either 1) to remobilization of ^{210}Pb and ^{210}Po out of sediments or 2) to the formation of complexes or stabilized colloids of organic compounds with ^{210}Pb and ^{210}Po in solution. Benninger's (1978) work also demonstrated rapid removal of ^{210}Pb from estuarine waters. His data showed a good correlation of ^{210}Pb in unfiltered water samples with the concentration of suspended particles. Within the uncertainties of the data, no dissolved ^{210}Pb could be discerned, indicating rapid removal onto particles. Benninger (1978) also constructed a mass balance for ^{210}Pb in Long Island Sound and confirmed that supply from rivers and the atmosphere were the dominant fluxes. Within the uncertainties of the calculations, all the ^{210}Pb added to the Sound was present in bottom sediments (making radioactive decay the primary removal pathway for ^{210}Pb in the Sound system).

As with ^{234}Th, excess ^{210}Pb (measured ^{210}Pb - measured ^{226}Ra) is redistributed in the bottom sediments as a function of sediment resuspension and the intensity of particle mixing by the benthic fauna (Turekian et al. 1980). Due to mixing of bottom sediments by organisms, the accumulation rates which can be calculated from the excess ^{210}Pb gradients in the sediment column are upper limits. Figure 10 shows the patterns of excess ^{210}Pb vs. depth which result from a combination of mixing and sedimentation in Long Island Sound sediments. Core DEEP 102375 is the same core for which excess ^{228}Th data are presented in Fig. 7, and the profiles demonstrate that mixing in the upper 5-10 cm produces gradients in short-lived tracers like ^{228}Th and tends to homogenize the activities of longer-lived tracers like ^{210}Pb. The ^{210}Pb activity decreases below the depth of rapid continuous mixing (Fig. 10). Even over this depth region, however, the apparent ^{210}Pb - derived accumulation rate is greater than that obtained from independent estimates based on sediment thickness or supply. In the case of core DEEP 102375, the accumulation rate derived from sediment thickness data is ∿ 0.12 cm/y (Bokuniewicz, Gebert and Gordon 1976) whereas that obtained from the ^{210}Pb profile is 0.6 cm/y. Deep burrows extending well below the surface mixed zone apparently serve to transport particles and excess ^{210}Pb to depth in the Long Island Sound sediment column. These burrowing events may be sufficiently frequent to produce smooth decreases in ^{210}Pb below the rapidly mixed zone or may produce discrete "spikes" of anomalously high excess ^{210}Pb activity at depth. Such tracer "spikes" have been linked to bioturbation by Benninger et al. (1979), who used X-radiographs as a guide to sampling discrete burrows in a sediment core and demonstrated that elevated ^{210}Pb activities were associated with burrow infilling. The effect of deep mixing on the calculation of sediment accumulation rates is not limited to estuarine sediments. Carpenter, Peterson and Bennett (1982) conclude from ^{210}Pb profiles in Washington (USA) continental slope sediments that neglect of deep particle mixing at a rate only 5 % of that in the surface rapidly-mixed zone produces sediment accumulation rates which are overestimated by a factor of 2 to 3. However, comparison of ^{210}Pb profiles with other tracers such as the artificial radionuclides $^{239,240}Pu$ or ^{137}Cs is useful in setting limits on deep mixing in sediment cores (Benninger et al. 1979; Nittrouer et al. 1984).

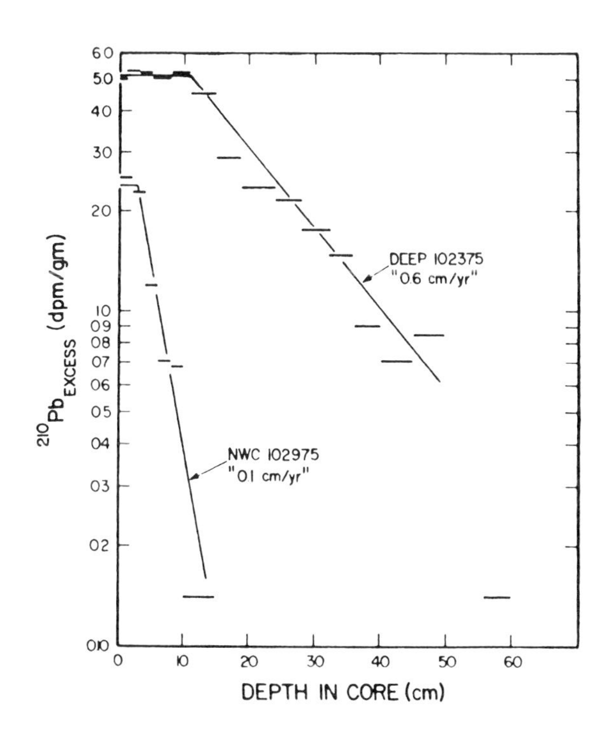

FIGURE 10. Excess ^{210}Pb activity as a function of depth at Stations NWC and DEEP in Long Island Sound. Excess ^{228}Th data for core 102375 are shown in Fig. 7. Values in quotation marks are the apparent sediment accumulation rates calculated from the decreasing part of the profiles (from Turekian et al. 1980).

Particle mixing by the benthic fauna is not the only process affecting excess ^{210}Pb profiles in estuarine sediments. Hirschberg and Schubel (1979) documented the effect of episodic input of sediment from the Susquehanna River on the ^{210}Pb profiles in northern Chesapeake Bay sediments. Decreasing activities of ^{210}Pb and ^{137}Cs over depth intervals inferred to be flood deposits resulting from a major storm were explained by the erosion and transport of progressively "older" material (hence lower in ^{210}Pb and ^{137}Cs) as the enhanced discharge continued. On a larger scale, Benninger and Krishnaswami (1981) have demonstrated that ^{210}Pb profiles in New York Bight sediments are explainable by rapid physical mixing of a phase carrying ^{210}Pb into existing sediment, with little net accumulation.

Thus both biological (i.e., bioturbation) and physical (i.e., sediment transport and redepositon) processes commonly affect ^{210}Pb profiles in estuarine sediments and complicate estimations of sediment accumulation rate. Successful reconstruction of the depositional history of an estuarine sediment column requires use of multiple tracers of different half-lives or independent information on the nature of the benthic community and sediment budgets for the area.

ACKNOWLEDGMENTS

The author's work in Long Island Sound and Narragansett Bay was supported by National Science Foundation grant OCE76-02-39 (Yale University) to Karl K. Turekian, Principal Investigator, and in Buzzards Bay by Department of Energy contract DE-AC02-81EV10694 (Woods Hole Oceanographic Institution). Laboratory assistance was provided by

Barbara Brockett and Peggy Delaney at Yale University and by Anne Carey and Lolita Surprenant at Woods Hole Oceanographic Institution. I am particularly grateful to S. Krishnaswami for his collaboration on the pore water ^{228}Ra analyses in Long Island Sound and Narragansett Bay and to Karl Turekian, Robert Aller, Larry Benninger and Edward Sholkovitz for sharing their insights with me. This is Contribution No. 427 of the Marine Sciences Research Center of the State University of New York. Permission to use Figures 8 and 10 was granted by Elsevier Science Publishers and by Academic Press, respectively.

REFERENCES CITED

Aller, R.C. 1977. The influence of macrobenthos on chemical diagenesis of marine sediments. Ph.D. Thesis, Yale University, New Haven, CT. 599pp.

Aller, R.C. 1980. Quantifying solute distributions in the bioturbated zone of marine sediments by defining an average microenvironment. *Geochim. Cosmochim. Acta 44*:1955-1965.

Aller, R.C., L.K. Benninger and J.K. Cochran. 1980. Tracking particle associated processes in nearshore environments by use of $^{234}Th/^{238}U$ disequilibrium. *Earth Planet. Sci. Lett. 47*:166-175.

Aller, R.C. and J.K. Cochran. 1976. $^{234}Th/^{238}U$ disequilibrium in nearshore sediment: particle reworking and diagenetic time scales, *Earth and Planet. Sci. Lett. 29*:37-50.

Benninger, L.K. 1978. ^{210}Pb balance in Long Island Sound. *Geochim. Cosmochim Acta 42:*1165-1174.

Benninger, L.K., R.C. Aller, J.K. Cochran and K.K. Turekian. 1979. Effects of biological sediment mixing on the ^{210}Pb chronology and trace metal distribution in a Long Island Sound sediment core. *Earth Planet. Sci. Lett. 43:*241-259.

Benninger, L.K. and S. Krishnaswami. 1981. Sedimentary processes in the inner New York Bight: evidence from excess ^{210}Pb and $^{239,240}Pu$. *Earth Planet. Sci. Lett. 53*:158-174.

Benninger, L.K., D.M. Lewis and K.K. Turekian. 1975. The use of natural Pb-210 as a heavy metal tracer in the river-estuarine system, pp. 202-210. *In:* T.M. Church (ed.), *Marine Chemistry in the Coastal Environment.* ACS Symposium Series 18, American Chemical Society.

Berelson, W.M., D.E. Hammond and C. Fuller. 1982. Radon-222 as a tracer for mixing in the water column and benthic exchange in the southern California borderland. *Earth Planet. Sci. Lett. 61:*41-54.

Berner, R.A. 1980. *Early Diagenesis, A Theoretical Approach.* Princeton Univ. Press, Princeton, NJ. 241pp.

Blanchard, R.L. and D. Oakes. 1965. Relationships between uranium and radium in coastal marine shells and their environment. *J. Geophys. Res. 70:*2911-2921.

Bokuniewicz, H.J., J. Gebert and R.B. Gordon. 1976. Sediment mass balance of a large estuary, Long Island Sound. *Estuar. Coastal Mar. Sci. 4:*523-536.

Borole, D.V., S. Krishnaswami and B.L. K. Somayajulu. 1977. Investigations on dissolved uranium, silicon and on particulate trace elements in estuaries. *Est. Coast. Mar. Sci.* *5*:743-754.

Borole, D.V., S. Krishnaswami and B.L.K. Somayajulu. 1982. Uranium isotopes in rivers, estuaries and adjacent coastal sediments of western India: their weathering, transport and oceanic budget. *Geochim. Cosmochim. Acta* *46*:125-137.

Broecker, W.S., A. Kaufman and R.M. Trier. 1973. The residence time of thorium in surface sea water and its implications regarding the fate of reactive pollutants. *Earth Planet. Sci. Lett.* *20*:35-44.

Carpenter, R., J.T. Bennett and M.L. Peterson. 1981. ^{210}Pb activities in and fluxes to sediments of the Washington continental slope and shelf. *Geochim. Cosmochim. Acta* *45*:1155-1172.

Carpenter, R., M.L. Peterson and J.T. Bennett. 1982. ^{210}Pb-derived sediment acumulation and mixing rates for the Washington continental slope. *Mar. Geol.* *48*:135-164.

Church, T.M., C.J. Lord III and B.L.K. Somayajulu. 1981. Uranium, thorium and lead nuclides in a Delaware salt marsh sediment. *Est. Coast. Shelf Sci.* *13*:267-275.

Cochran, J.K. 1979. The geochemistry of ^{226}Ra and ^{228}Ra in marine deposits. Ph.D. Thesis, Yale University, New Haven, CT. 260pp.

Cochran, J.K. and R.C. Aller. 1979. Particle reworking in sediments from the New York Bight Apex: Evidence from $^{234}Th/^{238}U$ disequilibrium. *Estuar. Coastal Mar. Sci.* *9*:739-747.

Cochran, J.K. and S. Krishnaswami. 1980. Radium, thorium, uranium and ^{210}Pb in deep-sea sediments and sediment pore waters from the North Equatorial Pacific. *Am. J. Sci.* *280*:849-889.

Elsinger, R.J. and W.S. Moore. 1980. ^{226}Ra behavior in the Pee Dee River-Winyah Bay estuary. *Earth Planet. Sci. Lett.* *48*:239-249.

Elsinger, R.J. and W.S. Moore. 1983. ^{224}Ra, ^{228}Ra, and ^{226}Ra in Winyah Bay and Delaware Bay. *Earth Planet. Sci. Lett.* *64*:430-436.

Feely, H.W., G.W. Kipphut, R.M. Trier and C. Kent. 1980. ^{228}Ra and ^{228}Th in coastal waters. *Est. Coast. Mar. Sci.* *11*:179-205.

Figuères, G., J.M. Martin and A.J. Thomas. 1982. Apport par les fleuves d'uranium dissous à l'océan: exemple du Zaire. *Ocean. Acta.* *5*:141-147.

Garrels, R.M. and F.T. Mackenzie. 1971. *Evolution of Sedimentary Rocks.* W.W. Norton & Co., New York. 397pp.

Goldberg, E.D., E. Gamble, J.J. Griffin and M. Koide. 1977. Pollution history in Narraganset Bay as recorded in its sediments. *Estuar. Coastal Mar. Sci.* *5*:549-561.

Goldberg, E.D., V. Hodge, M. Koide, J. Griffin, E. Gamble, O.P. Bricker, G. Matisoff, G.R. Holdren and R. Braun. 1978. A pollution history of Chesapeake Bay. *Geochim. Cosmochim. Acta.* *42*:1413-1425

Hammond, D.E., H.J. Simpson and G. Mathieu. 1975. Methane and radon-222 as tracers for mechanisms of exchange across the sediment-water interface in the Hudson River Estuary, pp. 119-132. *In:* T.M. Church (ed.), *Marine Chemistry in the Coastal Environment.* ACS Symposium Series 18, American Chemical Society.

Hammond, D.E., H.J. Simpson and G. Mathieu. 1977. Radon-222 distribution and transport across the sediment-water interface in the Hudson River Estuary. *J. Geophys. Res.* *82*:3913-3920.

Hirschberg, D.J. and J.R. Schubel. 1979. Recent geochemical history of flood deposits in the northern Chesapeake Bay. *Est. Coast. Mar. Sci.* *9*:771-784.

Ivanovich, M. 1982. Spectroscopic methods, pp. 107-144. *In:* M. Ivanovich and R. Harmon (eds.), *Uranium Series Disequilibrium: Applications to the Environmental Problems.* Clarendon Press, Oxford.

Kaufman, A., Y.H. Li and K.K. Turekian. 1981. The removal rates of ^{234}Th and ^{228}Th from waters of the New York Bight. *Earth Planet. Sci. Lett.* *54*:385-392.

Koczy, F.F., E. Picciotto, G. Poulaert and S. Wilgain. 1957. Mesure des isotopes du thorium dans l'eau de mer. *Geochim. Cosmochim. Acta.* *11*:103-129.

Koide, M. and K.W. Bruland. 1975. The electrodeposition and determination of radium by isotopic dilution in sea water and in sediment simultaneously with other natural radionuclides. *Anal. Chim. Acta.* *75*:1-19.

Koide, M., K.W. Bruland and E.D. Goldberg. 1973. Th-228/Th-232 and Pb-210 geochronologies in marine and lake sediments. *Geochimica et Cosmochimica Acta.* *37*:1171-1187.

Kolodny, Y. and I.R. Kaplan. 1973. Deposition of uranium in the sediment and interstitial water of an anoxic fjord, pp. 418-442. *In: Proceedings, Symposium on Hydrogeochemistry and Biogeochemistry, 1*, The Clark Company, Washington, DC.

Krishnaswami, S. and M.M. Sarin. 1976. The simultaneous determination of Th, Pu, Ra isotopes, ^{210}Pb, ^{55}Fe, ^{32}Si and ^{14}C in marine suspended phases. *Anal. Chim. Acta.* *83*:143-156.

Lewis, D.M. 1976. The geochemistry of manganese, iron, uranium, lead-210 and major ions in the Susquehanna River. Ph.D. Thesis, Yale University, New Haven, CT 272pp.

Lewis, D.M. 1977. The use of ^{210}Pb as a heavy metal tracer in the Susquehanna River system. *Geochim. Cosmochim. Acta.* *41*:1557-1564.

Li, Y.H. and L.H. Chan. 1979. Desorption of Ba and ^{226}Ra from riverborne sediments in the Hudson estuary. *Earth Planet Sci. Lett.* *43*:343-350.

Li, Y.H., H.W. Feely and P.H. Santschi, 1979. ^{228}Th - ^{228}Ra radioactive disequilibrium in the New York Bight and its implications for coastal pollution. *Earth Planet. Sci. Lett.* *42:*13-26.

Li, Y.H., G. Mathieu, P. Biscaye and H.J. Simpson. 1977. The flux of ^{226}Ra from estuarine and continental shelf sediments. *Earth Planet. Sci. Lett.* *57:*237-241.

Li, Y.H., P.H. Santschi, A. Kaufman, L.K. Benninger and H.W. Feely. 1981. Natural radionuclides in waters of the New York Bight. *Earth Planet. Sci. Lett.* *55:*217-228.

Maeda, M. and H.L. Windom. 1982. Behavior of uranium in two estuaries of the south-eastern United States. *Mar. Chem.* *11:*427-436.

Mangini, A., C. Sonntag, G. Bertsch and E. Muller. 1979. Evidence for a higher natural uranium content in world rivers. *Nature* *278:*337-339.

Martens, C.S., G.W. Kipphut and J.V. Klump. 1980. Sediment-water chemical exchange in the coastal zone traced by *in situ* radon-222 flux measurements. *Science* *208:*285-288.

Martin, J.M., V. Nijampurkar and F. Salvadori. 1978. Uranium and thorium isotope behavior in estuarine systems, pp. 111-126. *In: Biogeochemistry of estuarine sediments.* Proc. UNESCO/SCOR Workshop, Melreaux, Belgium. UNESCO Press, New York.

McKee, B.A., D.J. DeMaster and C.A. Nittrouer. 1983. Rates of particle scavenging on the Yangtze continental shelf based on measurements of ^{234}Th. *EOS, Trans. Am. Geophys.* *64*:76.

Moore, W.S. 1969. Measurement of Ra-228 and Th-228 in sea water. *J. Geophys. Res.* *74*:694-704

Moore, W.S. 1981. Radium isotopes in Chesapeake Bay. *Est. Coast. Shelf Sci.* *12*:713-723.

Myers, A.C. and D.K. Phelps. 1978. Criteria of benthic health: A transect study of Narragansett Bay, Rhode Island. Rept. US-EPA Contract No. P053203, 195pp.

Nittrouer, C.A., D.J. DeMaster, B.A. McKee, N.H. Cutshall and I.L. Larsen. 1984. The effect of sediment mixing on ^{210}Pb accumulation rates for the Washington continental shelf. *Mar. Geol.* *54*:201-222.

Osmond, J.K. and J.B. Cowart. 1976. The theory and uses of natural uranium isotopic variations in hydrology. *Atomic En. Rev. (IAEA)* *14*:621-679.

Rama, M. Koide and E.D. Goldberg. 1961. Lead-210 in natural waters. *Science* *134*:98-99.

Santschi, P.H., Y-H. Li and J. Bell. 1979. Natural radionuclides in the water of Narragansett Bay. *Earth Planet. Sci. Lett.* *45*:201-213.

Santschi, P.H., H.H. Li, J.J. Bell, R.M. Trier and K. Kawtaluk. 1980. Pu in coastal marine environments. *Earth Planet. Sci. Lett.* *51*:248-265.

Scott, M.R. 1982. The chemistry of U- and Th-series nuclides in rivers, pp. 181-201. *In:* M. Ivanovich and R. Harmon (eds.), *Uranium Series Disequilibrium: Application to Environmental Problems*. Clarendon Press, Oxford.

Sholkovitz, E.R., J.K. Cochran and A.E. Carey. 1983. Laboratory studies of radionuclide mobilization in coastal sediments, I: The artificial radionuclides $^{239,240}Pu$, ^{137}Cs and ^{55}Fe. *Geochim. Cosmochim. Acta.* *47:*1369-1380.

Smethie, W.M. Jr. 1981. Vertical mixing rates in fjords determined using radon and salinity as tracers. *Est. Coast. Shelf Sci.* *12:*131-153.

Smethie, W.M. Jr., C.A. Nittrouer and R.F.L. Self. 1981. The use of radon-222 as a tracer of sediment irrigation and mixing on the Washington continental shelf. *Mar. Geol.* *42:*173-200.

Thomson, J., K.K. Turekian and R.J. McCaffrey. 1975. The accumulation of metals in and release from sediments of Long Island Sound, pp. 28-43. *In:* L.E. Cronin (ed.), *Estuarine Research, Vol. 1.* Academic Press, New York.

Turekian, K.K. 1977. The fate of metals in the oceans. *Geochim. Cosmochim. Acta.* *41:*1139-1144.

Turekian, K.K. and J.K. Cochran. 1978. Determination of marine chronologics using natural radionuclides, pp. 313-361. *In:* J.P. Riley and R. Chester (eds.), *Chemical Oceanography Vol. 7,* Second Edition. Academic Press, New York.

Turekian, K.K. and J.K. Cochran. 1981. ^{210}Pb in surface air at Enewetak and the Asian dust flux to the Pacific. *Nature* *292*:522-524.

Turekian, K.K., J.K. Cochran, L.K. Benninger and R.C. Aller. 1980. The sources and sinks of nuclides in Long Island Sound, pp. 129-164. *In:* B. Saltzman (ed.), *Estuarine Physics and Chemistry: Studies in Long Island Sound.* Advances in Geophysics, Vol. 22, Academic Press, New York.

RESPONSE OF NORTHERN SAN FRANCISCO BAY TO RIVERINE INPUTS OF DISSOLVED INORGANIC CARBON, SILICON, NITROGEN AND PHOSPHORUS

Laurence E. Schemel
Dana D. Harmon
Stephen W. Hager
David H. Peterson

U. S. Geological Survey
Menlo Park, California

Abstract: Estuarine processes can be effective in modifying (filtering) distributions of dissolved inorganic forms of carbon (DIC), silicon (DIS), nitrogen (DIN), and phosphorus (DIP) in northern San Francisco Bay. During winter, high inflow from the Sacramento-San Joaquin river system supplied these nutrients to the estuary at rates that exceeded potential rates of estuarine supply and removal processes. During spring and summer, when inflow rates were lower, the estuary was an effective "filter" of the river inflow "signal" because rates of estuarine processes were high relative to river and other supply rates. At lower inflow rates, the river apparently influenced estuarine hydrodynamic features that controlled rates of phytoplankton nutrient removal. Largest biological removal effects were localized in San Pablo Bay during spring and Suisun Bay during summer, and they were generally more pronounced in shallow water areas of the bays. In San Pablo Bay, effects of biological removal appeared soon after river inflow decreased from high winter rates, but persisted for only a short time. During the following summer months, DIN and DIP distributions in San Pablo Bay indicated that estuarine sources contributed to higher concentrations of these nutrients.

ISBN 0-12-405070-0

INTRODUCTION

River inflow has a major effect on physical dynamics in northern San Francisco Bay and greatly influences biogeochemical features of this estuary. River inflow is simplified in this study as a two component "signal" to the estuary, namely inflow rate and chemical composition. Many responses of the estuary to these components of the river signal have been identified. River inflow controls non-tidal circulation, the location and structure of the salinity field (Conomos 1979), and some aspects of phytoplankton dynamics (Arthur and Ball 1979; Cloern et al. 1983). Chemical composition of the river inflow influences estuarine concentrations and distributions of many substances (Conomos et al. 1979), but our primary interests here are biologically reactive (nutrient) forms of dissolved inorganic carbon (DIC), silicon (DIS), nitrogen (DIN), and phosphorus (DIP). Biological removal and mineralization processes can modify estuarine distributions of these solutes (Conomos et al. 1979; Peterson 1979), which can be viewed as a "filtering" of the river chemical composition "signal" as it propagates seaward. The extent to which estuarine nutrient distributions are changed from that produced by mixing alone is a measure of the effectiveness of the estuary as a filter.

Nutrient substances are generally abundant in northern San Francisco Bay, owing in part to natural and anthropogenic sources (Conomos et al. 1979; Peterson 1979), but also to the turbid nature of the estuarine waters. Although phytoplankton nutrient uptake rates are controlled by a complex set of interrelated factors, light intensity (quantum flux density) appears to be the most limiting factor (Cloern and Cheng 1981). Light levels are primarily regulated by suspended particulate matter concentrations, which result from river supply and physical processes that suspend and transport particles in the estuary. Consequently, the photic zone is typically restricted to depths of less than 2m in the inner estuary and shallow areas and is deeper than 2m in the outer estuary (Peterson 1979). Effects of supply and removal processes in the photic zone, however, are averaged over the water column by vertical mixing induced primarily by tides and winds. The water column is mixed daily at most locations in the estuary (Peterson, Festa and Conomos 1978; Hammond and Fuller 1979).

Our objectives are to characterize the river as a source of DIC, DIS, DIN, and DIP to the estuary and then describe estuarine distributions and concentrations of these solutes on seasonal and shorter time scales. Lastly, with some knowledge of riverine/estuarine variability, we identify the locations, times, and some of the conditions for maximum "estuarine filtering."

METHODS

Measurements of DIC, DIS, DIN, and DIP were made at two-week intervals during neap tides at locations in deep (channel, >2m at MLLW) and shallow water (shoal, <2m at MLLW) areas of northern San Francisco Bay and the Sacramento and San Joaquin Rivers during 1980 (Fig. 1). Additional measurements were made in the Sacramento River at Rio Vista during 1980 and 1983. Results presented here are from the upper 2m of the water column. The two DIC components considered in this study are the river (total) alkalinity and the estuarine partial pressure of carbon dioxide (PCO_2). DIS and DIP are dissolved reactive forms of silica and phosphate. DIN is the sum of ammonium, nitrite and nitrate ion concentrations. Chlorophyll *a* concentrations were estimated from fluorescence measurements. Sampling techniques and analytical methods are detailed in Schemel and Dedini (1979), Smith, Herndon and Harmon (1979), and Schemel (1984a).

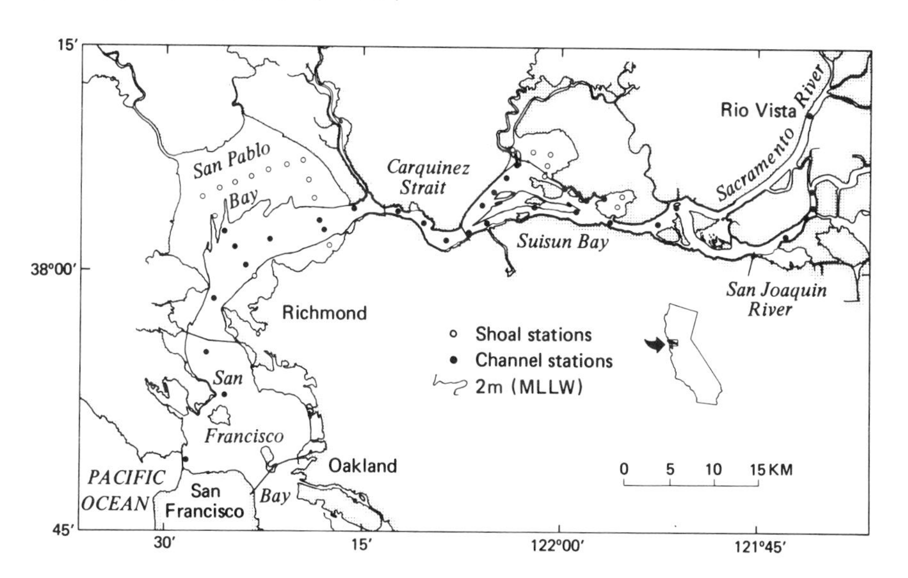

FIGURE 1. Northern San Francisco Bay, California, and the Sacramento and San Joaquin Rivers. Suisun Bay and San Pablo Bay are sub-embayments of San Francisco Bay. Sampling stations are indicated by open and solid circles.

FRESHWATER INFLOW AND SOLUTE TRANSPORT TO THE ESTUARY

The Sacramento-San Joaquin river system is the largest freshwater source to northern San Francisco Bay (Fig. 1). The Sacramento River is the larger of the two rivers, averaging 80% of their total inflow. Total annual inflow from other freshwater sources is typically a few percent of that from the Sacramento-San Joaquin river system (Conomos 1979).

Annual and seasonal freshwater inflow patterns can vary widely (California State Department of Water Resources 1982). For example, 1971 through 1980 mean inflow rate was 700 m^3 sec^{-1}, but annually-averaged inflow rates ranged from 110 m^3 sec^{-1} to 1300 m^3 sec^{-1}. Precipitation significantly affects inflow rates usually from late-fall to mid-spring. Monthly-averaged inflow rates from January through March are typically an order of magnitude higher than those during other months, accounting for a large fraction of the total annual inflow (eg., 74% during 1980). A major consequence of widely varying annual and seasonal river inflow patterns is that hydrologic conditions and the timing, locations, and development of phytoplankton blooms and associated biological nutrient removal change from year to year (Peterson et al. 1975a, b; Cloern et al. 1983).

River inflow rate affects input rates of riverine substances and (advective) freshwater replacement times in the estuary. Average freshwater replacement times can be estimated by comparing inflow volumes to the volume of river water in northern San Francisco Bay. The amount of river water in the bay varies with inflow rate and it appears to have been more than half of the total bay volume (4 x 10^9 m^3; Selleck et al. 1966) during January through March 1980 and less than half during the remainder of the year (Dedini, Schemel and Tembreull 1982). Total January through March inflow was 24 x 10^9 m^3, corresponding to an average freshwater replacement time of less than one week. In contrast, total April through December inflow was only 8 x 10^9 m^3, corresponding to an average freshwater replacement time of greater than ten weeks.

When evaluating estuarine supply and removal processes on the basis of constituent-salinity distributions, we must consider the effects of river concentration variability. In comparison to Sacramento River inflow, San Joaquin River inflow probably has little effect on solute distributions in the estuary during most times of year (Schemel 1984a,b). However, San Joaquin River inflow could have been important during April and May of 1980 because it contributed about 33% of the total freshwater inflow.

Alkalinity was measured more frequently than DIS, DIN, and DIP in the Sacramento River during 1980 and, therefore, alkalinity variability is better resolved (Fig. 2). During the high inflow period of January through March, major alkalinity variations corresponded to large changes in flow rate that resulted from major storms. During the lower inflow period of April through December, however, alkalinity variations were primarily the results of reservoir releases and inputs of agricultural waste waters (Schemel 1984a).

Sacramento River alkalinity during 1980 illustrates variations that can affect estuarine distributions (Schemel 1984b). Although alkalinity can be affected by estuarine supply and removal (Spiker and Schemel 1979), the 1980 alkalinity distributions (with respect to salinity) were primarily affected by river concentration variability. Large river alkalinity variations during high river inflow changed the slope of the alkalinity-salinity relation in the estuary, but resulted in linear or near-linear distributions (Schemel 1984b). Distributions in the estuary were non-linear from August through October in response to a large increase then decrease in river alkalinity. Non-linearity was primarily a result of the short period of the river variation relative to the flushing time (Dyer 1972) of the estuary. Linear and non-linear constituent-salinity distributions have been simulated numerically for similar conditions of river inflow and concentration variability (Loder and Reichard 1981).

Sacramento River DIS concentrations showed little variability during 1980, averaging about 250 μg-at L^{-1} (Fig. 2). DIN concentrations were highest (>25 μg-at L^{-1}) during January-February and November-December. With the exception of one measurement during May, DIN appears to vary over an annual cycle with the lowest concentrations (<15 μg-at L^{-1}) during July through September. DIP concentrations were lowest during March-April and highest during October-November. In general, Sacramento River DIS, DIN and DIP concentrations during April through December varied with longer periods than alkalinity, and their estuarine distributions probably were not as greatly affected by river concentration variability. When viewed on seasonal time scales, these river concentration variations broaden the ranges of estuarine concentrations, particularly in the inner estuary.

We examined Sacramento River concentration variability in greater detail during the 1983 high inflow period to determine why some riverine solutes were linearly distributed in the estuary, whereas others showed more concentration variability. Concentrations of riverine solutes can be related to river flow rate or changes in river flow rate (Kennedy and Malcolm 1978). Total monthly flow volumes varied by over a factor of

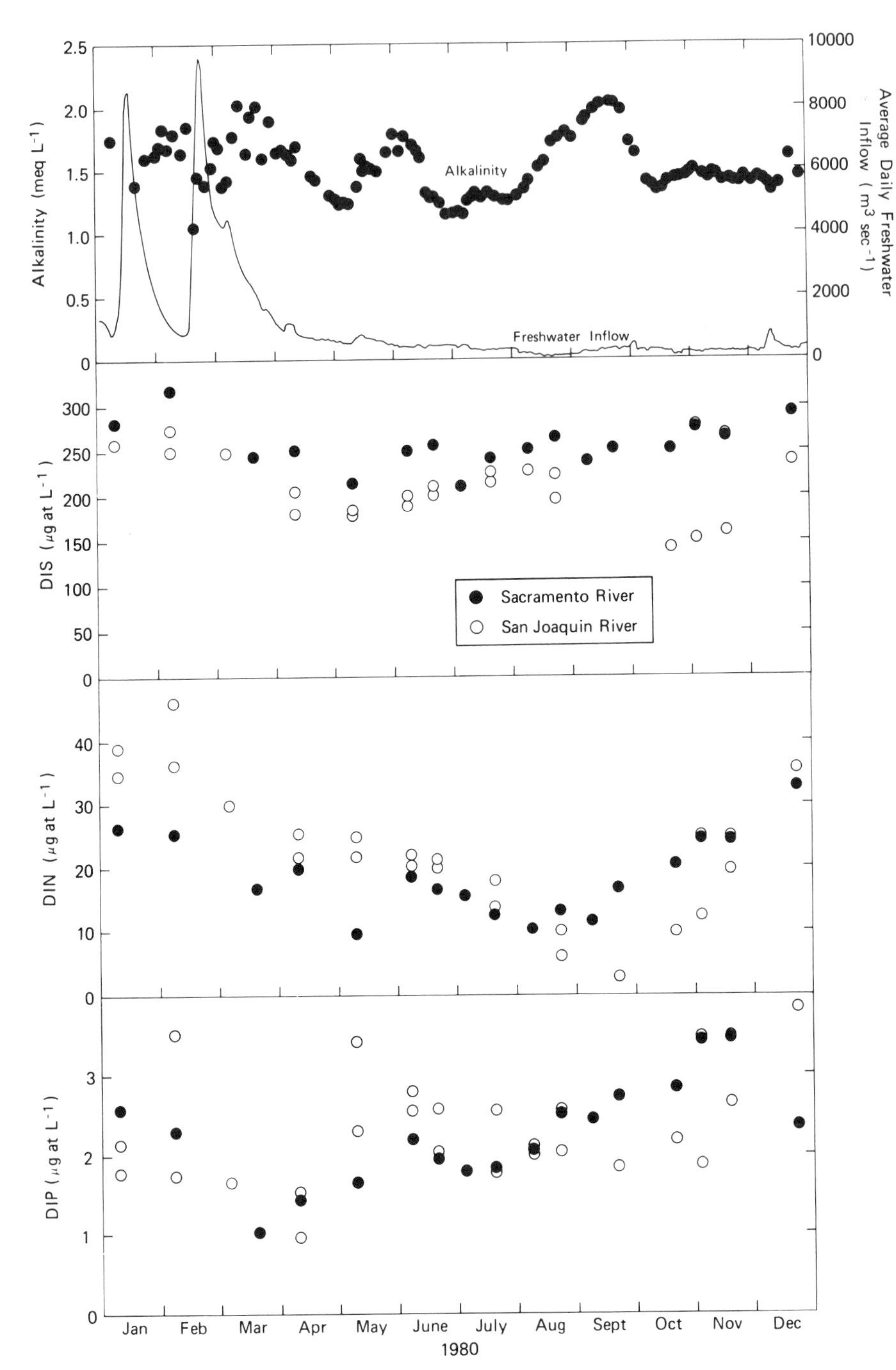

Alkalinity (meq L^{-1})
2.5
2.0
1.5
1.0
0.5
0
10000
8000
6000
4000
2000
0
Average Daily Freshwater Inflow (m^3 sec^{-1})
Alkalinity
Freshwater Inflow
DIS (μg at L^{-1})
300
250
200
150
100
50
0
Sacramento River
San Joaquin River
DIN (μg at L^{-1})
40
30
20
10
0
DIP (μg at L^{-1})
3
2
1
0
Jan
Feb
Mar
Apr
May
June
July
Aug
Sept
Oct
Nov
Dec
1980

three during the four-month 1983 study (Table 1). Monthly (discharge-volume weighted) average concentrations of alkalinity and DIS showed little variation over the study period and concentrations ranged less than DIN and DIP. Alkalinity and DIS distributions in the estuary are usually linear with respect to salinity during high inflow (Schemel 1984b; Peterson et al. 1975b), a feature that also indicates that their concentrations are relatively stable.

Average DIN and DIP concentrations during January were higher than those during the following three months, probably due to a washout of these substances from surface soils during the first major storms (Kennedy and Malcolm 1978). Furthermore, DIN and DIP were often more concentrated when flow rates were increasing with the early runoff from each major storm. River concentration variations were often large, as indicated by the concentration ranges (Table 1), and durations of these variations were short relative to estuary flushing times. Consequently, during high river inflow, we would expect DIN and DIP distributions in the estuary to be more irregular than DIS and alkalinity distributions.

Nutrient concentration variations in the river were small in comparison to the large seasonal variation of river inflow during 1980. Therefore, most of the annual transport of riverine DIC, DIS, DIN, and DIP occurred during January through March, when advective water replacement times were only about one week and solutes were rapidly moved through the estuary to the adjacent Pacific Ocean. Field observations from previous years and a numerical simulation model show that estuarine distributions of DIS are not significantly affected by biological removal when river inflow rates are high (Peterson et al. 1978). Consequently, on an annual time scale, the estuary may be relatively ineffective in modifying the transports of many riverine substances, and estuarine supply and removal processes need to be evaluated on shorter time scales.

FIGURE 2. Average daily freshwater inflow rate and concentrations of alkalinity and dissolved inorganic silicon (DIS), nitrogen (DIN), and phosphorus (DIP) in the Sacramento and San Joaquin Rivers during 1980.

TABLE 1. Volume weighted average concentrations and concentration ranges of alkalinity and dissolved inorganic silicon (DIS), nitrogen (DIN), and phosphorus (DIP) in the Sacramento River at Rio Vista during 1983. Samples were taken daily or more frequently during January through March, then every other day during April. Volumes are preliminary data courtesy of the U. S. Geological Survey District Office, Sacramento, California.

	January Concentrations		*February Concentrations*		*March Concentrations*		*April Concentrations*	
	Volume Weighted Average	*Range*	*Volume Weighted Average*	*Range*	*Volume Weighted Average*	*Range*	*Volume Weighted Average*	*Range*
Alkalinity[a]	1.25	0.90-1.60	1.28	1.08-1.39	1.28	1.12-1.47	1.33	0.97-1.77
DIS[b]	233	199-277	264	226-289	241	207-262	255	235-269
DIN[b]	21.1	12.1-36.1	13.7	11.8-18.4	13.3	9.9-16.8	14.0	9.8-16.8
DIP[b]	1.8	1.0-2.5	1.1	0.7-1.5	1.1	0.8-1.9	0.7	0.6-1.0
Total Volume	$5.0 \times 10^9 m^3$		$9.1 \times 19^9 m^3$		$15.8 \times 10^9 m^3$		$5.7 \times 10^9 m^3$	

[a] meq L^{-1}; [b] μg at L^{-1}

SEASONAL DISTRIBUTIONS OF DIS, DIN, AND DIP IN THE ESTUARY

Constituent-salinity distributions in the estuary can provide insights concerning the effects of competing estuarine supply and removal processes relative to the mixing of river and ocean source waters (Peterson et al. 1975b; 1978). Although this information does not directly tell us the dynamics of the substances involved, numerical models (Peterson et al. 1978; Officer 1980) can provide rate estimates.

Examination of our 1980 data revealed three time periods [winter (January to mid-April), spring (late April to mid-June) and summer (mid-June through October)] when constituent-salinity distributions showed similar patterns. The periods relate to hydrologic seasons that are expected to change from year to year due to inter-annual river inflow variability. During winter, river flow ranged from 650 to 9600 m^3 sec^{-1}, surface waters were often fresh in Carquinez Strait and San Pablo Bay, and the water column was stratified (Dedini et al. 1982). The salinity field shifted rapidly landward during spring. The water column was moderately stratified and river inflow ranged from 350 to 800 m^3 sec^{-1}. During summer, river inflow ranged from 60 to 500 m^3 sec^{-1} and the water column was partially to well mixed.

Individual distributions of DIS, DIN, and DIP during previous years have indicated the importance of estuarine supply and removal processes at channel locations in the estuary (e.g., Conomos et al. 1979; Peterson et al. 1975b; Peterson 1979). Our approach here is to combine all of the channel and shoal measurements during each of the three 1980 "seasons" and show concentration ranges for each solute over the estuarine salinity gradient (Fig. 3). In this way, major features of the distributions are identifiable along with their similarities and differences. When important differences exist between channel and adjacent shoal concentrations, they are detailed in the following text.

DIS was distributed linearly in the channels during winter (Fig. 3) and showed only minor departures from linearity in the shoals. The wider concentration range in the inner estuary indicates an effect of river concentration variability. Freshwater DIS concentrations were slightly lower during spring, in part due to the greater influence of the San Joaquin River (Fig. 2). The most distinctive feature during spring was the depression of concentrations at mid-salinities, corresponding to the shoals and to a lesser extent the channel of San Pablo Bay. The summer DIS distribution differed from those during winter and spring in that

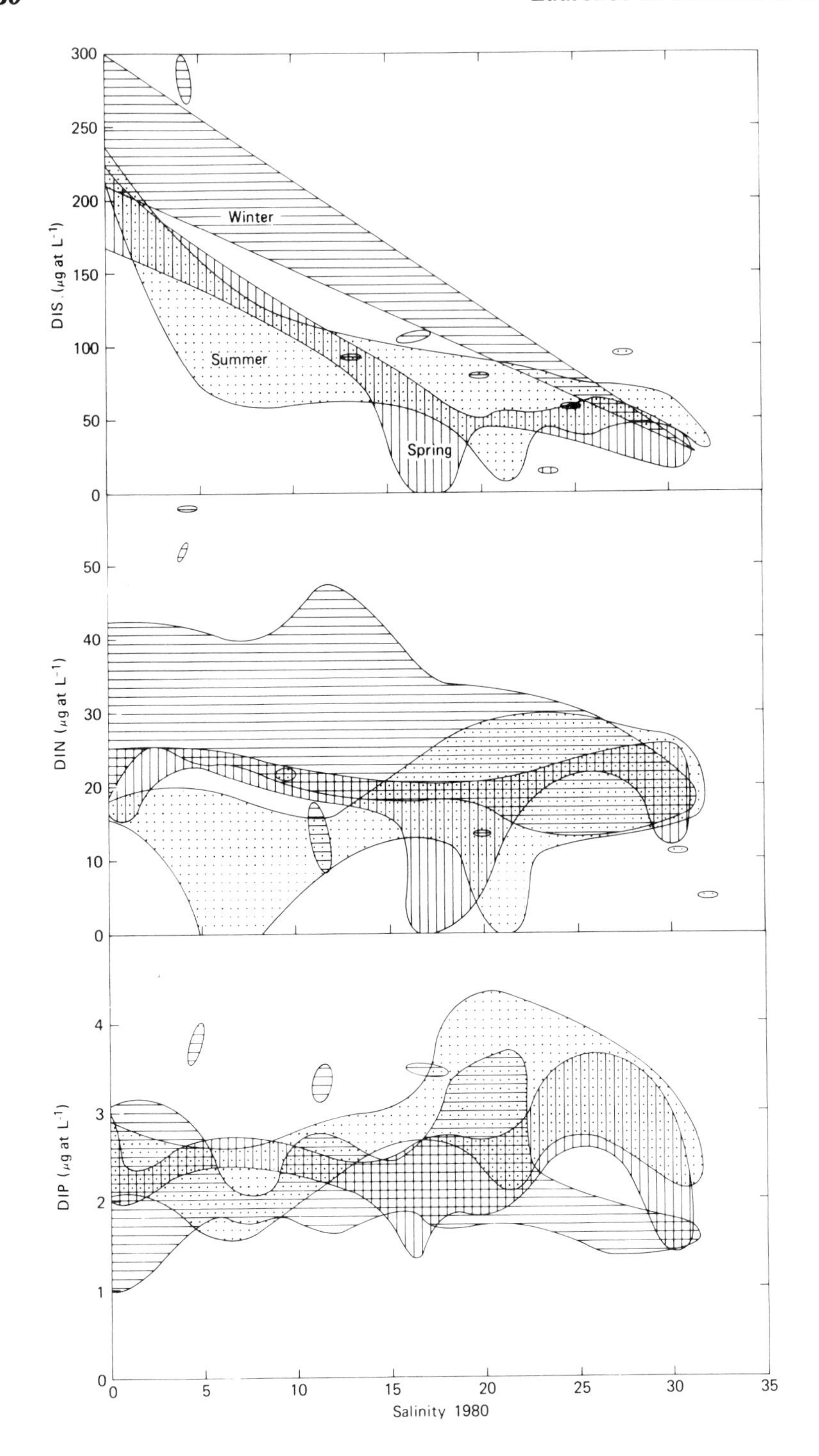
300
250
200
150
100
50
0
DIS. (μg at L⁻¹)
Winter
Summer
Spring
50
40
30
20
10
0
DIN (μg at L⁻¹)
4
3
2
1
0
DIP (μg at L⁻¹)
0
5
10
15
20
25
30
35
Salinity 1980

concentrations were much lower at low- to mid-salinities, corresponding to locations in Suisun Bay and Carquinez Strait. Low concentrations in the outer estuary were restricted to shoal locations in San Pablo Bay.

The wide range of DIN concentrations in the estuary during winter (Fig. 3) was presumably a consequence of river concentration variability. During spring, lower DIN concentrations were located in the channel and shoals of San Pablo Bay, with lowest concentrations in the shoals. Lowest river DIN concentrations occurred during summer, when DIN concentrations were lowest in the channels and shoals of Suisun Bay and the shoals of San Pablo Bay. Summer DIN concentrations at the outer estuary channel locations were generally higher than those of the freshwater inflow at that time. These higher concentrations probably resulted from a combination of estuarine, ocean and waste inputs.

DIP distributions during winter were often non-linear, without exhibiting a river to ocean gradient. River concentration variability probably contributed to the wide range of estuarine concentrations. During spring and summer DIP differed from DIN and DIS distributions in that DIP showed only slightly lower concentrations at the times and locations were DIS and DIN concentrations were strongly reduced. Summer DIP concentrations in the outer estuary were typically higher than those in the inner estuary.

The seasonal patterns (Fig. 3) indicated that estuarine supply and removal processes affected DIS, DIN and DIP concentrations at particular times and locations in the estuary. Suisun and San Pablo Bays were the sites of the largest net removals, but concentrations seaward of these locations appeared affected as water from these bays mixed downstream. DIS distributions showed net removal during spring and summer but did not show effects of estuarine supply processes. DIN distributions were affected by both estuarine supply and removal processes and, like DIS, they were most strongly affected by removal. DIP was primarily affected by supply processes in the outer estuary. DIP removal effects were small compared to those for DIS and DIN. Biological uptake stoichiometry (Redfield, Ketchum and Richards 1963) undoubtedly contributes to the smaller removal effects for DIP. In addition, compared to other estuaries, northern San Francisco Bay suspended sediment concentrations are generally higher

FIGURE 3. Seasonal concentration ranges of dissolved inorganic silicon (DIS), nitrogen (DIN), and phosphorus (DIP) with respect to salinity in northern San Francisco Bay during 1980.

and DIP concentrations are lower, when loading rates are normalized to water residence time (Nixon 1983). Therefore, it appears likely that DIP concentrations are buffered by interaction with suspended sediments (Pomeroy, Smith and Grant 1965).

BIOLOGICAL REMOVAL AS A MAJOR "FILTERING" PROCESS

Nutrient concentrations are determined by supply, removal and mixing rates, and it appears that the timing and magnitudes of net supply and removal effects varied seasonally during 1980, particularly in Suisun and San Pablo Bays. When PCO_2 decreases to below the atmospheric partial pressure (approx. 325 ppm by volume) in these bays, this indicates high rates of biological carbon dioxide removal (Spiker and Schemel 1979). During 1980, PCO_2 at the channel locations was lower than 325 ppm on only two occasions (Fig. 4), and these were localized in San Pablo Bay (early May) and Suisun Bay (mid-August). PCO_2 was higher in the bays two weeks before and after each minimum, illustrating the rapid kinetics of the processes controlling PCO_2 and possibly the short durations of the "maximum net rate" events. The minimum PCO_2 in both bays coincided with (channel- and shoal-wide averaged) maximum

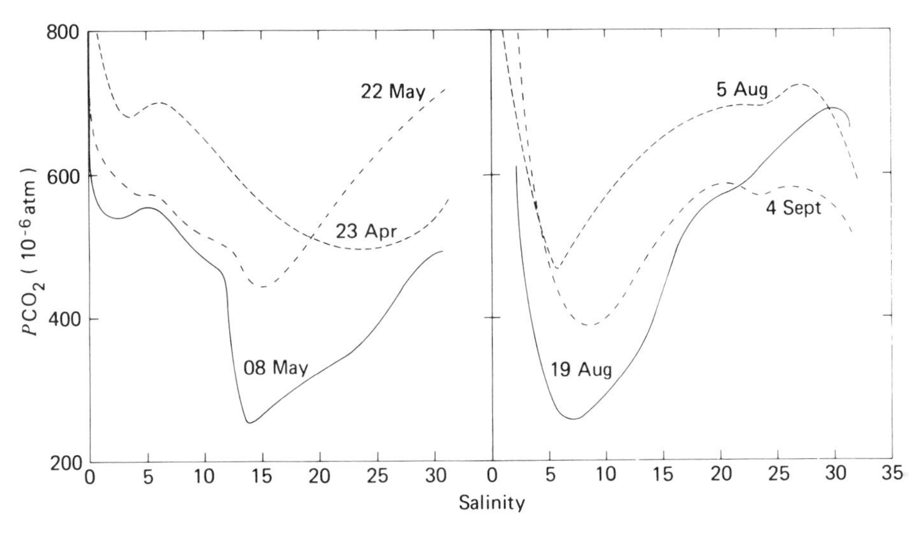

FIGURE 4. Distributions of PCO_2 with respect to salinity in northern San Francisco Bay during spring and summer 1980.

or near-maximum chlorophyll *a* concentrations, minimum concentrations of DIS and DIN, and minor decreases in DIP concentrations (Fig. 5). There were, however, distinctive differences in the hydrologic conditions during these two events, indicating that the biological removal of nutrients may have been controlled somewhat differently in San Pablo and Suisun Bays.

Factors controlling phytoplankton abundance and primary production in Suisun Bay appear to be better understood than those for San Pablo Bay (Cloern and Cheng 1981). Phytoplankton dynamics in Suisun Bay are controlled by hydrodynamic factors that are regulated by river inflow rate (Arthur and Ball 1979; Cloern et al. 1983). Minimum PCO_2 in Suisun Bay occurred during the four-week period when the average river inflow rate reached a seasonal minimum (125 $m^3 sec^{-1}$). Flow rates in the range of 100 to 350 $m^3 sec^{-1}$ place the non-tidal current null-zone in Suisun Bay (Peterson et al. 1975b; Cloern et al. 1983), a location with a shallow average cross-channel depth (Fig. 1). Higher and lower river inflow rates position the null zone seaward and landward, respectively, where average cross-channel water depths are deeper and, with all other factors equal, average water column light intensity is less. Therefore, it appears that a combination of higher average light levels and the particle-trapping ability of the estuarine circulation cell (Festa and Hansen 1978), which also traps diatoms with appropriate settling rates, contributed to increased phytoplankton biomass and nutrient removal in Suisun Bay during summer.

It appears unlikely that estuarine circulation influenced biological nutrient removal in San Pablo Bay in the same way that it did in Suisun Bay. The salinity range during the PCO_2 minimum in Suisun Bay (approx. 2 to 12; Fig. 5) corresponds to what would be expected in the null zone (Peterson et al. 1975a), whereas the salinity range in San Pablo Bay during the PCO_2 minimum was much higher (approx. 12 to 22). A physical feature that could have enhanced biological nutrient removal during the PCO_2 minimum in San Pablo Bay was salinity stratification (Dedini et al. 1982). Presumably, water column stratification can retain phytoplankton in the photic zone, thus increasing growth rates and population densities. A similar effect has been documented in the southern reach of San Francisco Bay (Cloern 1979). Stratification was greatly reduced by mid-June in San Pablo Bay, coinciding with the decline of chlorophyll *a* concentrations in the shoals and channels.

Chlorophyll *a* concentrations increased to a spring maximum then rapidly decreased in San Pablo Bay (Fig. 5). The chlorophyll *a* maximum occurred about three months later in Suisun Bay, where concentrations had increased regularly from spring

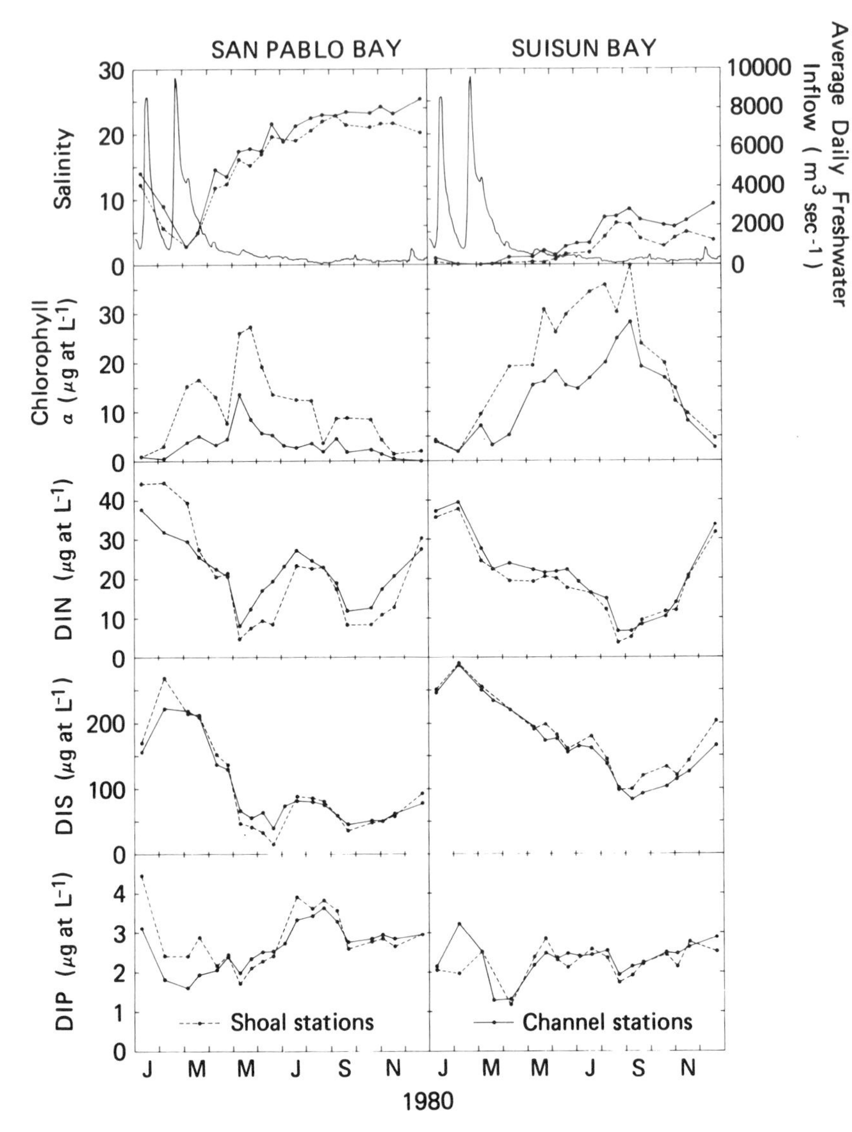

FIGURE 5. Average concentrations of salinity, chlorophyll a, and dissolved inorganic nitrogen (DIN), silicon (DIS), and phosphorus (DIP) in the channels and shoals of San Pablo and Suisun Bays during 1980.

to late summer. Therefore, it appears that biological removal and river inflow variability controlled the nutrient concentrations during summer in Suisun Bay, whereas mixing, estuarine supply, and other sources controlled concentrations in San Pablo Bay during most of the summer when biological removal rates were low.

The shoals are important for phytoplankton development in both Suisun and San Pablo Bays because a larger fraction of the water column is lighted (Cloern and Cheng 1981). When biological removal was most apparent, concentrations of chlorophyll *a* were generally higher and concentrations of nutrients were generally lower in the shoals relative to the channel in both bays (Fig. 5). Consequently, the shoals appeared to be the locations of maximum removal "filtering." Channel-shoal exchange of water was indicated by the similarities in the shoal and channel nutrient concentration variations in each bay (Fig. 5). Exchange could have enhanced estuarine filtering effects in the channel and diluted filtering effects to some extent in the shoals.

Magnitudes of nutrient inputs and removals were estimated for the San Pablo Bay channel locations during the spring period of maximum biological removal. Advective and exchange inputs were estimated from a one-layer box model (Officer 1980). Waste inputs are from Conomos (1979), and benthic mineralization is from Hammond (1981). Phytoplankton uptakes were estimated as net carbon production divided by the Redfield C:N and C:P ratios (Redfield, Ketchum and Richards (1963) and a C:Si ratio of 106:20 by atoms. Figure 6 summarizes DIN inputs and removals. The water column mineralization estimate for DIN is based on our unpublished observations, that indicate about 20% of the particulate nitrogen is mineralized per day. Denitrification was estimated from mass balance calculations for San Pablo Bay during late summer and is comparable to a rate measured under similar conditions of temperature and salinity in Narragansett Bay (Seitzinger et al. 1980). The key point of Figure 6 is that waste and mineralization inputs were large for DIN, amounting to about one-quarter and three-quarters, respectively, of the river input (advective plus exchange). The largest DIN loss was due to phytoplankton uptake, with denitrification and loss to the ocean being considerably less.

Similar estimates of the magnitudes of inputs and outputs were made for DIS and DIP for the same location and time period. Compared to river inputs, waste (2%) and mineralization (<20%) were relatively small inputs of DIS. DIS transport to the ocean was about 60% of the river inputs and four times the estimated phytoplankton uptake at that time. Waste was an important input for DIP, being 40% of the river

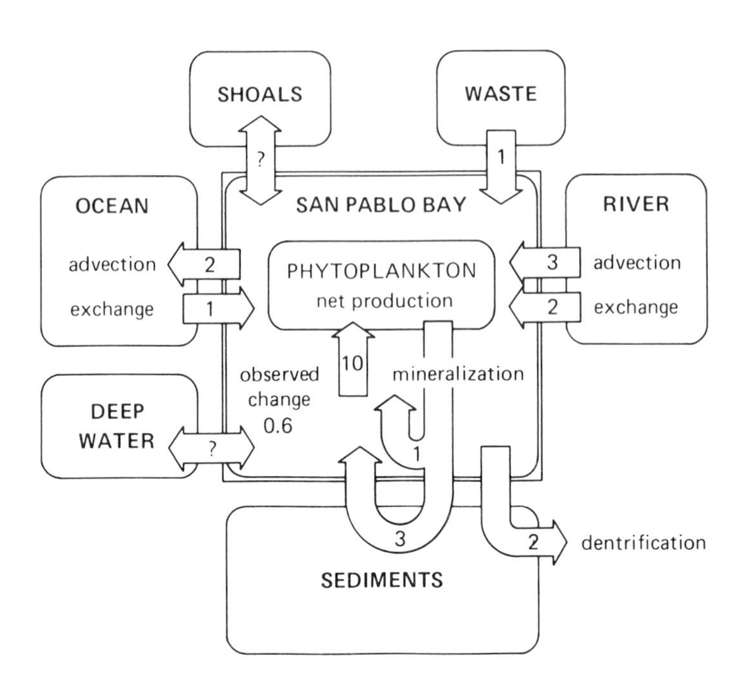

FIGURE 6. Average DIN supply and removal rates (mmoles m^{-2} d^{-1}) in the channel of San Pablo Bay during the six-week period following the $\underline{P}CO_2$ minimum.

inputs. Benthic mineralization rates for DIP were probably very low, as indicated by both mass balance calculations and benthic measurements (Hammond 1981). The largest DIP loss term was phytoplankton uptake, which appeared to be greater than river inputs at that time. DIP loss to the ocean was about equal to river inputs.

In conclusion, DIC, DIS, DIN, and DIP distributions in northern San Francisco Bay were affected by river inflow supply and estuarine supply and removal processes during 1980. River inflow supplied nutrients to the estuary during winter at rates that overwhelmed the potential effects of estuarine supply and removal processes. Estuarine processes were most effective in modifying estuarine distributions during lower river inflow periods, when river inflow controlled hydrodynamic factors that influenced phytoplankton dynamics and thus nutrient removal kinetics. Largest biological removal effects were localized in Suisun and San Pablo Bays. In Suisun Bay, biological removal of nutrients appeared effective over a longer period of time than in San Pablo Bay. During the summer period of low biological removal in San

Pablo Bay, estuarine sources appeared to contribute to increased DIN and DIP concentrations. A clearer understanding of estuarine "filtering" will require a better separation of the effects of river input variability from those of estuarine processes, which might be best accomplished by numerical simulation techniques.

ACKNOWLEDGMENTS

We wish to thank Jeanne Dileo-Stevens for technical assistance and Janet K. Thompson and T. John Conomos for their valuable ideas and criticisms of the manuscript.

REFERENCES CITED

Arthur, J. F. and M. D. Ball. 1979. Factors influencing the entrapment of suspended material in the San Francisco Bay-Delta Estuary, pp.143-174. *In:* T. J. Conomos (ed,), *San Francisco Bay, The Urbanized Estuary.* AAAS, San Francisco, California.

California State Department of Water Resources. 1982. Dayflow summary: water year 1955 through 1980. State of California Resources Agency, Sacramento, California. 599pp.

Cloern, J. E. 1979. Phytoplankton of the San Francisco Bay system: the status of our current understanding, pp.247-264. *In:* T. J. Conomos (ed.), *San Francisco Bay, The Urbanized Estuary.* AAAS, San Francisco, California.

Cloern, J. E. and R. T. Cheng. 1981. Simulation model of *Skeletonema costatum* population dynamics in northern San Francisco Bay, California. *Est. Coastal Shelf Sci. 12*:83-100.

Cloern, J. E., A. E. Alpine, B. E. Cole, R. L. Wong, J. F. Arthur and M. D. Ball. 1983. Discharge controls phytoplankton dynamics in the northern San Francisco Bay estuary. *Est. Coastal Shelf Sci. 16*:415-429.

Conomos, T. J. 1979. Properties and circulation of San Francisco Bay waters, pp.47-84. *In:* T. J. Conomos (ed.), *San Francisco Bay, The Urbanized Estuary*. AAAS, San Francisco, California.

Conomos, T. J., R. E. Smith, D. H. Peterson, S. W. Hager and L. E. Schemel. 1979. Processes affecting seasonal distributions of water properties in the San Francisco Bay Estuarine system, pp.115-142. *In:* T. J. Conomos (ed.), *San Francisco Bay, The Urbanized Estuary*. AAAS, San Francisco, California.

Dedini, L. A., L. E. Schemel and M. A. Trembreull. 1982. Salinity and temperature measurements in San Francisco Bay waters. U. S. Geological Survey Open-File Rpt. 82-125, Menlo Park, California. 130pp.

Dyer, K. R. 1972. *Estuaries: A Physical Introduction*. John Wiley and Sons, New York. 140pp.

Festa, J. F. and D. V. Hansen. 1978. Turbidity maxima in partially mixed estuaries: a two-dimensional numerical model. *Est. Coastal Shelf Sci. 7*:347-359.

Hammond, D. E. 1981. Nutrient exchange across the sediment-water interface in San Francisco Bay. *EOS 62*:925.

Hammond, D. E. and C. Fuller. 1979. The use of Radon-222 to estimate benthic exchange and atmospheric exchange rates in the San Francisco Bay, pp.213-230. *In:* T. J. Conomos (ed.), *San Francisco Bay, The Urbanized Estuary*. AAAS, San Francisco, California.

Kennedy, V. C. and R. L. Malcolm. 1978. Geochemistry of the Mattole River of Northern California. U. S. Geological Survey Open-File Rpt. 78-205, Menlo Park, California. 324pp.

Loder, T. C. and R. P. Reichard. 1981. The dynamics of conservative mixing in estuaries. *Estuaries 4*:64-69.

Nixon, S. W. 1983. Estuarine Ecology - a comparative and experimental analysis using 14 estuaries and the MERL microcosms. Final Report to the U. S. Environmental Protection Agency, Chesapeake Bay Program. Graduate School of Oceanography, Kingston, Rhode Island. 59pp.

Officer, C. B. 1980. Box models revisited, pp.65-114. *In:* P. Hamilton and K. B. Macdonald (eds.), *Estuarine and Wetland Processes*. Plenum Press, New York.

Peterson, D. H. 1979. Sources and sinks of biologically reactive oxygen, carbon, nitrogen, and silica in northern San Francisco Bay, pp.175-193. *In:* T. J. Conomos (ed.), *San Francisco Bay, The Urbanized Estuary*. AAAS, San Francisco, California.

Peterson, D. H., T. J. Conomos, W. W. Broenkow and P. C. Doherty. 1975a. Location of the nontidal current null zone in northern San Francisco Bay. *Est. Coastal Shelf Sci.* *3*:1-11.

Peterson, D. H., T. J. Conomos, W. W. Broenkow and E. P. Scrivani. 1975b. Processes controlling the dissolved silica distribution in San Francisco Bay, pp.153-187. *In:* L.E.Cronin (ed.), *Estuarine Research, Volume 1*. Academic Press, New York.

Peterson, D. H., J. F. Festa and T. J. Conomos. 1978. Numerical simulation of dissolved silica in the San Francisco Bay. *Est. Coastal Shelf Sci*. *7*:99-116.

Pomeroy, L. R., E. E. Smith and C. M. Grant. 1965. The exchange of phosphate between estuarine water and sediments. *Limnol. Oceangr.* *10*:167-172.

Redfield, A. C., B. H. Ketchum and F. A. Richards. 1963. The influence of organisms on the composition of sea water, pp.26-77. *In:* M. N. Hill (ed.) *The Sea, Volume 2*. Interscience, New York.

Schemel, L. E. 1984a. Salinity, alkalinity, and dissolved and particulate organic carbon in the Sacramento River at Rio Vista, California and at other locations in the Sacramento -San Joaquin Delta, 1980. U. S. Geological Survey Water Resources Investigation Rpt. 83-4059, Menlo Park, California. 45pp.

Schemel, L. E. 1984b. Seasonal patterns of alkalinity in the San Francisco Bay estuarine system, California, during 1980. U. S. Geological Survey Water Resources Investigation Rpt. 82-4102, Menlo Park, California. 80pp.

Schemel, L. E. and L. A. Dedini. 1979. A continuous water-sampling and multiparameter-measurement system for estuaries. U. S. Geological Survey Open-File Rpt. 79-273, Menlo Park, California. 92pp.

Seitzinger, S., S. Nixon, M. E. Pilson and S. Burke. 1980. Denitrification and N_2O production in near-shore marine sediments. *Geochimica et Cosmochimica Acta 44*:1853-1860.

Selleck, R. E., E. A. Pearson, B. Glenne and P. N. Storrs. 1966. Physical and hydrological characteristics of San Francisco Bay. Final report of a comprehensive study of San Francisco Bay. University of California, Sanitary Engineering Research Laboratory Rpt. 65-10, Berkeley, California. 99pp.

Smith, R. E., R. E. Herndon and D. D. Harmon. 1979. Physical and chemical properties of San Francisco Bay waters, 1969-1976. U. S. Geological Survey Open-File Rpt. 79-511, Menlo Park, California. 607pp.

Spiker, E. C. and L. E. Schemel. 1979. Distribution and stable-isotope composition of carbon in San Francisco Bay, pp.195-212. *In:* T. J. Conomos (ed.), *San Francisco Bay, The Urbanized Estuary*. AAAS, San Francisco, California.

THE ESTUARINE INTERACTION OF NUTRIENTS, ORGANICS, AND METALS: A CASE STUDY IN THE DELAWARE ESTUARY

Jonathan H. Sharp
Jonathan R. Pennock
Thomas M. Church
John M. Tramontano
Luis A. Cifuentes

College of Marine Studies
University of Delaware
Lewes, Delaware

Abstract: In the estuarine environment, biogeochemical processes alter concentrations of soluble nutrients, organic matter, and trace metals. Some constituents show geochemical reactivity and are filtered out by "flocculation" type reactions; these may be considered as a geochemical "filter". Other constituents show biochemical reactivity and are filtered out by organismic processes; these may be considered as a biochemical "filter". Through use of data from the Delaware Estuary, the geochemical filter is illustrated as it affects humic acids, phosphate, and iron; the biochemical filter as it affects ammonium, phosphate, silicate, and urea. Contrasting examples are presented for the transition elements copper and nickel which show little filtration, despite the potential for bioreactivity. Cadmium and phosphate are used to illustrate a combined biogeochemical filter.

INTRODUCTION

Estuaries serve as corridors for the transit of terrigenous materials to the sea. Much of the dissolved material carried into the estuary is simply diluted upon

ISBN 0-12-405070-0

mixing and delivered to the sea. In contrast, some dissolved chemicals react within the estuary and are removed from the dissolved phase, thus making the estuary a "filter". The major mechanism for removal of dissolved chemicals is through reactions with particulate phases. These reactions potentially include geochemical processes

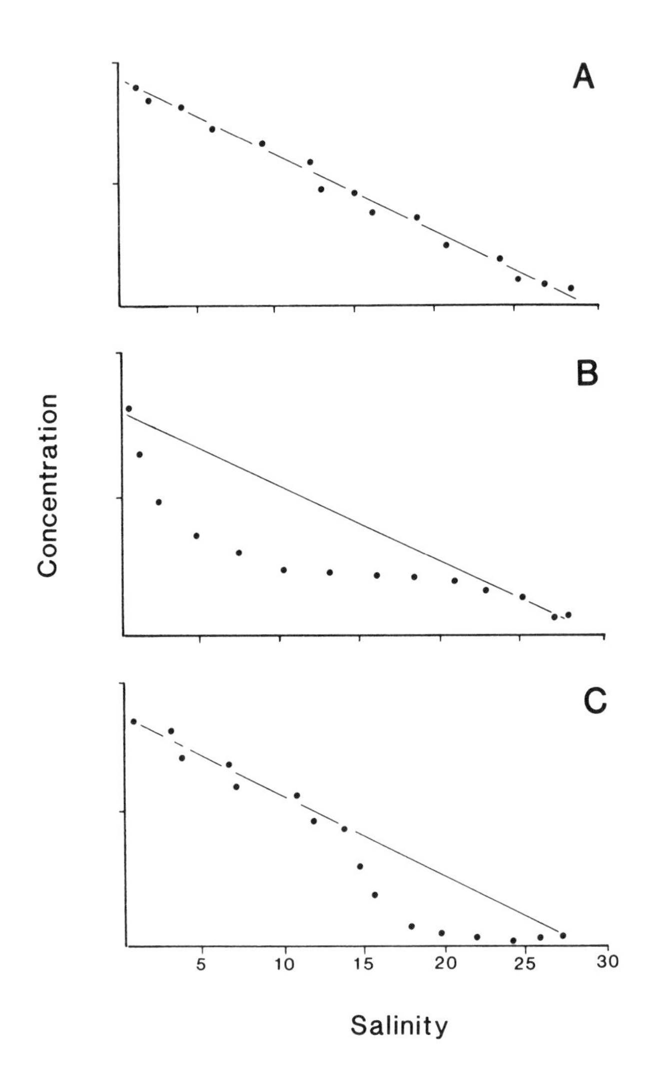

FIGURE 1. Hypothetical estuarine mixing curves for the Delaware Estuary. Curve A illustrates apparent conservative mixing. Curve B indicates a geochemical estuarine sink (filter). Curve C indicates a biochemical estuarine sink (filter).

of adsorption, flocculation, and precipitation, and biochemical processes such as uptake. All of these reactions serve as filters for dissolved chemicals; however, only when particulates are buried in the sediment does the estuary act as a true sink for terrigenous materials.

The concept of conservative and non-conservative mixing has been applied in recent years to nutrients (e.g. Liss 1976), organic matter (e.g. Sholkovitz 1976), and metals (e.g. Boyle et al. 1974) as one method of assessing the reactivity of chemical constitutents in estuaries. Using this concept, one can assess whether or not an estuary serves as a filter for a dissolved material by plotting the concentrations of any material against the corresponding salinities (property-salinity plot). A straight line is interpreted to indicate conservative mixing (Fig. 1A) while a downward curvilinearity indicates an estuarine sink, i.e. filtration (Fig. 1B-C). Of course, non-conservative behavior with an upward curve can also occur and is considered to represent the presence of estuarine sources, but the emphasis here is on the estuary as a filter.

An understanding of estuarine hydrodynamics and chemical reaction rates is necessary to interpret property-salinity plots. Inherent assumptions in these plots are that salt is mixed conservatively and that the freshwater and sea water endmember concentrations of the parameter are non-varying. However, endmember concentrations do vary, and have been shown to result in a curvilinear pattern indicating apparent removal, when in reality, conservative mixing was occurring (Loder and Reichard 1981; Officer and Lynch 1981). We interpret curvilinear patterns of nitrate in the upper Delaware Estuary as being primarily caused by varying endmembers (L.A. Cifuentes, unpublished data). In contrast, observed linearity may also be misleading when materials are consumed and produced within the estuary at the same average rate (which is rapid compared to estuarine flushing).

Examples of nutrient, organic, and metal behavior in the Delaware Estuary are used to illustrate the concept of the estuary as a filter. A conceptual model is presented that differentiates between geochemical and biochemical filtration in the estuary (Fig. 2). Geochemical filtration occurs primarily in the low salinity upper region. This is the area of highest particulate concentration, due partially to flocculation but primarily to resuspension of bottom sediments (Biggs et al. 1983). Biochemical filtration occurs predominately in the lower estuary, the area of highest biological productivity (Fig. 2). The spatial separation between these regions affords us the ability to distinguish between these processes.

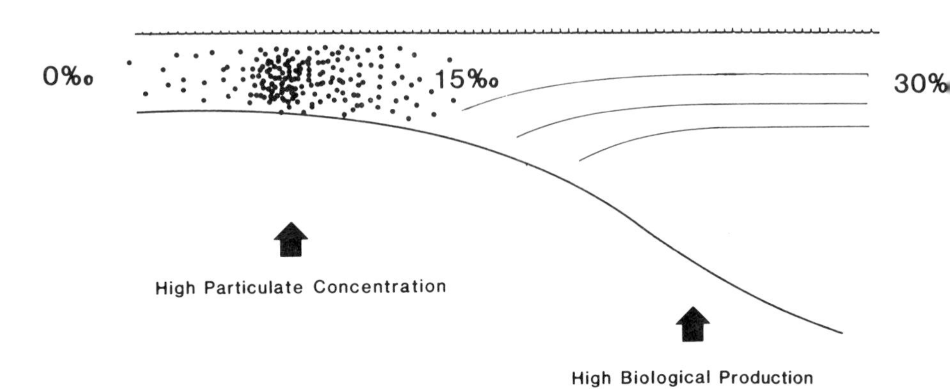

FIGURE 2. Schematic longitudinal cross section of the Delaware Estuary showing locations where the geochemical and biochemical filters are important (^{o}/oo = salinity).

STUDY AREA

The Delaware Estuary is a large shallow turbid estuary in the middle Atlantic region of the United States. A discussion of its geographic and physical setting, as well as a general chemical description, has been given by Sharp, Culberson and Church (1982). There is a pronounced turbidity maximum region in the upper estuary with clearer waters at higher salinities (Biggs et al. 1983). We have sampled the full salinity gradient (0-31°/oo salinity) of the estuary on 27 separate occasions from May 1978 through July 1983 during all seasons, and covering extremes in temperature and flow. Data from specific cruises have been chosen for use in this paper to illustrate estuarine filtering activity. Analyses for salinity, nutrients, dissolved organic carbon, metals, primary productivity, and soluble-particulate separations have been performed with routine methods that have been modified as described in Sharp et al. (1982). Analysis for total dissolved amino acids and urea are described in Cifuentes (1982).

Surface and deep water sampling was performed in the channel down the salinity gradient and in the shoals. Emphasis here is on surface samples through the salinity gradient that extends 95-130 km between Philadelphia and the mouth of the estuary. The flushing time under average flow conditions for the entire estuary has been estimated at about 100 days (Ketchum 1952). However, flow conditions and thus flushing time vary considerably over the year. River flow displays a maximum in March-May (510 m^3 sec^{-1}), moderate flow from November-February (334 m^3 sec^{-1}), and low flow from June to October (195 m^3 sec^{-1}) (Smullen et al. 1983). Under any conditions, the flushing time through the salinity gradient in the Delaware Estuary represents a relatively long time period when one considers the shorter time scales often associated with chemical kinetics and biological rate processes. Thus, one would expect to be able to observe the effects of biogeochemical processes on chemical distributions in the system.

THE GEOCHEMICAL FILTER

Geochemical filtration in the Delaware Estuary appears as an abrupt drop in the concentrations of some dissolved materials in the low salinity region of the estuary (Fig. 1B), due primarily to "flocculation" (aggregation, precipitation, adsorption, etc.) processes. Estuarine flocculation is largely a physico-chemical process associated with colloidal instability (Stuum and Morgan 1981). However, the carrier phases for such colloidal formation may be specific such that the estuarine geochemical filter for trace metals (e.g. iron and manganese) may be distinctly different both kinetically and geographically from those of nutrients (e.g. phosphate) and humic substances.

In the Delaware, classical examples of geochemical removal are found for humic acids (Fig. 3), iron (Fig. 4A), and phosphate in winter (Fig. 4B). However, misinterpretation of such property-salinity plots is likely without a knowledge of the chemical rate processes. Several important points are discussed below.

Although a number of chemicals may show grossly similar shapes in property-salinity plots, the processes causing the patterns may be different. For example, the coincident flocculation of humic acid and iron in an estuary has been interpreted in the past as being due to a single process (Sholkovitz, Boyle and Price 1978). From work in the Delaware Estuary and its subtributaries, it has been demonstrated that removal of humic acid and iron from flocculation is the result of two separate and semi-independent processes (Fox 1981; Fox and Wofsy 1983).

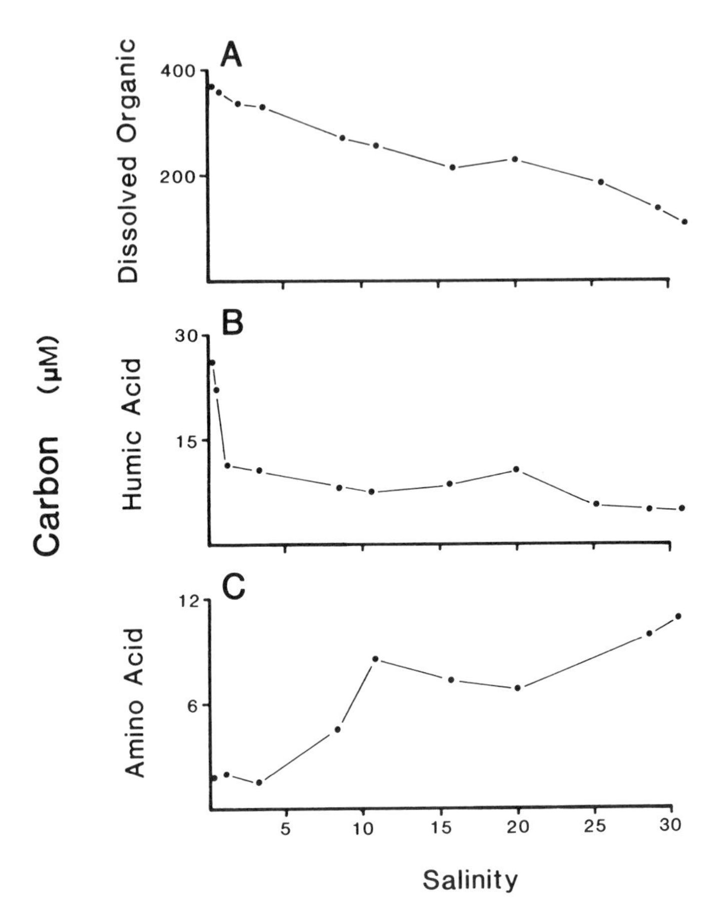

FIGURE 3. Total dissolved organic carbon, humic acids and amino acids versus salinity (micromolar carbon) from the Delaware Estuary in March 1982.

Similarly, by plotting iron concentration against phosphate concentration (Fig. 5), one can see a correlation between the two parameters throughout the estuary during winter suggesting that both parameters may be affected by the same process. However, the slope of this line (0.21) is considerably below the value of 1.0-1.5 that one would expect if the removal were due to stoichiometric precipitation of the iron phosphorus compounds strengite [$FePO_4$] or vivianite [$Fe_3(PO_4)_2$]. In addition, iron reaches an apparent steady-state lower limit at

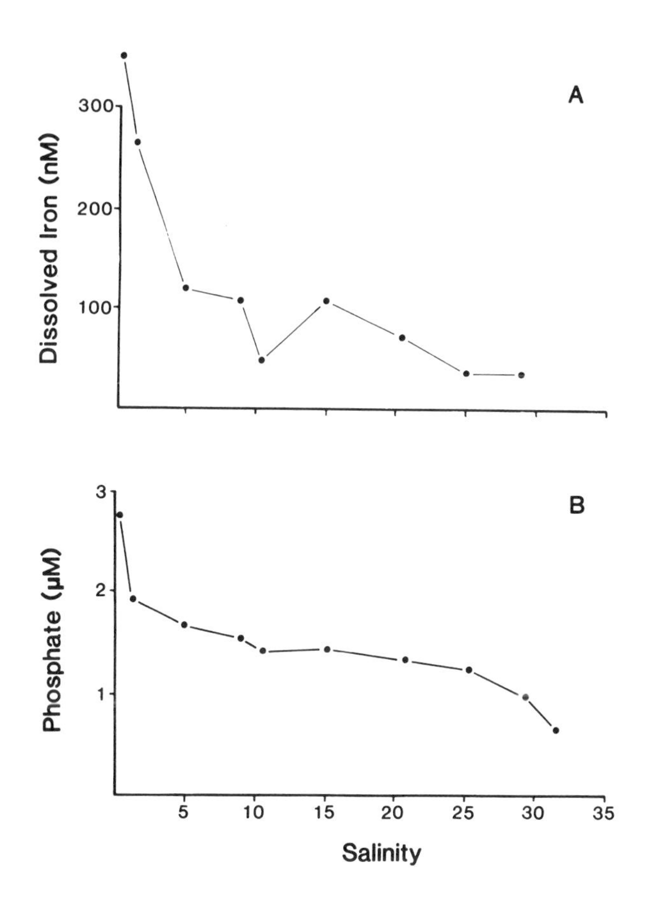

FIGURE 4. (A.) Dissolved iron concentration and (B.) phosphate concentration versus salinity in the Delaware Estuary in January 1982.

approximately 30 nM while the phosphate minimum is around 1 μM. This indicates that whatever process is removing iron in this case leaves excess phosphate. We believe that some of the excess phosphate and the relatively uniform mid-estuary phosphate concentrations are due to the estuarine phosphorus buffer effect. This is the phenomenon in which phosphate maintains a relatively steady concentration along the salinity gradient due to equilibrium adsorption-desorption reactions despite dilution by mixing (Pomeroy, Smith and Grant 1965; Morris, Bale and Howland 1981).

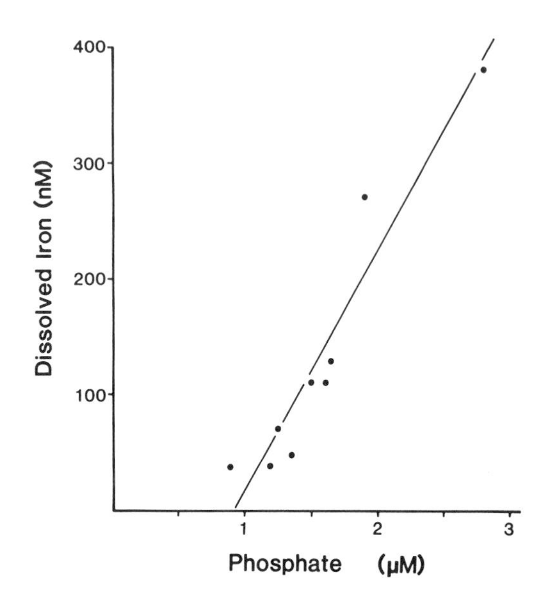

FIGURE 5. Dissolved iron concentration versus phosphate concentration in the Delaware Estuary in January 1982. The mole to mole slope of the line is 0.21 (R^2 = 0.91).

Finally, it is possible for apparent conservative mixing of a broadly defined class of materials (e.g. dissolved organic carbon) to occur while a number of its components (e.g. humic acids, amino acids) are markedly non-conservative (Fox 1981). Although humic acid carbon shows a strong sink in the upper estuary, and amino acid carbon, another component of the dissolved organic carbon pool, shows a source in the lower estuary, dissolved organic carbon displays conservative behavior (Fig. 3). In this example, the scale of the total dissolved organic carbon plot is too large to indicate variations brought out by components making up only about 10% of the pool.

THE BIOCHEMICAL FILTER

Biochemical filtration in the Delaware Estuary is the result of biological fixation of dissolved materials into particulate form followed by removal during food chain transfer, transport out of the estuary, or sedimentation.

The net result of biological filtration in the Delaware Estuary appears as a non-linear decrease in constituent concentration in the lower estuary (Fig. 1C).

The abrupt decline in ammonium, phosphate, and silicate concentrations in mid-estuary (Fig. 6) that occurs during the spring phytoplankton bloom provides the best example of biological filtration in the Delaware Estuary. This pattern is the result of phytoplankton uptake just downstream from the turbidity maximum region (Pennock, Sharp and Canzonier 1983). This clear interrelationship between non-conservative nutrient behavior and phytoplankton production during spring is due to several factors. First, high flushing rates in the upper estuary decrease the influence of geochemical processes in the river during this period (residence time is relatively short in comparison to chemical rate kinetics). Second, high river flow causes vertical stratification in mid-estuary which increases the amount of light available for phytoplankton growth, thus allowing the bloom to occur (Pennock 1983). During this period, phytoplankton nutrient demands in mid-estuary are met primarily by depletion of nutrient standing stock (ambient levels); this gives the appearance of strong non-conservative behavior in property-salinity plots.

This mid-estuarine decline in nutrient concentration during spring is quite different from the patterns for other times of year which do not show marked depletion of the nutrients at high salinities (Sharp et al. 1982). However, the apparent conservative behavior in the summer should not be considered *a priori* as evidence of low primary productivity. In this case, the apparent conservative pattern is due to utilization of regenerated nutrients during summer (Pennock 1983). In contrast to utilization of ambient nutrient standing-stocks during spring, production that is based on rapid regeneration of nutrients, as occurred during summer, is not observable in property-salinity plots.

The influence of nutrient recycling on property-salinity diagrams is also evident for urea in the Delaware Estuary. During winter, urea and ammonium display similar property-salinity patterns (Fig. 7A). During spring, the abrupt mid-estuary decline in ammonium due to phytoplankton uptake (Pennock 1983) presumably also affects urea (Fig. 7B). However, during this period, deviation from otherwise parallel behavior of urea and ammonium occurs in the lower estuary (Fig. 7B) and is probably due to urea excretion that is not rapidly utilized. In this case, removal of nitrogen as ammonium, urea, and nitrate does not result in

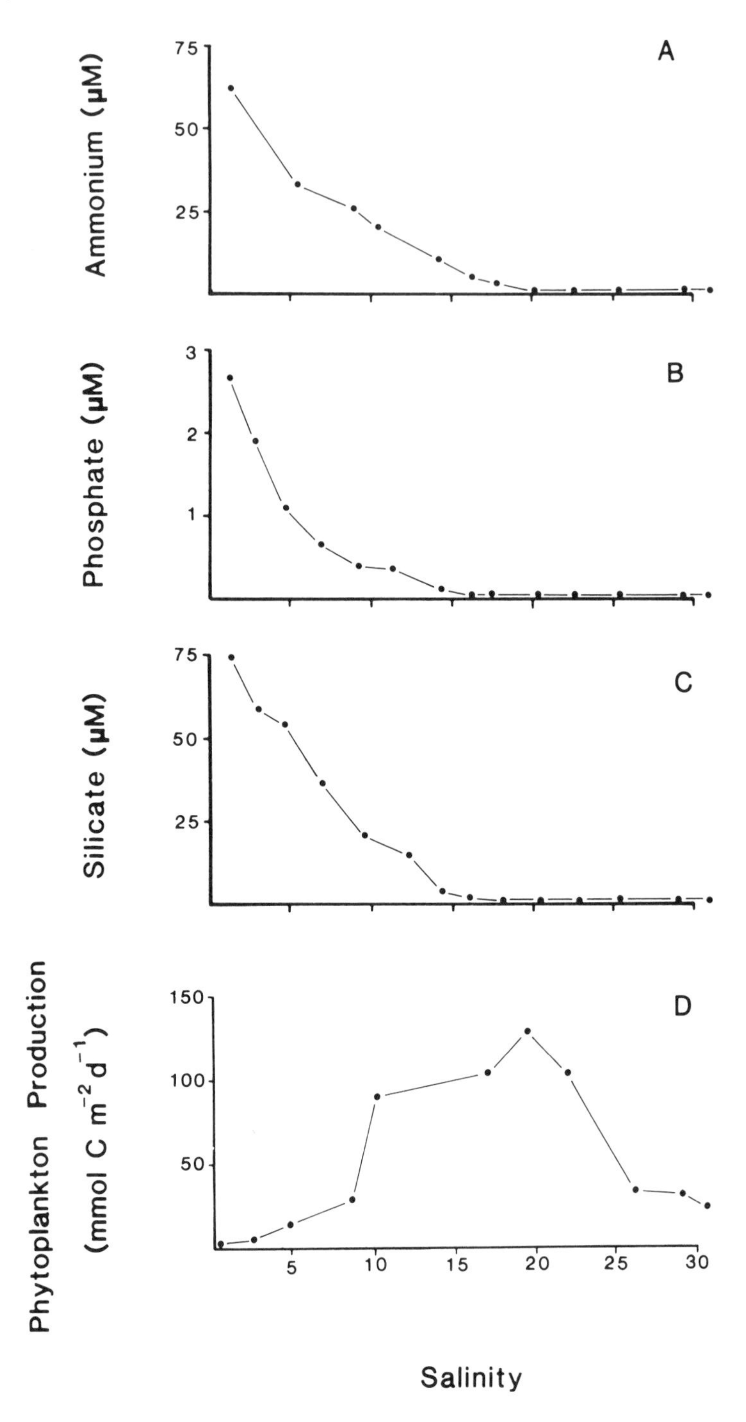
Ammonium (μM)
75
50
25
A
Phosphate (μM)
3
2
1
B
Silicate (μM)
75
50
25
C
Phytoplankton Production
(mmol C m⁻² d⁻¹)
150
100
50
D
5
10
15
20
25
30
Salinity

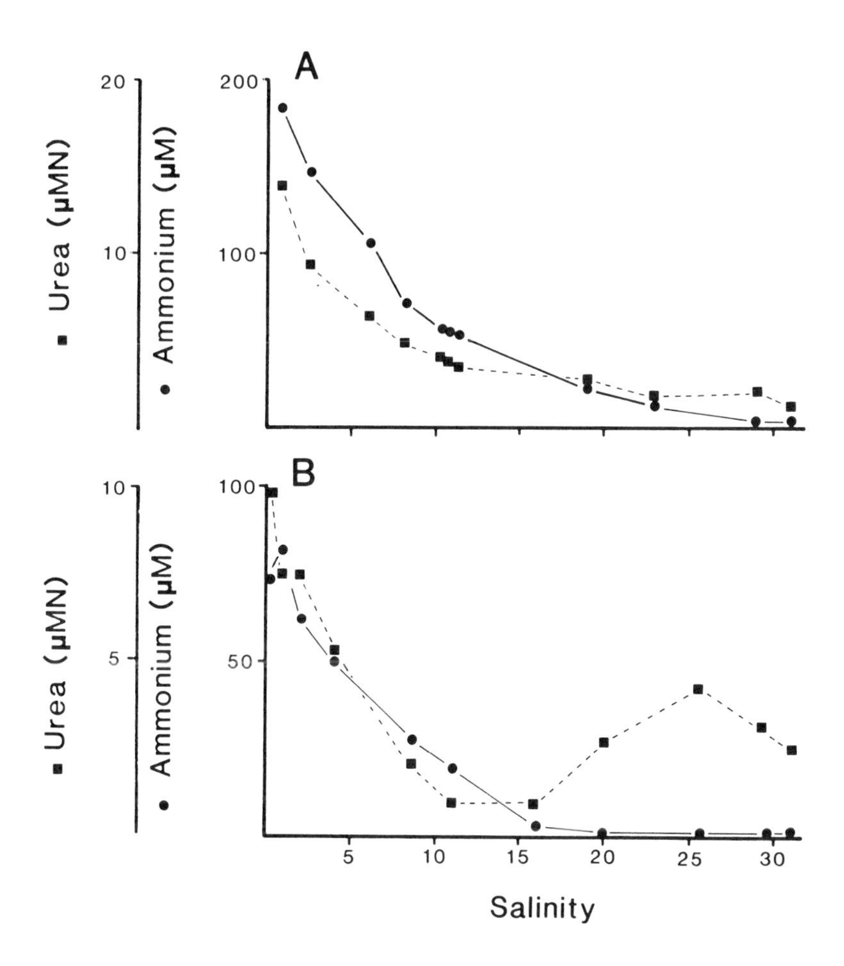

FIGURE 7. Ammonium and urea concentrations versus salinity in the Delaware Estuary in (A.) January and (B.) March 1981. Note that the y axis scales are not the same for the two frames.

FIGURE 6 (opposite). Concentrations of (A.) Ammonium, (B.) phosphate, and (C.) silicate, and (D.) areal primary productivity versus salinity in the Delaware Estuary in March 1982.

complete removal (filtration) from the estuary because portions are remineralized by other trophic levels.

The group of trace metals that include copper, cadmium, and nickel are often biologically active. One might therefore expect to see a close association between such trace metals and nutrients during periods of high productivity. However, plots of copper and nickel concentration versus salinity for the end of the spring bloom period display no pronounced curvilinearity during high river flow and intense nutrient utilization (Fig. 8). In fact, the requirements for trace metals by phytoplankton, while finite, are almost negligible compared to estuarine abundances. While plankton in oligotrophic waters may be limited sometimes by iron, coastal and estuarine waters seem to have trace metal abundances far in excess of demands even during bloom conditions (J. Martin, Moss Landing Marine Laboratory, personal communication). Another explanation of such apparent conservancy of copper and nickel may be related to metal speciation and availability since only the free ion may be bioactive and much of the metal pool is probably tied up in organic and inorganic complexes (Stuum and Morgan 1981). Many organisms can actively excrete trace metal chelators which may be mechanisms for assimilation and detoxification. Chelation by phytoplankton has been demonstrated for iron (Murphy, Lean and Nalewajko 1976) and for copper (McKnight and Morel 1979). It is possible that during productive periods, estuarine trace metals are subjected to excess complexation rendering them far less reactive. In the Delaware, such periods are often coincident with times of greatest flushing (e.g. May) and may thus be significant periods of trace metal export to shelf waters.

The trace metal cadmium may show a combination of geochemical and biochemical behaviors. Although cadmium is neither an essential element nor actively taken up by phytoplankton cultures, it is one of the few trace metals that is anomalously enriched in natural plankton populations and appears to be regenerated with phosphate (Martin, Bruland and Broenkow 1976). At the end of the spring bloom period, a drop can be seen in cadmium concentration in the area of the geochemical filter with a second drop in the area of the biochemical filter (Fig. 9B); this is similar to the pattern for phosphate at the same time (Fig. 9A). The plot of cadmium versus phosphate concentrations for this period (Fig. 10) is linear. The slope of this line, 3.9×10^{-4}, approximates the molar regeneration ratio of cadmium to phosphate (3.1×10^{-4}) in the water column and is similar to that found in microplankton (3.9×10^{-4}) of the northeast Pacific Ocean

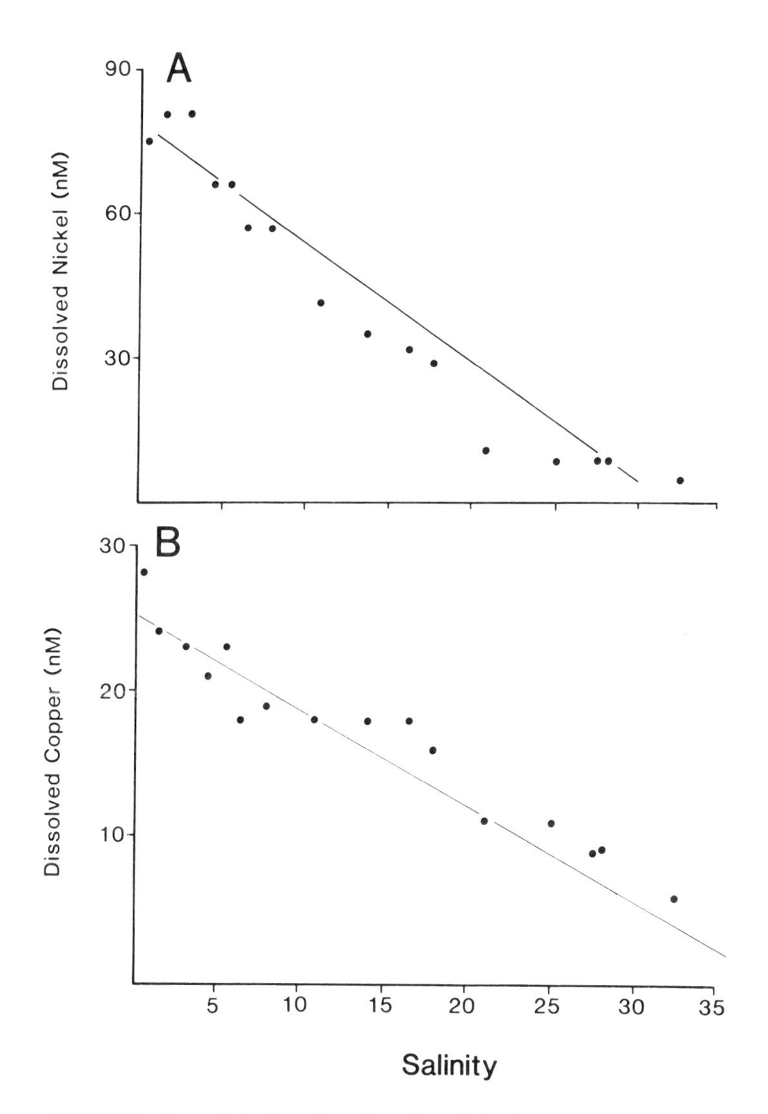

FIGURE 8. (A.) Dissolved nickel concentration and (B.) dissolved copper concentration versus salinity for the Delaware Estuary in May 1981.

(Bruland, Knauer and Martin 1978). This biochemical regeneration process appears to be a universal one controlling the cycling of cadmium by microplankton and associated organic matter, operating not only in oceanic waters (Boyle, Slater and Edmond 1976), but, as we have shown, in estuaries as well. Apparently the estuarine

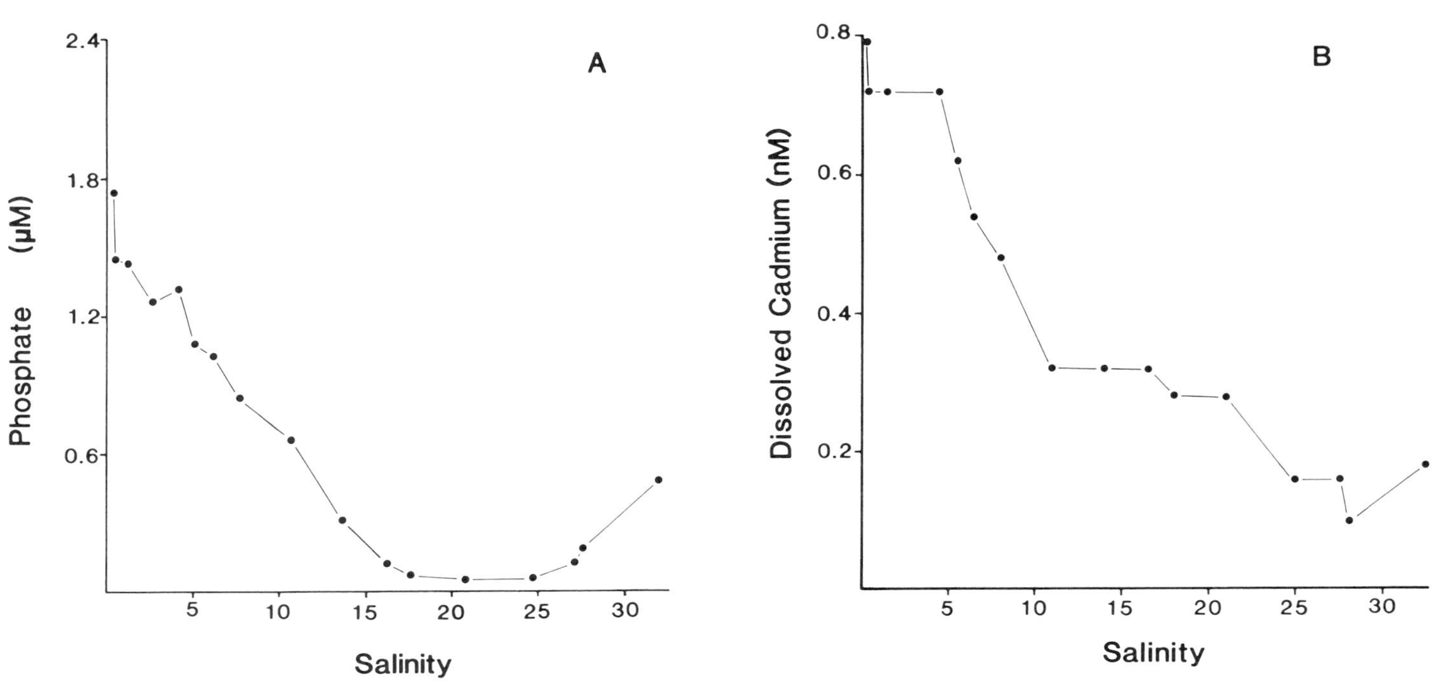

FIGURE 9. (A.) Dissolved phosphate concentration and (B.) dissolved cadmium concentration versus salinity in the Delaware Estuary in May 1981.

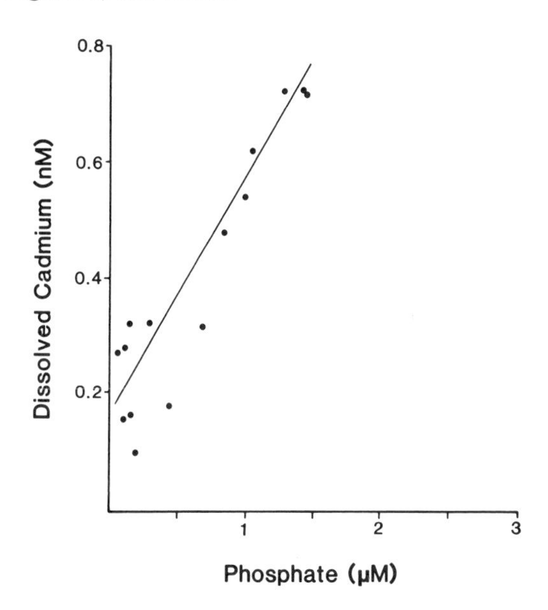

FIGURE 10. Dissolved cadmium concentration versus phosphate concentration in the Delaware Estuary in May 1981. The mole to mole slope of the line is 3.9 x 10^{-4} (R^2 = 0.88).

process involves a combined biogeochemical filter with geochemical exchange dominant at lower salinities and biochemical uptake and regeneration dominant at higher salinities.

CONCLUSIONS

Examples of removal (filtration) of dissolved constituents in estuaries are found in the Delaware Estuary. Under certain conditions, it is possible to distinguish conceptually between geochemical and biochemical filtration in the estuary with property-salinity diagrams. However, this interpretation is influenced by the physics of the system and chemical rate kinetics. Geochemical removal can be seen during winter when primary production is low and river flow is average. Biochemical removal can be seen during the spring bloom when primary production is high and geochemical activity is affected by the flushing rate.

These same biogeochemical processes probably occur at other times of the year and in other estuaries.

However, they may not be as readily demonstrable since the two processes are often difficult to distinguish. In the Delaware Estuary, the regions of geochemical (high turbidity zone) and biochemical (high productivity zone) filtration are geographically separate much of the year and are of comparable length. Furthermore, both regions have sufficient salinity gradients for representative sampling; this is often not the case in many estuaries. It is valuable to address the mechanisms by which estuaries act as chemical filters and the illustrations given here attempt to aid in that understanding.

An important next step is to analyze whether the estuary acts as a true sink for dissolved constituents. Alternatively, materials could be exported from the estuary as particulate matter, especially during storm periods. This will be best addressed by budgeting total elemental inventories with particulate measurements and estimates of depositional removal. We are currently working on this extension.

ACKNOWLEDGMENTS

This research was supported by a grant from the Delaware River and Bay Authority and by the Office of Sea Grant (NA80AA-D-00106) of the U.S. National Oceanic and Atmospheric Administration. Data for the exercise done here were provided by a number of people in our research group. We wish especially to acknowledge Andrew C. Frake and Sharon E. Pike for analytical aid and Shui-long Huang for computer aid.

REFERENCES CITED

Biggs, R.B., J.H. Sharp, T.M. Church and J.M. Tramontano. 1983. Optical properties, suspended sediments, and chemistry associated with the turbidity maxima of the Delaware Estuary. *Can. J. Fish. Aquat. Sci.* 40:172-179.

Boyle, E.A., R. Collier, A.T. Dengler, J.M. Edmond, A.C. Ng and R.F. Stallard. 1974. On the chemical mass-balance in estuaries. *Geochim. Cosmochim. Acta* 38:1719-1728.

Boyle, E.A., F. Slater and J.M. Edmond. 1976. On the marine geochemistry of cadmium. *Nature* 263:42-44.

Bruland, K.W., G.W. Knauer and J.H. Martin. 1978. Cadmium in northeast Pacific waters. *Limnol. Oceanogr.* 23:618-625.

Cifuentes, L.A. 1982. The character and behavior of organic nitrogen in the Delaware Estuary salinity gradient. M.S. Thesis, University of Delaware, College of Marine Studies. 81pp.

Fox, L.E. 1981. The geochemistry of humic acid and iron during estuarine mixing. Ph.D. Dissertation, University of Delaware, College of Marine Studies. 219pp.

Fox, L.E. and S.C. Wofsy. 1983. Kinetics of removal of iron colloids from estuaries. *Geochim. Cosmochim. Acta.* 47:211-216.

Ketchum, B.H. 1952. The distribution of salinity in the estuary of the Delaware River. Woods Hole Oceanogr. Inst. Rep. 52-103. 52p.

Liss, P.S. 1976. Conservative and non-conservatiave behavior of dissolved constitutents during estuarine mixing, pp.93-130. In: J.D. Burton and P.S. Liss (eds.), *Estuarine Chemistry*. Academic Press, New York.

Loder, T.C. and R.P. Reichard. 1981. The dynamics of conservative mixing in estuaries. *Estuaries* 4:64-69.

Martin, J.H., K.W. Bruland and W.W. Broenkow. 1976. Cadmium transport in the Caifornia current, pp.159-184. In: H.L. Windom & R.A. Duce (eds.). *Rain Pollutant Transfer*. Lexington Press, Lexington, MA.

McKnight, D.A. and F.M.M. Morel. 1979. Release of weak and strong copper-complexing agents by algae. *Limnol. Oceanogr.* 24:823-837.

Morris, A.W., A.J. Bale and R.J. Howland. 1981. Nutrient distributions in an estuary: Evidence of chemical precipitation of dissolved silicate and phosphate. *Estuarine Coastal Mar. Sci.* 12:205-216.

Murphy, T.P., D.R. Leon and C. Nalewajko. 1976. Blue-green algae: Their excretion of iron selective chelators enables them to dominate other algae. *Science* 192:900-902.

Officer, C.B. and D.R. Lynch. 1981. Dynamics of mixing in estuaries. *Estuarine Coastal Mar. Sci.* 12:525-534.

Pennock, J.R. 1983. Regulation of phytoplankton carbon and nitrogen production in the Delaware Estuary. Ph.D. Dissertation, University of Delaware, College of Marine Studies. 289pp.

Pennock, J.R., J.H. Sharp and W.J. Canzonier. 1983. Phytoplankton, pp.133-155. In: J.H. Sharp (ed.). *The Delaware Estuary*. University of Delaware and New Jersey Marine Sciences Consortium, Lewes, DE.

Pomeroy, L.R., E.E. Smith and C.M. Grant. 1965. The exchange of phosphate between estuarine waters and sediments. *Limnol. Oceanogr.* 10:167-172.

Sharp, J.H., C.H. Culberson and T.M. Church. 1982. The chemistry of the Delaware Estuary: General considerations. *Limnol. Oceanogr.* 27:1015-1028.

Sholkovitz, E.R. 1976. Flocculation of dissolved organic and inorganic matter during the mixing of river water and seawater. *Geochim. Cosmochim. Acta* 40:831-845.

Sholkovitz, E.R., E.A. Boyle and N.B. Price. 1978. The removal of dissolved humic acids and iron during estuarine mixing. *Earth Planet. Sci. Lett.* 40:130-136.

Smullen, J.T., J.H. Sharp, R.W. Garvine and H.H. Haskin. 1983. River flow and salinity, pp.9-25. In: J.H. Sharp (ed.). *The Delaware Estuary*. University of Delaware and New Jersey Marine Sciences Consortium, Lewes, DE.

Stumm, W. and J.J. Morgan. 1981. *Aquatic Chemistry*. 2nd Edition. John Wiley and Sons, New York. 780pp.

BIOLOGICAL PROCESSES

ESTUARINE TOTAL SYSTEM METABOLISM AND ORGANIC EXCHANGE CALCULATED FROM NUTRIENT RATIOS: AN EXAMPLE FROM NARRAGANSETT BAY

Scott W. Nixon
Michael E. Q. Pilson

Graduate School of Oceanography
University of Rhode Island
Narragansett, Rhode Island

Abstract. The average ratio of the concentration of dissolved inorganic nitrogen (DIN) to phosphorus (DIP) found in many estuaries differs from that in the rivers, sewage, rain, and offshore waters that contribute nutrients to these systems. These differences are largely the result of various biological processes which take place in the estuary, including denitrification, the net production of organic matter, and the consumption and remineralization of organic matter carried into the estuary. Net production in this case represents the total amount of organic matter exported, harvested, and removed through long-term burial that is in excess of the amount imported and consumed within the estuary. It is not the same as the commonly reported net production by the autotrophic components of the system. Together with physical mixing and chemical exchange processes which take place in the system, these biological transformations provide a dynamic "filter" which often makes the quantity and form of nutrients exported from an estuary quite different from those it imports. These changes can themselves provide insight into the internal processes of the estuary.

For Narragansett Bay (RI, USA) we have combined measurements of the annual nutrient input with the ratio of the annual time and volume-weighted mean concentrations of DIN and DIP found in the Bay, measurements of the annual rate of denitrification, and data on the chemical composition of the

ISBN 0-12-405070-0

plankton to calculate the total system metabolism of the Bay. The result suggests that the system is autotrophic, with net production exceeding consumption by about 80 g C m^{-2} y^{-1}. Since only a small portion of this organic matter accumulates in the sediments (∿ 6 g C m^{-2} y^{-1}) or is removed in the fishery (∿ 1 g C m^{-2} y^{-1}), some 70-75 g C m^{-2} y^{-1} may be exported from the Bay, an amount equal to 22-24% of the reported production by the phytoplankton.

INTRODUCTION

With certain notable exceptions, such as the Wadden Sea, Delaware Bay, and a number of other very turbid areas, there exists in our minds a close association between estuaries and high productivity. Odum (1971) emphasized the fact for countless students in Fundamentals of Ecology, writing that, "Characteristically, estuaries tend to be more productive than either the sea on one side or the freshwater drainage on the other." While this impression may be traced back to early work on estuarine systems dominated by sea grasses (Peterson 1915) and salt marshes (Teal 1962), numerous recent measurements of phytoplankton production have also extended the association to open water, plankton-based estuaries (see reviews by Nixon 1981a, 1982; Boynton, Kemp and Keefe 1982; Boynton et al. 1983).

In the early work on the salt marshes of the Georgia coast, it was emphasized that the high net production of the emergent grasses should not be confused with the production of the marsh system as a whole, and pioneering efforts were made to balance the heterotrophic demands of other components of the marsh against the surplus production of the plants (Teal 1962). The result suggested that the emergent marsh could support all of its own respiratory demands and still export almost half of the net production of the grasses to adjacent systems. The validity of this assessment and the extent and importance of "outwelling" from other marshes is now being debated (see reviews by Odum 1980 and Nixon 1980), but the point to be made is that those working in marshes have maintained a clear distinction between the net production by the primary producers and the net production of the total marsh system. With few exceptions, such as Riley's (1956) analysis of Long Island Sound, the same distinction has received considerably less attention in open water estuaries, and it is often assumed that the finding of high rates of phytoplankton production implies a correspondingly large export of organic matter. As Odum

(1971) put it, "...estuaries often generate more fertility than they can use (P exceeds R), resulting in the export or outwelling of nutrient and organic detritus into the ocean... The export of meroplankton and detritus from estuaries can particularly enhance the secondary productivity of coastal waters."

While it might seem likely that the export of living and dead organic matter from open water, plankton-based systems would be more effective than the flushing of macrophyte detritus from the winding creeks and intertidal areas of marshes, we have found no reports of long-term measurements of such transport. The reason for this is simple. As a practical matter, the direct measurement of net material fluxes across the mouth of a large estuary over a significant period of time appears to be impossible with reasonable resources. The difficulties described by Boon (1975, 1978), Kjerfve and Proehl (1979), and Kjerfve et al. (1981, 1982) in working with relatively small tidal creeks just tens or a few hundreds of meters across seem insurmountable for transects thousands of meters long and 10 to 50 m deep.

A knowledge of the exchange of organic matter between estuaries and offshore waters (and the input to the estuaries from land drainage) is nevertheless important for understanding both the economy of the coastal waters and processes within the estuary. Without evaluating these terms in the material budgets of an estuary it is difficult to assess the efficiency with which the estuary acts as a "filter", by retaining or removing organic carbon, nitrogen, phosphorus, and various pollutants that may be of interest. The growing body of information on the net productivity of estuarine phytoplankton (as measured by ^{14}C uptake) is useful in addressing such questions, but we also need some way to estimate the net production (or consumption) of estuarine systems as a whole.

OTHER APPROACHES

Mass Balances

Six years before Teal (1962) published his estimate of the organic budget for a Georgia salt marsh, Riley (1956) reported similar calculations for the open waters of Long Island Sound. The results of these efforts were quite different, since the sum of various rate measurements suggested that Long Island Sound was essentially in metabolic balance over an annual cycle (annual net phytoplankton production ≅ annual respiration of pelagic and benthic heterotrophs) while net plant production on the Georgia marsh appeared to exceed respiration by almost a factor of two. For a number of reasons, Riley's early effort was seldom repeated in other plankton-based estuaries. First, the Long Island Sound study had measured pelagic metabolism using changes in dissolved oxygen in light and dark bottles, but the increasingly popular ^{14}C uptake technique for measuring primary production provides no data on pelagic respiration or net production by the plankton community (including zooplankton and microheterotrophs). Second, until quite recently there were remarkably few measurements of the consumption of organic matter over an annual cycle by estuarine bottom communities (see reviews by Zeitzschel 1980; Nixon 1981b). With the notable exception of Carey's (1967) studies in Long Island Sound, and the work of Pamatmat and Banse (1969) in Puget Sound and Smith (1973) off the Georgia coast, studies of subtidal bottom communities tended to focus on aspects of community structure rather than metabolism (Mills 1975).

A third factor may be a suspicion held by many ecologists that measurements of the parts of a complex system can not simply be summed to give an accurate picture of the whole. Cumulative errors become very large when many measurements, each with its own variance, are combined. And, on a more fundamental level, the very act of isolating parts for controlled measurement may alter their behavior. Odum (1968) touched on the problem, writing that, "It should be mentioned in passing that carbon-14 and other measurements made within closed containers have an extremely high variance so that averages must be based on a very large number of samples. The challenge...is to replace present small sample 'closed system' measurements with large sample 'open system' ones."

Total System Metabolism

In many ways, E. P. Odum's challenge had already been met with the diel free water oxygen curves that H. T. Odum and his coworkers had been measuring in the bays and lagoons along the coast of Texas (summarized by Odum 1967). Unfortunately, the attempt to measure total system production and respiration over a 24-h period by following oxygen changes at a fixed station was confounded by not knowing the history of each water parcel sampled and by variations in the rates of air-sea gas exchange caused by changes in wind and currents. Mixing and air-sea exchange also limit attempts to follow a particular water mass as it moves about an estuary. In spite of these difficulties, however, Odum's efforts provided many of the first direct estimates of total estuarine net production or, in some cases, consumption. For the reasons noted, few estuaries have been studied using the diel curve technique, though in cases where there is a large water mass where biological and chemical conditions are homogeneous or where advection is small or can be measured, the technique has been used effectively. For example, in a small salt marsh embayment where tidal flow could be blocked for 24 h periods, it was possible to show that the waters and subtidal system adjacent to the marsh were heterotrophic over an annual cycle and presumably relied on organic matter exported from the emergent marsh (Nixon and Oviatt 1973).

In a few areas, attempts were also made to measure total system metabolism using longer-term changes in dissolved oxygen, nutrients, or carbon dioxide. A part of Riley's (1956) calculations for Long Island Sound relied on an analysis of changes in oxygen and phosphate over periods of several weeks to several months to show an annual balance in production and consumption, and in his studies of biweekly to monthly changes in pCO_2 in the plankton-dominated waters off Woods Hole, MA, Teal (1967) found that total system respiration exceeded total system production by about 33% over an 11-month period. In the latter case, Teal attributed the difference to an inflow of organic matter from Nantucket Sound. Unfortunately, a number of problems make it difficult to know how much confidence to place in either of these calculations, and the findings apparently did not do a great deal to stimulate similar work in other estuaries.

A Calculation Using Nutrient Ratios

A characteristic feature of many estuaries is that the average ratio of dissolved inorganic nitrogen (DIN) to inorganic phosphorus (DIP) in their waters is maintained at a value which is low relative to the 16:1 ratio found in the open sea or in the particulate organic matter formed in the estuary (Redfield 1934; Harris and Riley 1956; Goldman, McCarthy and Peavey 1979; Nixon 1981b; Boynton et al. 1982; Copin-Montegut and Copin-Montegut 1983). In Narragansett Bay, R.I., USA, and a number of other estuaries, particularly near their seaward ends, the DIN:DIP ratio is also markedly lower than that found in the anthropogenic and riverine inputs to the system, though it is not uncommon to find systems which maintain a higher ratio than their inputs (Fig. 1). It is difficult to draw a general conclusion about which condition is more common because there are surprisingly few reliable measurements of nutrient inputs to estuaries and the concentration data for nutrients in estuaries are seldom volume-weighted to provide a meaningful average for such a comparison. Nevertheless, it seems evident that the DIN:DIP ratios of many estuaries result from processes within the estuaries themselves and do not simply reflect the ratios of their inputs.

As Banse (1974) has emphasized, the DIN:DIP ratio of a body of water reflects the net result of all of the uptake and regeneration processes involving each nutrient, not just the uptake by the phytoplankton. It is possible to take advantage of this fact and use the ratio to calculate the total metabolism of an estuary, bearing in mind that three basic processes act to change the DIN/DIP ratio of the system from that of its inputs. First, denitrification will lower the ratio. Second, the net production of organic matter will do the same if, as is usually the case, nitrogen is more abundant relative to phosphorus in the organic matter than it is in inorganic form in the input waters. Third, the consumption of organic matter carried into the estuary from outside sources will increase the ratio if the N/P ratio of the material is high relative to the DIN/DIP ratio of the water. Nitrogen fixation would also increase the ratio, but has not been shown to be an important process in open water estuarine areas. It is important to emphasize that the net production referred to here is the total amount of organic matter exported, harvested, and removed through long-term burial that is in excess of any organic matter imported and consumed. It is not simply the net production of the autotrophic components of the system.

The purpose of this paper is to describe the calculation of total system metabolism using nutrient ratios and to apply it to the example of Narragansett Bay, a system with which we are familiar and for which the

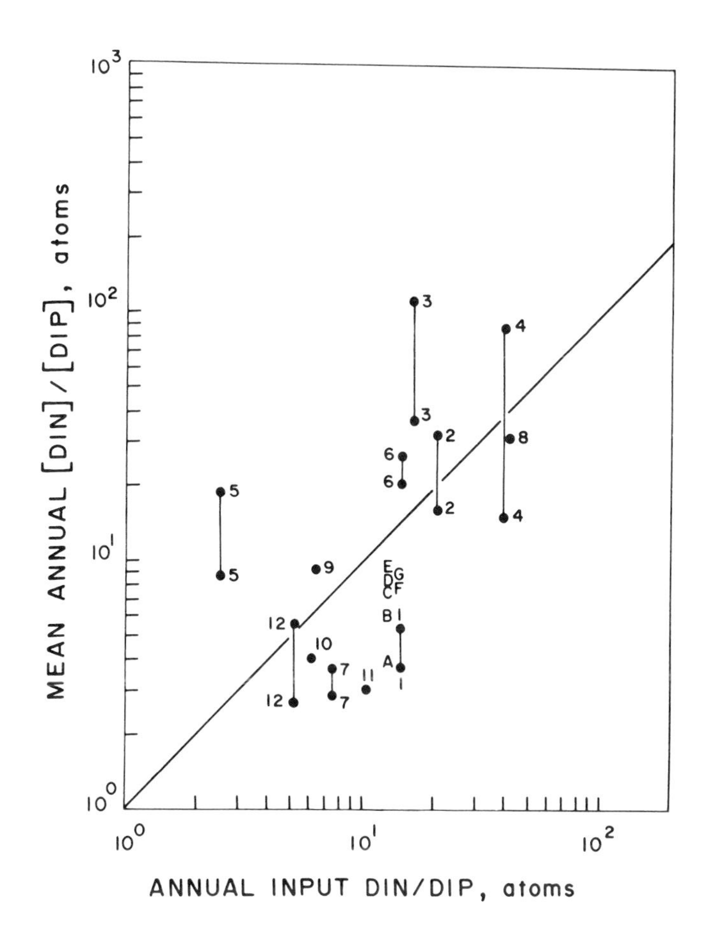

FIGURE 1. Ratio of mean annual concentration of dissolved inorganic nitrogen to dissolved inorganic phosphorus in various estuaries as a function of estimated DIN/DIP ratio of nutrients input to each system from land drainage and anthropogenic sources but not from adjacent ocean waters. Data from estuaries have not been volume-weighted, a correction which would decrease the ratio. Bars show ranges between upper (top) and lower reaches of the estuaries. The line shows a 1:1 correspondence. Data from various sources summarized in Nixon (1983). 1 = Narragansett Bay, 2 = New York Bay, 3 = Delaware Bay, 4 = Chesapeake Bay, 5 = Patuxent River Estuary, 6 = Potomac River Estuary (mid and lower reaches), 7 = Pamlico River Estuary (mid and lower reaches), 8 = Apalachicola Bay, 9 = North San Francisco Bay, 10 = South San Francisco Bay, 11 = Kaneohe Bay, 12 = Mobile Bay. A-G = MERL experimental tanks.

necessary supporting data are, with varying degrees of reliability, in hand. While the arithmetic is straightforward, it does not appear that the implications of the stoichiometry of an estuary have been considered in this way before.

We begin by assuming a year-to-year steady state, and defining the parameter, DIN_o, which represents the dissolved inorganic nitrogen in the estuary which is available for export. DIN_o is equal to the difference between the annual input of dissolved inorganic nitrogen from all sources (DIN_{in}) and the annual losses of DIN due to denitrification (DNF) and the net annual uptake or release of DIN into or from organic matter (PON). Again, it is important to remember that this net uptake (PON) represents organic matter that remains after all recycling; it is not the same as net uptake measured in short-term ^{14}C or ^{15}N uptake studies. Assuming that the concentration of organic matter is not changing from year-to-year, it is this material (PON) which is lost from the estuary in long-term burial, through fisheries, or through export to offshore waters ("outwelling").

$$DIN_o = DIN_{in} - DNF - PON \tag{1}$$

According to the "Redfield model," the amount of nitrogen in the organic matter is, on average, 16 x the amount of organic phosphorus (POP), though this may be varied if another ratio is found more appropriate. Again, POP is the net annual production (or consumption) of organic phosphorus by the estuarine system.

$$PON = 16 \bullet POP \tag{2}$$

Substituting:

$$DIN_o = DIN_{in} - DNF - (16 \bullet POP) \tag{3}$$

Rearranging:

$$16 \bullet POP = DIN_{in} - DNF - DIN_o \tag{4}$$

The value of DIN_o can be expressed as the mean annual volume-weighted DIN:DIP ratio (R) multiplied by the amount of dissolved inorganic phosphorus exported from the estuary (DIP_o).

$$DIN_o = R \bullet DIP_o \quad (5)$$

But DIP_o can also be set equal to the difference between the annual input of dissolved inorganic phosphorus (DIP_{in}) and the annual net uptake into or release from organic matter (POP) as in Eq. (1) for nitrogen.

$$DIP_o = DIP_{in} - POP \quad (6)$$

Substituting back into Eq. (5):

$$DIN_o = R(DIP_{in} - POP) \quad (7)$$

Substituting for DIN_o in Eq.(4) gives:

$$16 \bullet POP = DIN_{in} - DNF - R(DIP_{in} - POP) \quad (8)$$

The annual net production (or consumption if the sign is negative) of organic phosphorus by the system (POP) can be calculated by substituting empirical data for the annual inputs of inorganic nitrogen and phosphorus, the average DIN:DIP ratio in the estuary, and an estimate of the annual loss of nitrogen in denitrification. Units of DIN_{in}, DIP_{in}, and DNF should be the same (for example, mmol m^{-2} y^{-1}) and set the units for production or consumption. The annual net production or consumption of organic nitrogen and organic carbon (POC) can then be calculated by ratio using the Redfield model (PON = 16 POP; POC = 106 POP) or an empirical regression for the composition of organic matter in the estuary. The export of organic matter from the estuary (if any) must then be the difference between the calculated net annual total system production and the amount buried in the sediments and removed through harvest.

TOTAL SYSTEM METABOLISM OF NARRAGANSETT BAY

Narragansett Bay (Fig. 2) has been described frequently in the literature (see, for example, Hicks 1959; Nixon and Kremer 1977; Kremer and Nixon 1978) and numerous studies have been devoted to various aspects of its ecology (Dunn, Hale and Bucci 1979 list over 1700 entries in a recent bibliography). Nevertheless, additional work would increase greatly the level of confidence that could be

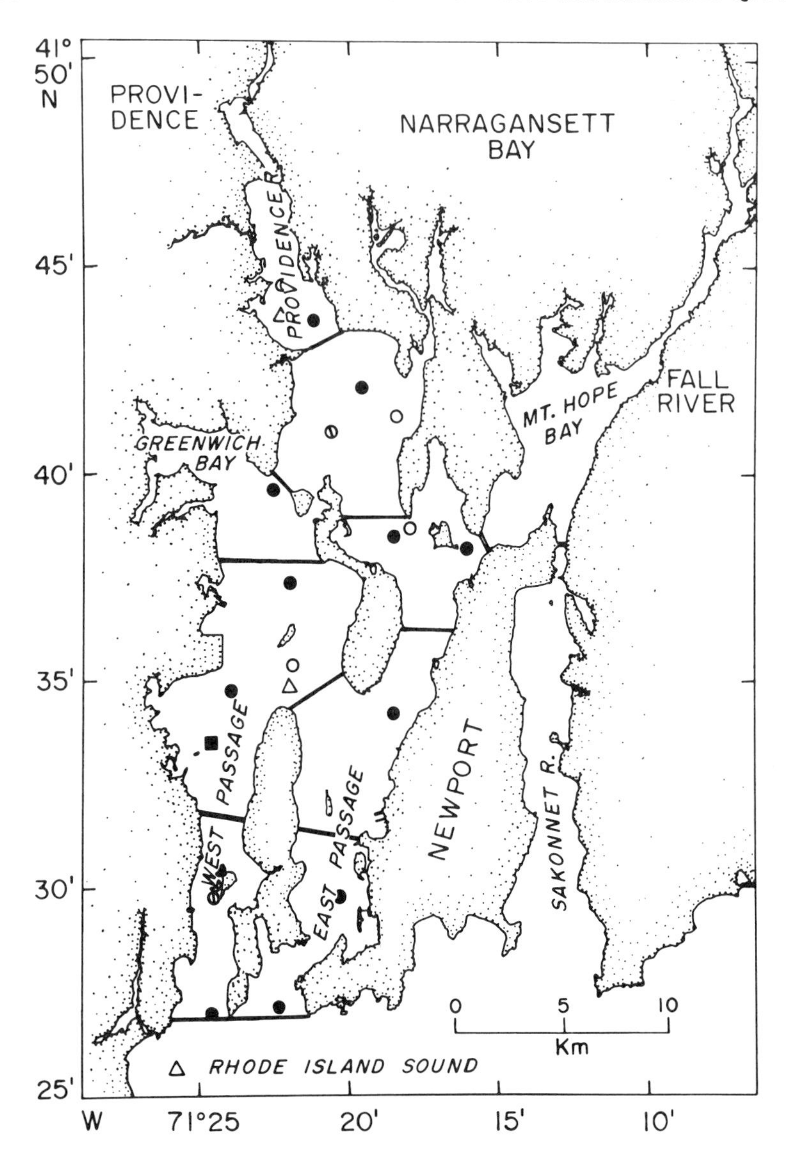

FIGURE 2. Narragansett Bay, RI showing sampling stations for measurements of DIN and DIP concentrations (●), the station where water was taken by Furnas and Smayda (see Table 1) for ^{14}C uptake measurements and the analysis of the chemical composition of phytoplankton (■), sites where denitrification was measured (Δ), and locations where sediment cores were taken to estimate sediment accretion rates (⦸) and chemical composition (⦶). Segments of the Bay used for volume-weighting DIN and DIP data are shown by heavy lines.

placed in the estimates that follow. Our impression, however, is that each term in the calculation is constrained to some degree by several of the others, with the result that the numerical syntheses of individual measurements may be more credible than its individual parts.

General Description of the Bay

The Bay is a plankton-based system covering some 312 km^2 with a mean depth of about 9 m. Water temperatures range from about -1 C to 24 C. There is a relatively modest fresh water input with a long-term average according to the most recent analysis (Pilson, unpublished) of about 105 m^3 s^{-1}. As a result, Narragansett Bay has a high volume-weighted mean salinity that may vary during a year from about 27 o/oo to 31 o/oo (Pilson, unpublished). The Bay also shows only weak vertical stratification except in its upper reaches. The Providence metropolitan district lies at the head of Narragansett Bay and the rivers entering the system drain large urban and industrial areas. As a result, there are substantial inputs of nutrients and various pollutants in the upper Bay with marked concentration gradients decreasing toward the offshore waters of Rhode Island and Block Island Sounds at the seaward end. The residence time of water in the Bay appears to average about 30 days (Kremer and Nixon 1978).

The Stoichiometric Calculation

The various sources of data used to solve Eq. (8) for Narragansett Bay are summarized in Table 1 with sampling locations shown in Fig. 2. The average DIN/DIP ratio in the system ranges from 0.2 to 12.4 over an annual cycle, with the lowest values during periods when biological activity is greatest (Fig. 3). As noted earlier, the time-averaged and volume-weighted mean of 5.00 is considerably lower than the DIN/DIP of 9.92 found in the estimated total input from rivers, sewage, rain, and offshore. A summary of the stoichiometric ratios measured in the Bay shows the potential importance of organic production and storage in maintaining the low DIN/DIP ratio of the Bay waters (Fig. 4). The relative composition of the plankton varies during an annual cycle, but a

TABLE 1. Summary of data sources used in the stoichiometric calculation of annual net total system production and organic export from Narragansett Bay.

DIN/DIP in the water column	-Biweekly survey of 13 stations around the Bay during 1972-73 (Fig. 2); near-surface and near-bottom samples averaged and volume-weighted using segments from Kremer and Nixon (1978).
DIN and DIP inputs from sewage and rivers	-Biweekly survey of concentrations in effluents from major treatment plants and at the dams of tributaries to Providence River during 1975-76. Taunton River assumed equal to Blackstone and 70% assigned to the Bay. Massachusetts sewage input calculated by flow relative to R.I. measurements. (Modified from Nixon 1981b; Nixon, Oviatt and Nowicki, unpublished).
DIN and DIP input from atmosphere	-Deposition assumed equal to that measured near Woods Hole, MA (Valiela et al. 1978).
DIN and DIP input from offshore	-Calculated using flushing rate of Bay as a function of mean monthly fresh water input (Pilson, unpublished) for 1972-73. Flushing rate (day^{-1}) was multiplied by volume of the Bay and mean monthly DIN and DIP concentration of bottom water in the lower East Passage (Fig. 2).
Sediment accumulation rate	-Six coring sites (Fig. 2) analyzed for various radionuclides ($^{234}Th_{xs}$, $^{210}Pb_{xs}$, $^{239,240}Pu$) with known

(Continued)

TABLE I (continued)

	input functions. Accretion rates averaged by sediment type over the Bay (Santschi et al., in press).
C, N, P content of sediments	-One upper and one lower Bay core (Fig. 2) analyzed at 1 cm intervals to 1 m and averaged (Fig. 6; Nixon, Nowicki and Northby, unpublished).
Denitrification rate in sediments	-Upper, mid, and lower Bay sites (Fig. 2) measured directly for N_2 and N_2O flux five times during an annual cycle (1978-79), data averaged and integrated for the year (Seitzinger 1982; Seitzinger et al. 1984).
C:N:P in Bay phytoplankton	-Weekly surface, mid, and near-bottom samples pooled, filtered through 158 μm screen anu analyzed over an annual cycle during 1975-76. Data from near mid-Bay, West Passage (Fig. 2). From M. Furnas and T. Smayda, Univ. of R.I., unpublished.

functional regression through the data (Ricker 1973) provides a C:N:P ratio of 110:13:1 by atoms which is not very different from the Redfield model of 106:16:1 (Fig. 5). During mineralization in the water column and surface sediments, this organic matter loses nitrogen relative to phosphorus as shown by the composition of the sediments (Figs. 4 and 6). However, the flux of DIN/DIP returned from the sediments to the overlying water is low in nitrogen relative to phosphorus (Nixon, Oviatt and Hale 1976; Nixon et al. 1980). The missing nitrogen appears to be accounted for in denitrification which removes some 515 mmol N m^{-2} y^{-1} from the Bay (Seitzinger, Nixon and Pilson 1984). The remaining values needed for solving Eq. (8) are the absolute amounts of DIN and DIP entering the Bay.

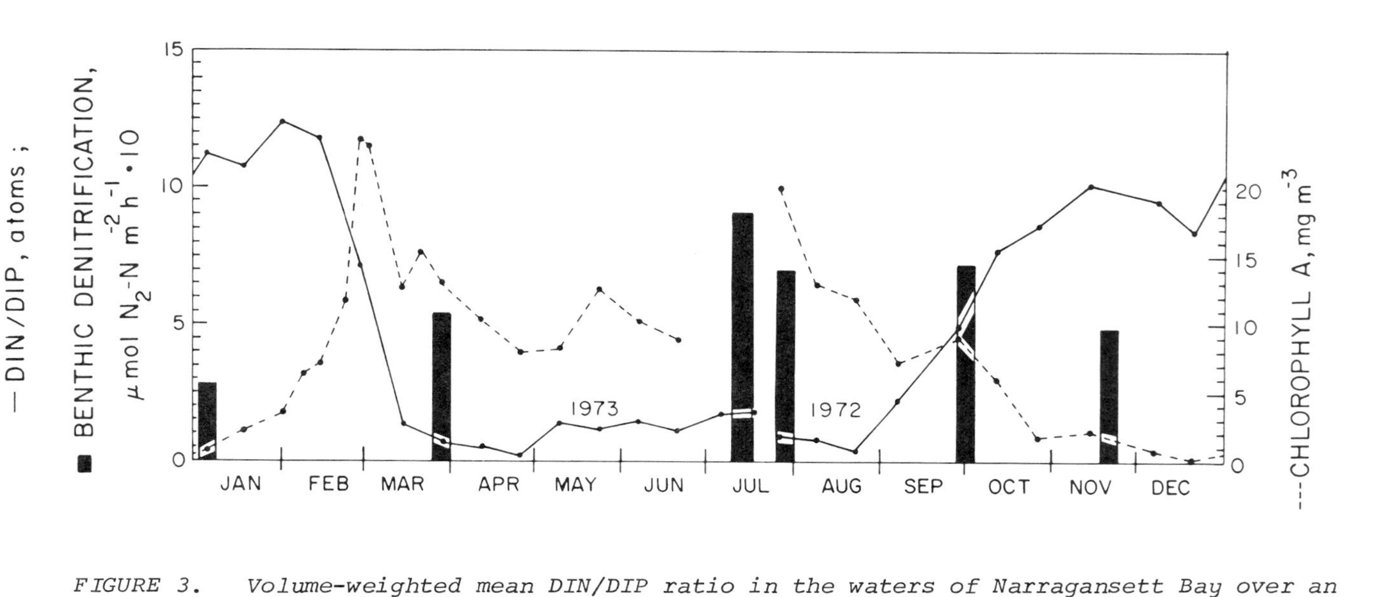

FIGURE 3. Volume-weighted mean DIN/DIP ratio in the waters of Narragansett Bay over an annual cycle, July 1972 to Aug. 1973. Chlorophyll data are for the same year at a station near mid-Bay. Denitrification data are averaged from Seitzinger, Nixon and Pilson (1984) for a more recent year (1978-1979).

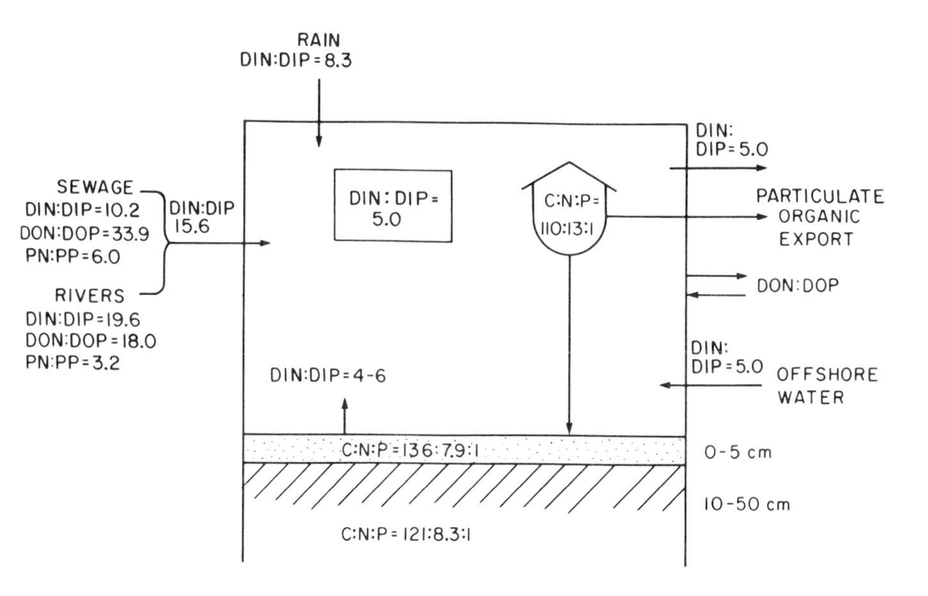

FIGURE 4. An estimate of the annual mean stoichiometric conditions in Narragansett Bay. Inputs and water include DIN and DIP; particulate matter and sediments are "total" C, N, and P. Data from various sources noted in Table 1.

Estimated as in Table 1, the major sources appear to amount to at least 2019 mmol N m^{-2} y^{-1} and 204 mmol P m^{-2} y^{-1} distributed as follows:

	DIN_{in}, mmol N m^{-2} y^{-1}	DIP_{in}, mmol P m^{-2} y^{-1}
Atmospheric deposition	29	3.5
Rivers	1102	56
Sewage effluent	429	42
Offshore waters	459	102
TOTAL	2019	204

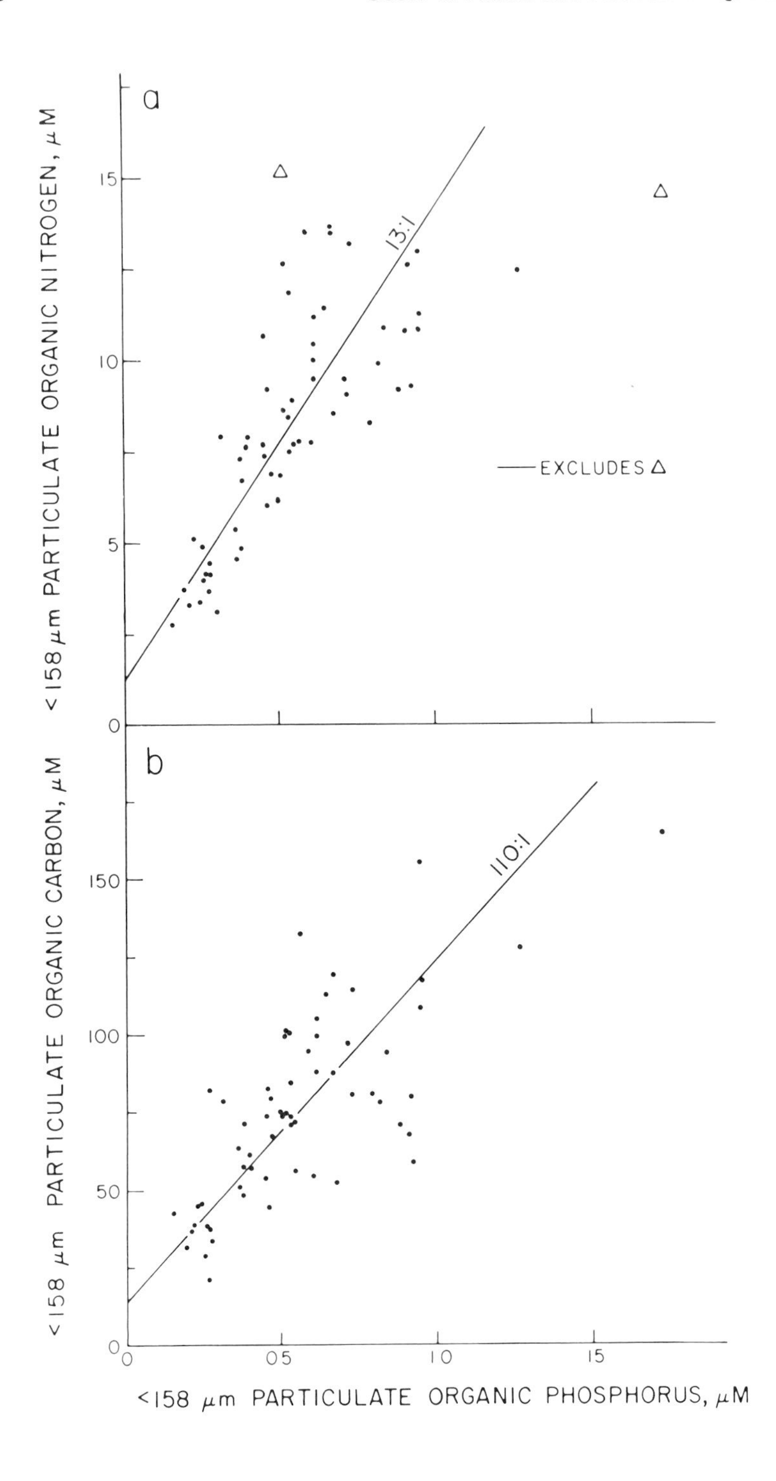
a
13:1
—— EXCLUDES Δ
<158 μm PARTICULATE ORGANIC NITROGEN, μM
b
110:1
<158 μm PARTICULATE ORGANIC CARBON, μM
<158 μm PARTICULATE ORGANIC PHOSPHORUS, μM

The numbers of significant figures given above and carried through the following calculations are certainly greater than can be justified from the actual measurements. They are carried in order to avoid progressive accumulation of rounding errors, to help in following the calculation, to avoid loss of some of the smaller terms which are less than the rounding errors of the larger terms, and in some cases because we don't know how to assess the uncertainty.

With these values substituted in Eq. (8), the calculated annual total system net production in Narragansett Bay is:

$$POP = 60.5 \text{ mmol P m}^{-2} \text{ y}^{-1} = 1.9 \text{ g P m}^{-2} \text{ y}^{-1}$$

$$PON = 13 \text{ POP} = 787 \text{ mmol N m}^{-2} \text{ y}^{-1} = 11 \text{ g N m}^{-2} \text{ y}^{-1}$$

$$POC = 110 \text{ POP} = 6655 \text{ mmol C m}^{-2} \text{ y}^{-1} = 80 \text{ g C m}^{-2} \text{ y}^{-1}$$

This amounts to about 26% of the 310 g C m^{-2} y^{-1} of net phytoplankton production measured during 1974 using water from one station near mid-Bay (Furnas, Hitchcock and Smayda 1976), or 30% of the 269 g C m^{-2} y^{-1} estimated for the whole Bay by Oviatt, Buckley and Nixon (1981) from a variety of types of oxygen measurements.

Using the remarkably constant C, N, and P compositions of the sediments below 10 cm (the surface sediments are actively bioturbated and mixed with freshly deposited material) (Fig. 6) and the average sediment deposition rate for the Bay of 250 g m^{-2} y^{-1} (Santschi et al., in press), it is possible to estimate the amount of the net production that is removed through long-term burial in the sediments. The result suggests that about 4 mmol P m^{-2} y^{-1}, 33 mmol N

FIGURE 5 (opposite). Composition of the phytoplankton (< 158 μm) measured weekly over an annual cycle in the mid-West Passage of Narragansett Bay. Lines show functional regressions (Ricker 1973) through the data, excluding the two points noted. Data from M. Furnas and T. J. Smayda (unpublished).

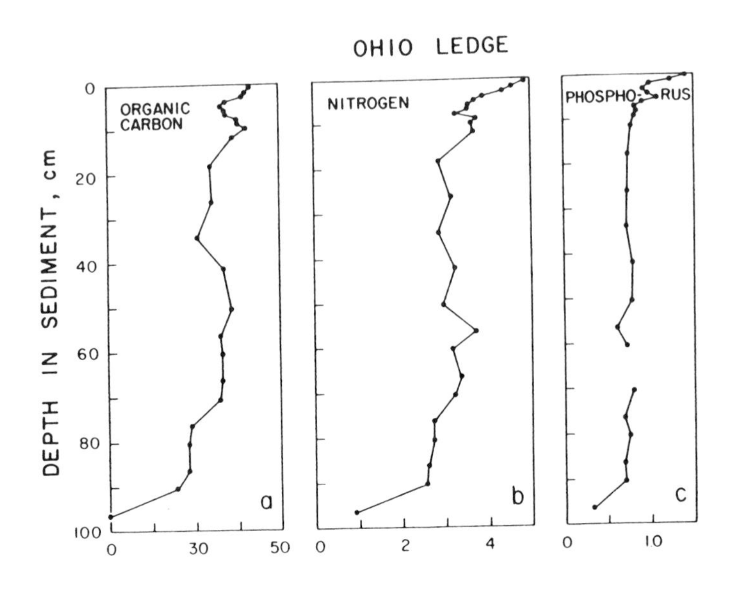

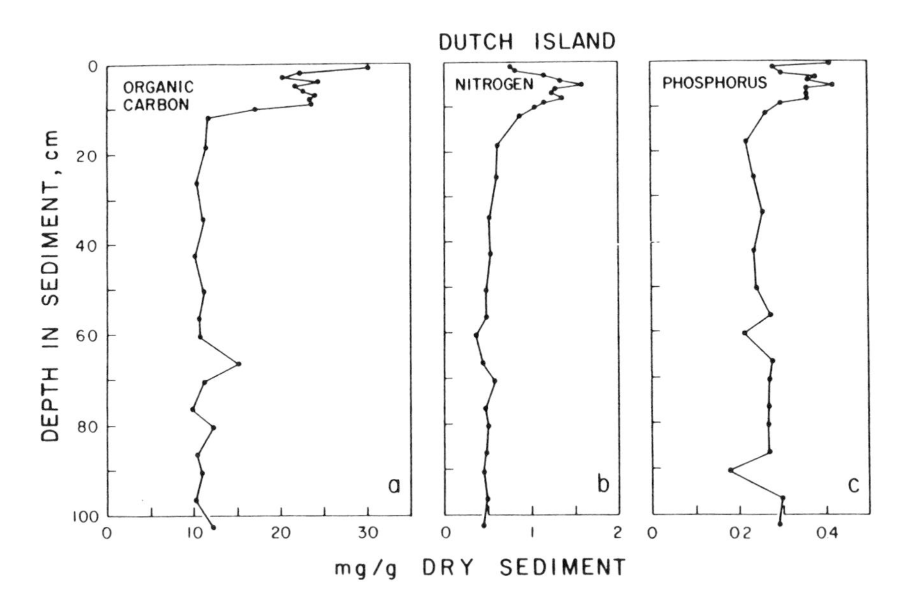

FIGURE 6. Concentrations of carbon, nitrogen and phosphorus at various depths in the sediment from upper (Ohio Ledge) and lower (Dutch Island) Narragansett Bay. Data from Nixon, Nowicki and Northby (unpublished).

m^{-2} y^{-1}, and 500 mmol C m^{-2} y^{-1} are stored in this way, amounting to 7%, 4%, and 7% of the total system net production of organic phosphorus, nitrogen, and carbon, respectively.

The amount of total system net production removed in fisheries is not well known, but it must certainly be small. If landings were 100 kg ha^{-1} y^{-1} (Nixon 1982) they would account for 0.5 to 1 g C m^{-2} y^{-1} or some 0.5 to 1% of the total, and this is likely an overestimate.

If we neglect the small and uncertain influence of the fishery, the amount of organic matter exported or "outwelled" from Narragansett Bay to the offshore waters of Block Island and Rhode Island Sounds must be on the order of 6655 mmol C m^{-2} y^{-1} net production - 500 mmol C m^{-2} y^{-1} buried = 6155 mmol C m^{-2} y^{-1} or 74 g C m^{-2} y^{-1} exported. This is about 24 or 28% of the reported net phytoplankton production.

TOTAL SYSTEM NET PRODUCTION AND ORGANIC EXPORT IN THE CONTEXT OF A MASS BALANCE FOR C, P, AND N IN NARRAGANSETT BAY

The stoichiometric calculation suggests that about 75% of the organic matter fixed in Narragansett Bay is consumed or retained within the system. With this information it is possible to develop in some detail preliminary pictures of the approximate annual budgets of carbon, phosphorus and nitrogen in the Bay (Figs. 7, 8, 9), though the contribution of dissolved and particulate organic matter brought into the system remains unknown.

While the exact value of any term in the budgets is subject to a considerable uncertainty (for example, Taunton River inputs have not been measured directly, ^{14}C uptake has only been measured using water from one station, sediment composition is only taken from two locations, no benthic measurements are available from the East Passage of the Bay, etc.), it seems necessary from other evidence that organic export from the Bay must be on the order of the amount calculated or less. Earlier measurements of benthic oxygen uptake showed that an amount of organic matter equal to 25-50% of that fixed by the phytoplankton is decomposed by the benthos (Nixon et al. 1976). In addition, the net input of inorganic nitrogen from outside sources plus benthic recycling only appears large enough to support some 60% of the net phytoplankton production (Nixon 1981), with the obvious implication that pelagic recycling must be important. Measurements of excretion by the meso- and macrozooplankton (Kremer 1975; Vargo 1979) showed that their contribution fell considerably short of the amount required for water column regeneration, so that there must also be a significant regeneration by microheterotrophs.

The measured and estimated input and removal rates also provide a minimum estimate of the export of inorganic nutrients from Narragansett Bay (Table 2). Available data

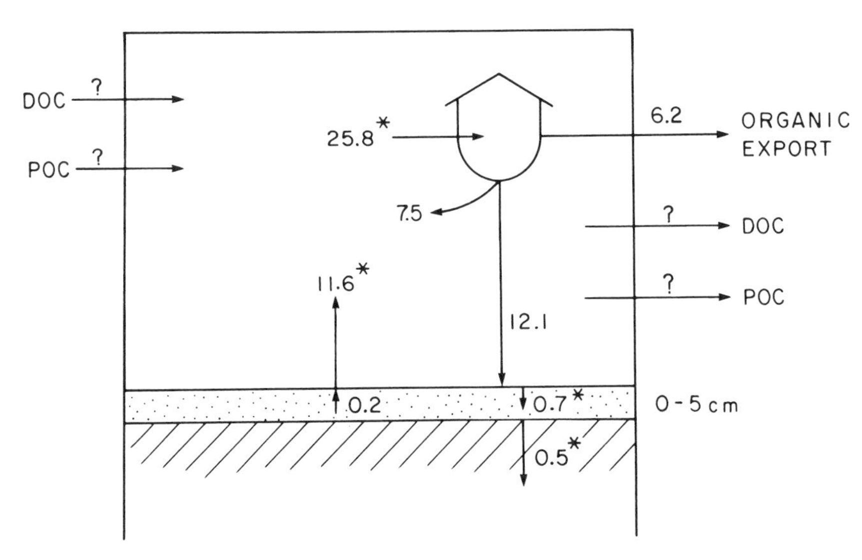

FIGURE 7. Present status of the annual carbon budget for Narragansett Bay. Values marked by (*) have been measured; others were calculated by difference or (in the case of organic export) indirectly.

do not allow a direct measurement of this export, which in any case is most likely greater than given in Table 2. The total input of dissolved and particulate organic carbon from offshore, rivers, and sewage cannot be reliably estimated with the data at hand (although we believe it to be large and significant, and some partial fluxes are entered into Figs. 7, 8 and 9). The remineralization of any of this input would provide an additional input of DIN and DIP, according to the stoichiometry of the remineralization process. The effect of this extra input on the fluxes shown in Figures 7, 8 and 9 would be accommodated by increasing the fluxes through the microheterotrophs. This unmeasured pathway is apparently a major one in Narragansett Bay.

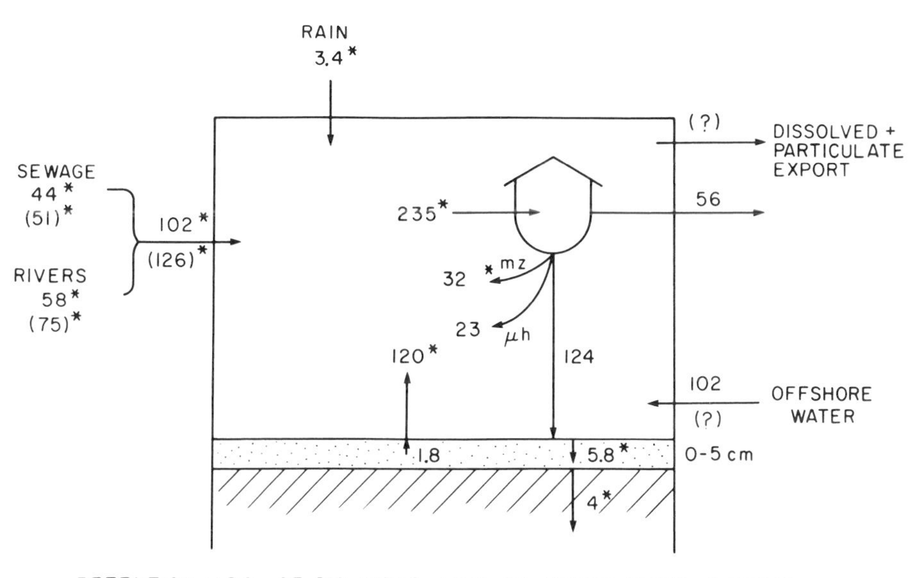

FIGURE 8. Present status of the annual phosphorus budget for Narragansett Bay. Values marked by (*) have been measured; others were calculated by difference or (in the case of organic export) indirectly. Values in ( ) are for total phosphorus in inputs and outputs assuming that the organic inputs do not become remineralized in the Bay. mz is pelagic excretion by meso and macrozooplankton, μh is pelagic remineralization by microheterotrophs.

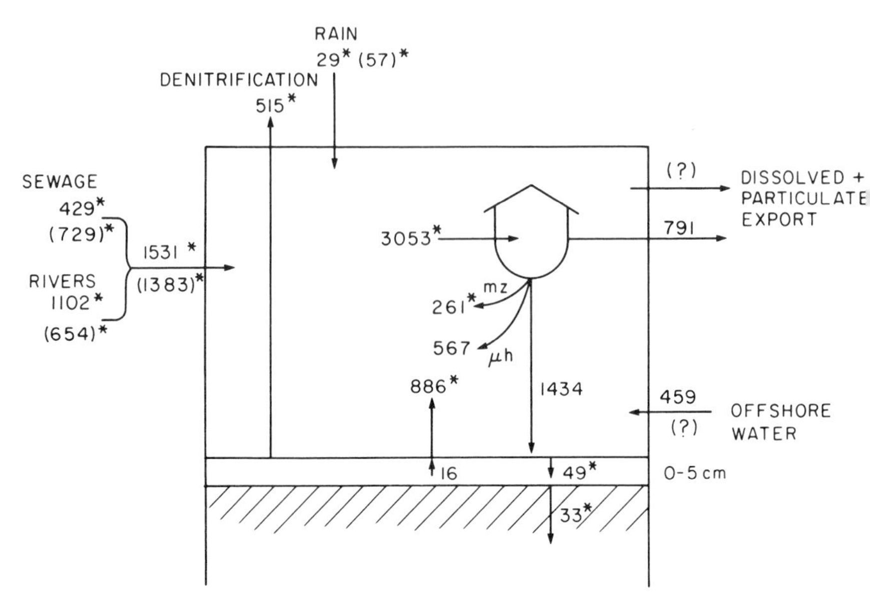

FIGURE 9. Present status of the annual nitrogen budget for Narragansett Bay (as Fig. 8).

TABLE 2. Comparison of estimated inputs and outputs of inorganic nitrogen and phosphate in Narragansett Bay.

	Annual flux, mmol m^{-2}	
	Nitrogen	*Phosphorus*
Input of inorganic forms	2019	204
Export of autochthonous organic matter (calculated)	-791	-60
Denitrification	-515	--
Burial	- 33	-4
Input available for output	680	140

Comparison with Other Systems

It is difficult to assess the total system net production of most of the marshes that have been studied because most of the studies of organic export have not adequately taken account of organic accumulation in the marsh sediments. On the basis of the data available, it appears that the total of particulate organic export and accumulation may commonly amount to some 100-200 g C m^{-2} y^{-1} (Nixon 1980). In rapidly accreting areas, burial may appear quite large, but it is necessary to correct for carbon entering the marsh associated with the sediment that is being deposited. Strictly speaking, this is also true for deposition in systems like Narragansett Bay, where some component of the small amount of organic matter accumulating in the sediments must come from terrestrial or offshore sources.

The only plankton-based systems for which we could find long-term data on total system metabolism were the nearby waters of Long Island Sound and the Woods Hole-Vinyard Sound-Buzzards Bay area. In the former, Riley's most recent (1972) analysis showed a small (probably insignificant) net consumption of 5 g C m^{-2} y^{-1}, and for the latter, Teal (1967) reported a net consumption of 22 g C m^{-2} over an 11-month period. Oviatt et al. (1981) combined a variety of oxygen-based measurements in light bottles, benthic chambers, and microcosms to provide a partially indirect estimate of total system metabolism of Narragansett Bay, and concluded that the Bay is essentially in balance with regard to organic carbon, integrated over the annual cycle. For the turbid Wadden Sea, where there are large intertidal mudflat areas and almost half the primary production is contributed by benthic microflora, Postma (1981) calculated that decomposition in the estuary exceeded production by about 115 g C m^{-2} y^{-1}, emphasizing the importance of organic inputs from the North Sea. None of these results (uncertain as they are) supports the notion that plankton-based estuaries are marked by high net production or export. The stoichiometric calculation for Narragansett Bay (imperfect as it is) appears to be the first indication of a significant net production and export from a plankton-based estuary.

In spite of the more intimate hydrodynamic linkage between most plankton-based estuaries and offshore waters compared with intertidal marshes, it does not appear that the pelagic systems export a proportionally larger fraction of their primary production. The high respiration: biomass costs of small plankton and the better nutritional quality of plankton and the detritus they produce may lead to systems where production and consumption remain closely in balance over an annual cycle.

SOME ADVANTAGES AND LIMITATIONS OF THE STOICHIOMETRIC CALCULATION

The approach described here takes advantage of the fact that all of the various processes contributing to the total metabolism of an estuary leave their imprint on the DIN/DIP ratio observed in the water. Yet, only one biological rate measurement, denitrification, is required for the calculation of total system net production. Other terms (DIN/DIP ratio, and inputs of DIN and DIP) require standing crop and water transport measurements. The various difficulties of bottle effects, enrichment effects, patchiness, and the large variance often associated with biological rate measurements are, to a large extent, avoided or minimized. Instead of measuring a highly variable rate-of-change (as with free water O_2, CO_2, or nutrient calculations), the stoichiometric approach uses the much less "noisy" integral of many autotrophic and heterotrophic processes. Recent advances in measuring denitrification (e.g., Seitzinger et al. 1980) make it likely that assessments of the loss of fixed nitrogen will become increasingly common and reliable. Measurements of the input of nitrogen and phosphorus from rivers, atmosphere, and anthropogenic sources are reasonably straightforward, though estimating inputs from offshore waters will continue to provide a challenge.

In order to account for seasonal variations in the input of DIN and DIP as well as variations in the DIN/DIP ratio of the Bay and the C:N:P composition of the plankton, we have calculated mean monthly values for the terms in Eq. (8), solved the equation using these, and summed the results for an annual total. This value is not significantly different from that obtained using annual means for each term as described in the text.

Major limitations with the approach stem from the simplifying assumptions necessary to arrive at an estimate of the inputs from offshore and from the potential role of the inputs of particulate and dissolved organic nitrogen and phosphorus. There is also no substitute for the information which is yet lacking regarding conditions in the water and sediments of the Providence River, Mt. Hope Bay, and the East Passage of Narragansett Bay. Doubtless our knowledge of many other estuaries must suffer from a similar lack of information on geographical and year-to-year variability. Nevertheless, an attempt to consider and calculate the total metabolism of an estuary provides a powerful synthesis of the information at hand, sets some limits on the magnitude of organic export from the system, and focuses attention on the kinds and quality of measurements that will be required to go further in constraining various mass balances.

ACKNOWLEDGMENTS

The stoichiometric calculation draws on several sets of data, and we are grateful to those who made it possible for us to go through this exercise, especially Miles Furnas and Ted Smayda for their information on the composition of the phytoplankton in the Bay, Candace Oviatt and Jim Kremer for their role in the 1972-73 Bay survey, Barbara Nowicki and Sharon Northby for data on the composition of the sediments, and Barbara Nowicki for her help with the measurement of nutrient inputs to the Bay. Sybil Seitzinger and Peter Santschi provided access to data which are still in press at the time of writing. In addition, a number of past and present graduate students and staff helped with various chemical analyses. The work reported here was supported largely by grants from the Office of Sea Grant, NOAA and from the National Science Foundation.

REFERENCES CITED

Banse, K. 1974. The nitrogen-to-phosphorus ratio in the photic zone of the sea and the elemental composition of the plankton. *Deep-Sea Res. 21*:767-771.

Boon, J., III. 1975. Tidal discharge asymmetry in a salt marsh drainage system. *Limnol. Oceanogr. 20*:71-80.

Boon, J., III. 1978. Suspended solids transport in a salt marsh creek--an analysis of errors, pp. 147-159. *In*: B. Kjerfve (ed.), *Estuarine Transport Processes*. Univ. South Carolina Press, Columbia, SC.

Boynton, W. R., C. A. Hall, P. G. Falkowski, C. W. Keefe and W. M. Kemp. 1983. Phytoplankton productivity in aquatic systems, pp. 305-327. *In:* O. L. Lange, P. S. Nobel, C. V. Osmond and H. Ziegler (eds.), *Encyclopedia of Plant Physiology*, New Series, Vol. *12D: Physiological Plant Ecology IV*. Springer-Verlag, New York.

Boynton, W. R., W. M. Kemp and C. W. Keefe. 1982. A comparative analysis of nutrients and other factors influencing estuarine phytoplankton production, pp. 69-90. *In:* V. S. Kennedy (ed.), *Estuarine Comparisons*. Academic Press, New York.

Carey, A. G. 1967. Energetics of the benthos of Long Island Sound. I. Oxygen utilization of sediment. *Bull. Bingham Oceanogr. Coll. 19:*136-144.

Copin-Montegut, C. and G. Copin-Montegut. 1983. Stoichiometry of carbon, nitrogen and phosphorus in marine particulate matter. *Deep-Sea Res. 30:*31-46.

Dunn, C. Q., L. Z. Hale and A. Bucci. 1979. *The Bay Bib: Rhode Island Marine Bibliography*, revised edition. Vols. *I* and *II*. University of Rhode Island Marine Technical Report Nos. 70 and 71. Kingston, RI.

Furnas, M. J., G. L. Hitchcock and T. J. Smayda. 1976. Nutrient-phytoplankton relationships in Narragansett Bay during the 1974 summer bloom, pp. 118-134. *In:* M. L. Wiley (ed.), *Estuarine Processes: Uses, Stresses and Adaptation to the Estuary*, Vol. *1*. Academic Press, New York.

Goldman, J. C., J. J. McCarthy and D. G. Peavey. 1979. Growth rate influence on the chemical composition of phytoplankton in oceanic waters. *Nature 279:*210-214.

Harris, E. and G. A. Riley. 1956. Oceanography of Long Island Sound, 1952-54. VIII. Chemical composition of the plankton. *Bull. Bingham Oceanogr. Coll. 15:*315-323.

Hicks, S. D. 1959. The physical oceanography of Narragansett Bay. *Limnol. Oceanogr. 4:*316-327.

Kjerfve, B. and J. A. Proehl. 1979. Velocity variability in a cross-section of a well-mixed estuary. *J. Mar. Res. 37*:409-418.

Kjerfve, B., J. A. Proehl, F. B. Schwing, H. E. Seim and M. Marozas. 1982. Temporal and spatial considerations in measuring estuarine water fluxes, pp. 37-51. *In:* V. S. Kennedy (ed.), *Estuarine Comparisons*. Academic Press, New York.

Kjerfve, B., L. H. Stevenson, J. A. Proehl, T. H. Chrzanowski and W. M. Kitchens. 1981. Estimation of material fluxes in an estuarine cross section: A critical analysis of spatial measurement density and errors. *Limnol. Oceanogr. 26*:325-335.

Kremer, J. N. and S. W. Nixon. 1978. *A Coastal Marine Ecosystem, Simulation and Analysis*. Ecological Studies *24*. Springer-Verlag, New York. 217 pp.

Kremer, P. 1975. The ecology of the ctenophore *Mnemiopsis leidyi* in Narragansett Bay. Ph.D. Thesis, Univ. Rhode Island, Kingston, RI. 311 pp.

Mills, E. 1975. Benthic organisms and the structure of marine ecosystems. *J. Fish. Res. Board Can. 32*:1657-1663.

Nixon, S. W. 1980. Between coastal marshes and coastal waters--A review of twenty years of speculation and research on the role of salt marshes in estuarine productivity and water chemistry, pp. 437-525. *In*: P. Hamilton and K. B. MacDonald (eds.), *Estuarine and Wetland Processes*. Plenum Publishing Corporation, New York.

Nixon, S. W. 1981a. Freshwater inputs and estuarine productivity, pp. 31-57. *In*: R. D. Cross and D. L. Williams (eds.), *Proceedings of the National Symposium on Freshwater Inflow to Estuaries*, Vol. *1*. U. S. Fish and Wildlife Service, Office of Biological Services. FWS/OBS-81/04.

Nixon, S. W. 1981b. Remineralization and nutrient cycling in coastal marine ecosystems, pp. 111-138. *In*: B. J. Neilson and L. E. Cronin (eds.), *Estuaries and Nutrients*. Humana Press, Clifton, New Jersey.

Nixon, S. W. 1982. Nutrient dynamics, Primary production and fisheries yields of lagoons. Proceedings of the International Symposium on Coastal Lagoons, Bordeaux, France, September, 1981. *Oceanologica Acta*. Special Edition, pp. 357-371.

Nixon, S. W. 1983. Estuarine ecology--A comparative and experimental analysis using 14 estuaries and the MERL microcosms. Final report to the U. S. Environmental Protection Agency, Chesapeake Bay Program, Annapolis, MD.

Nixon, S. W., J. R. Kelly, B. N. Furnas, C. A. Oviatt and S. S. Hale. 1980. Phosphorus regeneration and the metabolism of coastal marine bottom communities, pp. 219-243. *In:* K. R. Tenore and B. C. Coull (eds.), *Marine Benthic Dynamics*. Univ. South Carolina Press, Columbia, SC.

Nixon, S. W. and J. N. Kremer. 1977. Narragansett Bay--the development of a composite simulation model for a New England estuary, pp. 622-673. *In:* C. Hall and J. Day (eds.), *Ecosystem Modeling in Theory and Practice*. Wiley Interscience, New York.

Nixon, S. W. and C. A. Oviatt. 1973. Ecology of a New England Salt Marsh. *Ecol. Monog.* *43*:463-498.

Nixon, S. W., C. A. Oviatt and S. S. Hale. 1976. Nitrogen regeneration and the metabolism of coastal marine bottom communities, pp. 269-283. *In*: J. M. Anderson and A. Macfadyen (eds.), *The Role of Terrestrial and Aquatic Organisms in Decomposition Processes*. Blackwell Scientific Publications, London.

Odum, E. P. 1968. A research challenge: evaluating the productivity of coastal and estuarine water. Proceedings of the Second Sea Grant Conference, pp. 63-64. Graduate School of Oceanography, Univ. Rhode Island, Kingston, RI.

Odum, E. P. 1971. *Fundamentals of Ecology*. W. B. Saunders Company, Philadelphia, PA. 574 pp.

Odum, E. P. 1980. The status of three ecosystem-level hypothesis regarding salt marsh estuaries: Tidal subsidy, outwelling and detritus-based food chains, pp. 485-495. *In:* V. S. Kennedy (ed.), *Estuarine Perspectives*. Academic Press, New York.

Odum, H. T. 1967. Biological circuits and the marine systems of Texas, pp. 99-157. *In:* T. A. Olson and F. J. Burgess (eds.), *Pollution and Marine Ecology*. Interscience Publishers, New York.

Oviatt, C., B. Buckley and S. Nixon. 1981. Annual phytoplankton metabolism in Narragansett Bay calculated from survey field measurements and microcosm observations. *Estuaries 4:*167-175.

Pamatmat, M. M. and K. Banse. 1969. Oxygen consumption by the seabed. 2. *In situ* measurement to a depth of 180 m. *Limnol. Oceanogr. 14:*250-259.

Petersen, C. J. G. 1915. A preliminary result of the investigation on the valuation of the sea. *Rep. Danish Biol. Sta. 23:*29-33.

Postma, H. 1981. Exchange of materials between the North Sea and the Wadden Sea. *Mar. Geol. 40:*199-213.

Redfield, A. C. 1934. On the proportions of organic derivatives in sea water and their relation to the composition of plankton, pp. 176-192. *In: James Johnstone Memorial Volume*. Liverpool University Press, Liverpool.

Ricker, W. E. 1973. Linear regressions in fishery research. *J. Fish. Res. Board Can. 30:*409-434.

Riley, G. A. 1956. Oceanography of Long Island Sound, 1952-54. IX. Production and utilization of organic matter. *Bull. Bingham. Oceanogr. Coll. 15:*324-344.

Riley, G. A. 1972. Patterns of production in marine ecosystems, pp. 91-112. *In*: J. A. Weins (ed.), *Ecosystems, Structure and Functions*. Oregon State University Press, Corvallis, OR.

Santschi, P. H., S. W. Nixon, M. E. Q. Pilson and C. Hunt. In press. Accumulation of sediments, trace metals (Pb, Cu) and hydrocarbons in Narragansett Bay, Rhode Island. *Est. Coast. Shelf Sci.*

Seitzinger, S. P. 1982. The importance of denitrification and nitrous oxide production in the nitrogen dynamics and ecology of Narragansett Bay, Rhode Island. Ph.D. Thesis, Univ. Rhode Island, Kingston, RI. 145 pp.

Seitzinger, S. P., S. W. Nixon and M. E. Q. Pilson. 1984. The importance of denitrification and nitrous oxide production in the ecology and nitrogen dynamics of a coastal marine ecosystem. *Limnol. Oceanogr. 29:*73-83.

Seitzinger, S. P., S. W. Nixon, M. E. Q. Pilson and Suzanne Burke. 1980. Denitrification and N_2O production in near-shore marine sediments. *Geochem. Cosmochem. Acta. 44:* 1853-1860.

Smith, K. L. 1973. Respiration of a sublittoral community. *Ecology 54:*1065-1075.

Teal, J. M. 1962. Energy flow in the salt marsh ecosystem of Georgia. *Ecology 43:*614-624.

Teal, J. M. 1967. Biological production and distribution of pCO_2 in Woods Hole waters, pp. 336-340. *In:* G. H. Lauff (ed.), *Estuaries*. Amer. Assoc. Adv. Sci., Pub. No. 83, Washington, DC.

Valiela, I., J. M. Teal, S. Volkman, D. Shafer and E. J. Carpenter. 1978. Nutrient and particulate fluxes in a salt marsh ecosystem: Tidal exchanges and inputs by precipitation and ground water. *Limnol. Oceanogr. 23:*798-812.

Vargo, G. A. 1976. The influence of grazing and nutrient excretion by zooplankton on the growth and production of the marine diatom, *Skeletonema costatum* (Greville) Cleve, in Narragansett Bay. Ph.D. Thesis, Univ. Rhode Island, Kingston, RI. 216 pp.

Vargo, G. A. 1979. The contribution of ammonia excreted by zooplankton to phytoplankton production in Narragansett Bay. *J. Plankton Res. 1:*75-84.

Zeitzchel, B. F. 1980. Sediment-water interactions in nutrient dynamics, pp. 195-212. *In*: K. R. Tenore and B. C. Coull (eds.), *Marine Benthic Dynamics*. Univ. South Carolina Press, Columbia, SC.

ANTHROPOGENIC NITROGEN LOADING AND ASSIMILATION CAPACITY OF THE HUDSON RIVER ESTUARINE SYSTEM, USA

Thomas C. Malone

Horn Point Environmental Laboratories
Center for Environmental and Estuarine Studies
University of Maryland
Cambridge, Maryland

Abstract: The coastal plume of the Hudson River estuary receives inputs of new nitrogen of sewage origin from the lower estuary and of offshore origin from adjacent coastal water. As a consequence of these inputs, the plume is one of the most productive coastal systems in the world's ocean. The extent to which such high production reflects the input of sewage-nitrogen depends on the capacity of phytoplankton to assimilate new nitrogen input to the plume and on the magnitude of sewage-nitrogen input relative to coastal inputs, i.e., on the capacity of the plume to "filter" inputs of new nitrogen.

The importance of new nitrogen varies seasonally relative to nitrogen recycled within the plume. Regenerated production increases from 10% of total phytoplankton production during winter to a maximum of 80% during summer. Sewage-nitrogen supports an average of 54% of new production during the spring bloom period (March-May) compared to 121% during the subsequent period of stratification (June-October) and 221% during winter (November-February). On an annual basis, these estimates indicate that phytoplankton production has increased by ca. 30% in response to the input of sewage-nitrogen.

These patterns have important implications in terms of cross-shelf transport and exchange. The supply of new nitrogen from offshore is highest during March-May and,

ISBN 0-12-405070-0

along with sewage-nitrogen, appears to support a large seaward export of nitrogen in the form of phytoplankton biomass. In contrast, most new nitrogen is exported via plankton food webs during summer. Consequently, a relatively small fraction of phytoplankton production is available to fuel oxygen demand below the pycnocline during summer when assimilation capacity is greatest.

INTRODUCTION

Anthropogenic nutrient inputs to estuaries have increased rapidly as sewage production, agricultural fertilization, and urbanization have increased (e.g., Milliman 1981; van Bennekom and Salomons 1981). For example, the input of anthropogenic nitrogen, the nutrient which most frequently limits phytoplankton production in coastal waters (Eppley et al. 1971; Ryther and Dunstan 1971; Goldman 1976), has been estimated to have increased by an order of magnitude over the past three decades (Walsh et al. 1981). Such an increase will have pronounced effects on water quality and fisheries depending on the capacity of estuarine communities to assimilate (or "filter") nitrogen inputs and on the pathways by which assimilation occurs.

Phytoplankton, because of their importance as primary producers and their high growth rates, play a central role in governing the capacity of estuarine systems to assimilate inorganic nitrogen inputs. In this context, phytoplankton production can be partitioned into new and regenerated production, the former being dependent on imported nitrogen and the latter on nitrogen recycled within the system (Dugdale and Goering 1967). When regenerated production is high relative to new production, phytoplankton production as a whole is usually low and nitrogen limited (Thomas 1970a; Eppley et al. 1973; Harrison 1980). Such systems develop when phytoplankton production, heterotrophic consumption, and nitrogen regeneration are closely coupled in time and space (Eppley et al. 1973; Harrison 1978; Eppley and Peterson 1979). As the proportion of new production increases in response to new nitrogen input, phytoplankton become less nitrogen deficient and total production increases (Thomas 1970b; Eppley, Renger and Harrison 1979; Harrison 1980; Malone et al. 1983b). The increase in phytoplankton production reflects both an increase in the supply of new nitrogen and a decrease in the degree of coupling between primary producers and consumers (e.g., Walsh 1976). This separation in time or space of phytoplankton production and heterotrophic consumption results in accumulations of phytoplankton biomass and increases the susceptibility of the system to episodes of oxygen depletion (cf. Officer and Ryther 1977). Thus, the capacity to assimilate new nitrogen inputs with a minimal increase in oxygen demand should be greatest when regenerated production is high and new production is low, i.e., the balance between new and regenerated production should be an index of assimilation capacity.

Nitrogen of sewage origin is new nitrogen, and the capacity for its assimilation should vary depending on the extent to which phytoplankton utilize recycled nitrogen and on the magnitude of other inputs of new nitrogen. This report is concerned with seasonal variations in the importance of sewage-derived nitrogen relative to other sources of nitrogen and with the relationship between new nitrogen supply and assimilation of nitrogen by phytoplankton in the Hudson estuary.

THE HUDSON ESTUARINE SYSTEM

The salt-intruded reach of the Hudson estuary and its coastal plume is a partially mixed estuarine system characterized by vertical and horizontal salinity gradients throughout the year on a seasonal time scale (Figs. 1 and 2). The distribution of salt reflects a two-layered circulation system with a net upstream flow of salt water in the bottom layer and a net seaward flow of fresher water in the surface layer, the latter being driven by fresh water flow of the Hudson River (Fig. 2). Input of new nitrogen to the estuary proper is independent of fresh water flow and dominated by sewage-nitrogen (Garside et al. 1976; Malone 1982), most of which is discharged into the inner harbor region (Fig. 3). Dissolved inorganic nitrogen (DIN, nitrate and nitrite and ammonium) fluctuates around 60 μg-at liter^{-1} in this region (Fig. 4) and, to a first approximation, distributes conservatively with salt in both upstream and downstream directions within the estuary (Garside et al. 1976; Deck 1981).

Assimilation of DIN by phytoplankton in the estuary is small compared to sewage-nitrogen input, and most sewage-nitrogen is transported to the plume (Malone 1982). In addition to sewage-nitrogen, the plume receives inputs of new nitrogen in the form of nitrate from offshore (Fig. 5). These inputs support a large crop of phytoplankton in the plume relative to the estuary (Fig. 4) and to adjacent coastal water (Malone et al., 1983b). Phytoplankton biomass in the plume, expressed as chlorophyll *a* ,tends to fluctuate

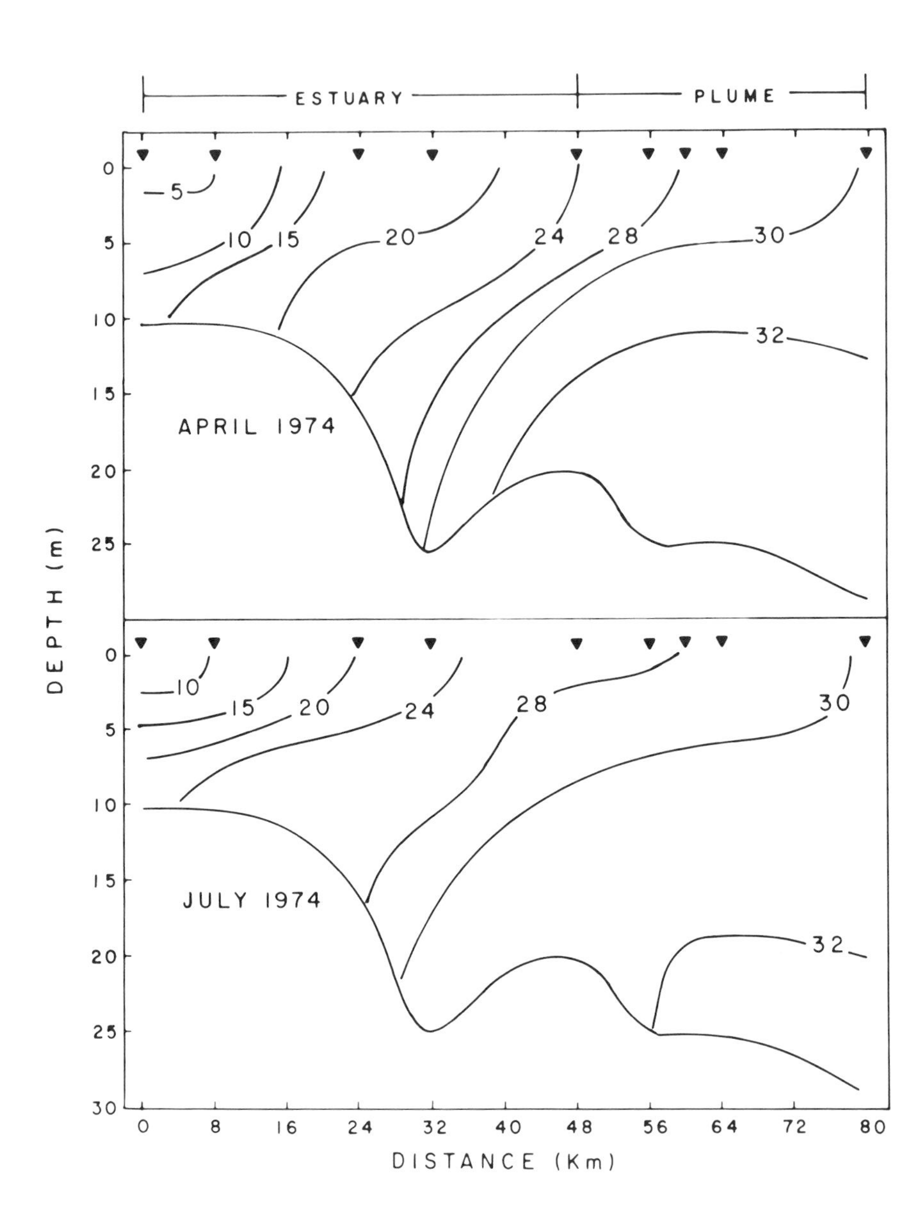

FIGURE 1. Salinity distributions down the axis of the lower Hudson estuary (midchannel from 40° 53' N, 73° 58' W) and coastal plume (40° 30' N, 73° 58' W to 40° 10' N, 73° 50'W) under high (April) and low (July) flow conditions (▼, station locations).

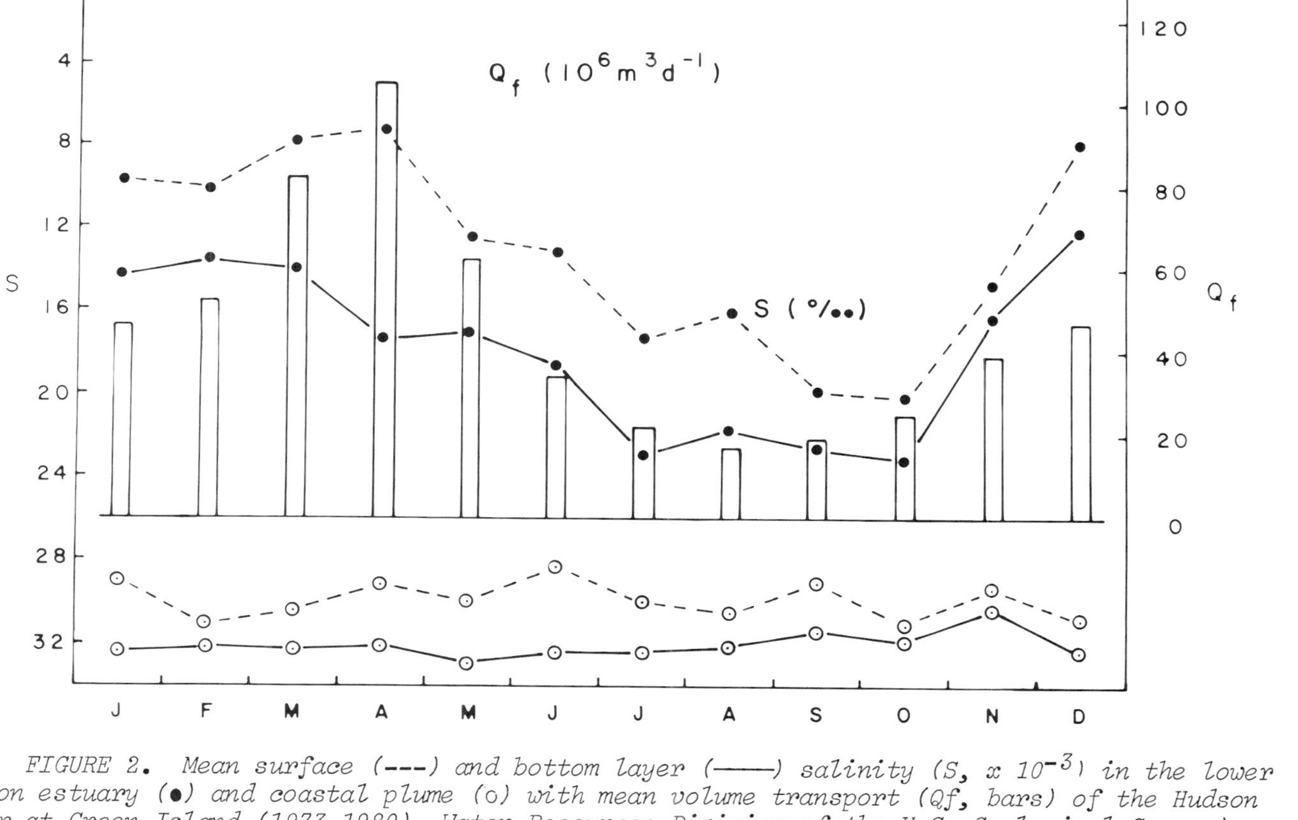

FIGURE 2. Mean surface (---) and bottom layer (——) salinity (S, x 10^{-3}) in the lower Hudson estuary (●) and coastal plume (○) with mean volume transport (Qf, bars) of the Hudson River at Green Island (1973-1980), Water Resources Division of the U.S. Geological Survey).

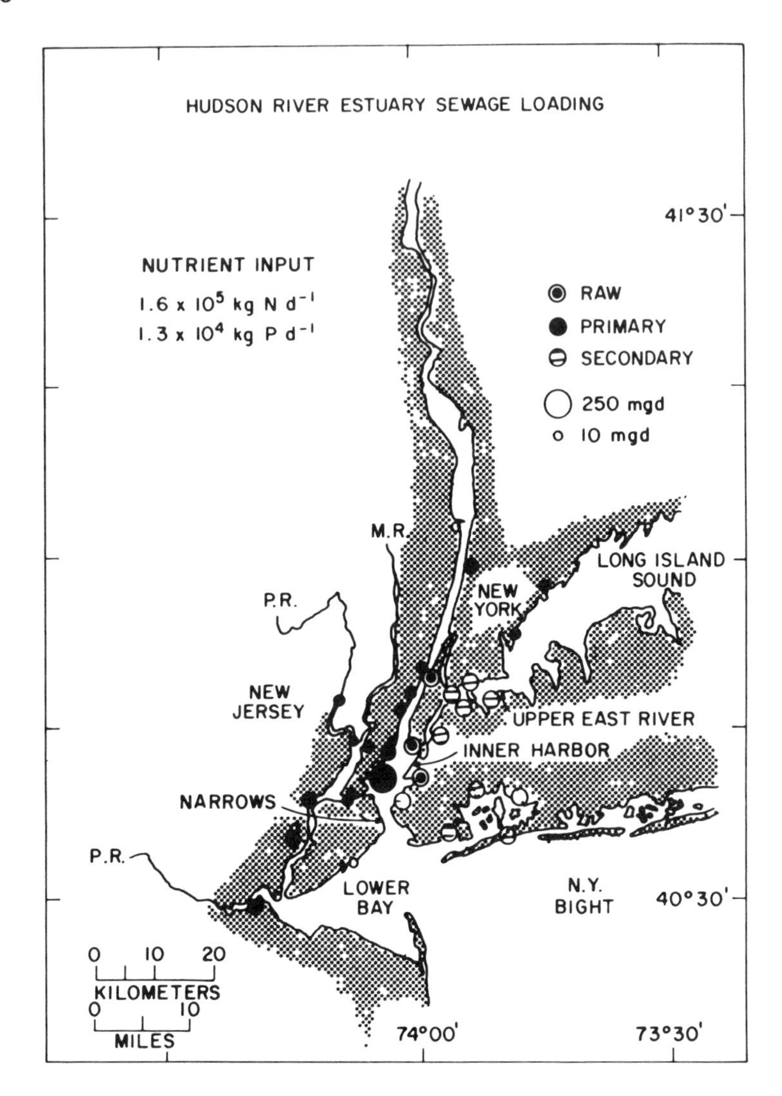

FIGURE 3. Location of sewage outfalls and treatment plants discharging into the Hudson estuary with nutrient loading in terms of nitrogen and phosphorus.

between 40 and 60 mg m^{-2} except during March-April when the monthly mean approaches 200 mg m^{-2} (Fig. 4). Chain forming diatoms dominate this spring bloom period, and integral chlorophyll a often exceeds 500 mg m^{-2} (e.g. Malone et al. 1983a). In contrast, small solitary chlorophytes dominate during summer when chlorophyll a content is lower and less variable on a time scale of days-weeks (e.g. Malone and Chervin 1979). High event scale variability

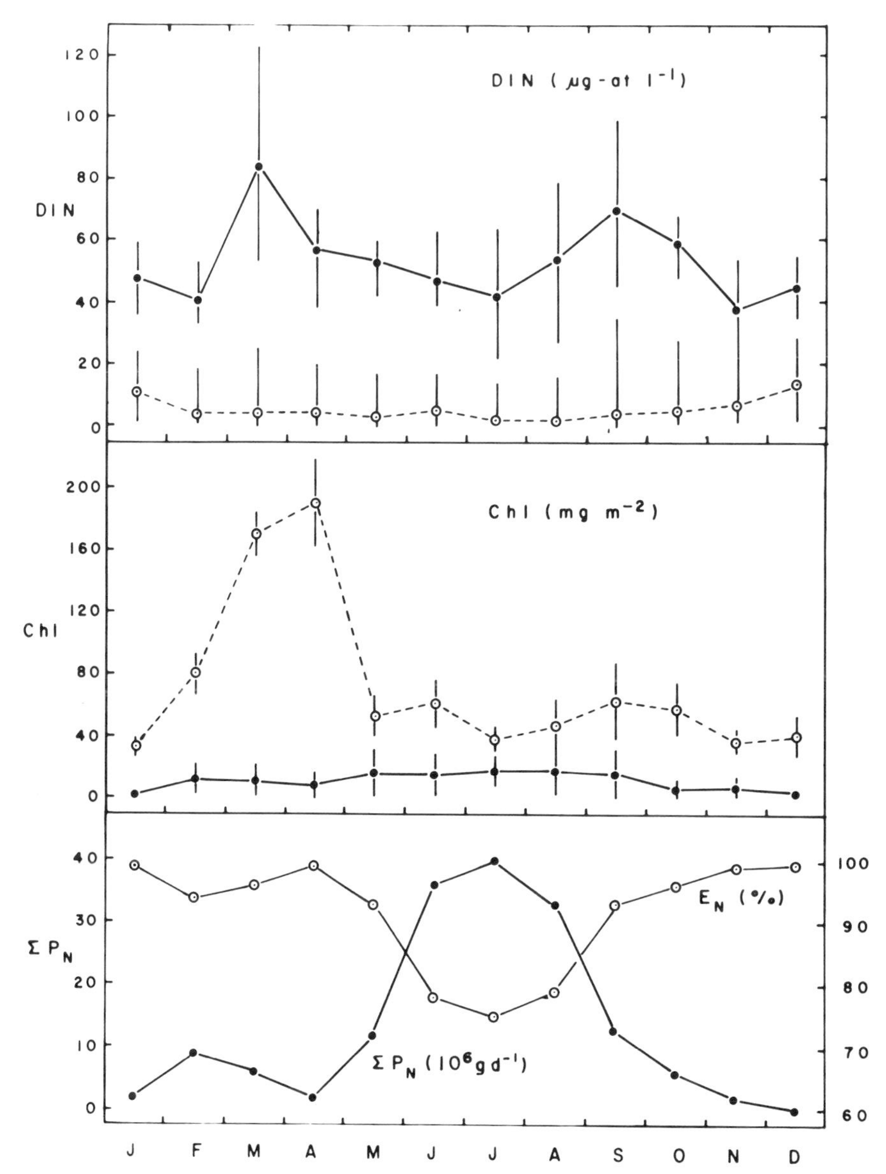

FIGURE 4. Monthly means for the surface layer of the lower estuary and for the coastal plumes: DIN concentration (● -estuary, ○ -plume; range indicated by vertical bars), chlorophyll a content (● -estuary, ○ -plume; ± 2SE), nitrogen assimilation by phytoplankton over the salt-intruded reach of the estuary ($\Sigma \bar{P}_n$, 10^6 g N d^{-1} calculated from area and depth-integrated phytoplankton production using the Redfield N/C of 0.18 w/w), and transport of sewage-nitrogen into the plume as a percent of input (E_n, % on right axis).

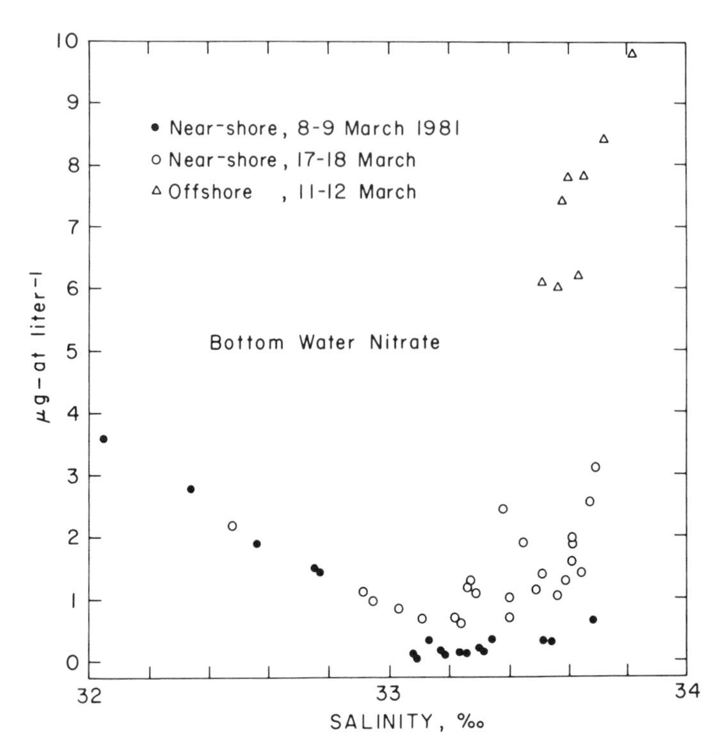

FIGURE 5. Relationship between nitrate concentration and salinity in water below the plume (near-shore) before and after a southwest wind event that caused onshore transport of nitrate-rich bottom water from offshore; note that the distribution is dominated by nitrate from the estuary prior to the storm and by nitrate from offshore after the storm.

during the spring bloom period is a consequence of high storm frequency (0.2 d^{-1} on average) and low grazing mortality. Relatively low biomass and the lack of such variability during summer is related to low storm frequency (0.1 d^{-1} on average) and to high grazing mortality.

Phytoplankton production (P_C) varies with chlorophyll *a*, ranging from a monthly mean of 0.2 g C $m^{-2}d^{-1}$ during November-January to nearly 5 g C $m^{-2}d^{-1}$ in April (Fig. 6). Annual phytoplankton production is on the order of 590 g C $m^{-2}yr^{-1}$ or about that expected for estuarine systems given an input of sewage-nitrogen of 43 g N $m^{-2}yr^{-1}$ (Boynton, Kemp and Keefe 1982; Malone 1982). Phytoplankton production of this magnitude places the Hudson plume among the most productive coastal systems in the world's ocean (cf. Leith and Whittaker 1975) and accounts for 75% of total particulate organic carbon inputs to the plume (Malone 1982)

NEW AND REGENERATED PRODUCTION

The relative importance of new and regenerated nitrogen in sustaining such high phytoplankton production can be calculated from variations in the turnover rate of DIN and phytoplankton biomass within the plume. Assuming that phytoplankton account for all DIN assimilation and that input and assimilation of DIN approach steady-state over monthly time intervals, the ratio of DIN turnover due to

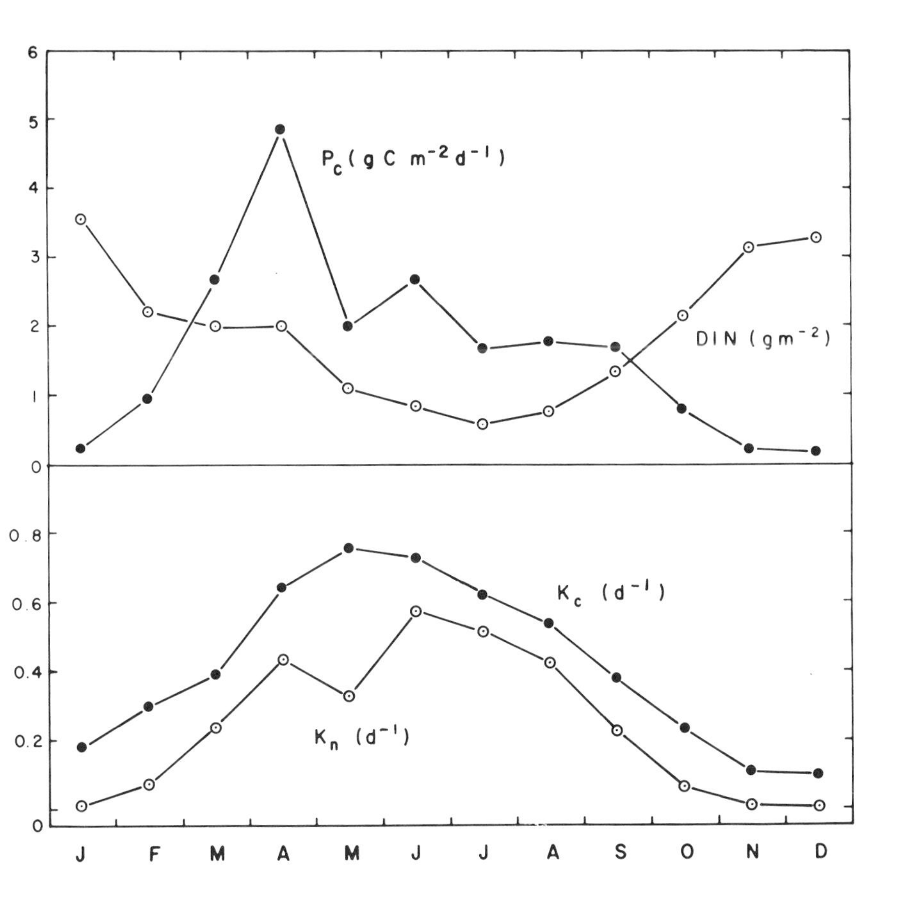

FIGURE 6. Depth-integrated monthly means for the coastal plume: DIN content, phytoplankton production (P_c), turnover rate of phytoplankton biomass due to photosynthetic growth (K_c), and turnover rate of DIN due to phytoplankton assimilation (K_n); units for dependent variables scaled on the ordinate are indicated on the respective graphs.

phytoplankton assimilation (K_n, d^{-1}) to the turnover of phytoplankton biomass due to photosynthesis (K_c, d^{-1}) is an estimate of the proportion of production supported by regenerated nitrogen, i.e., as K_n/K_c approaches 1.0, regenerated production approaches total production; and as K_n/K_c approaches 0, new production approaches total production.

Turnover rates are calculated from depth integrated (over the euphotic zone which is roughly equal to the depth of the plume) and area weighted (plume) rates and pool sizes as follows:

$K_c = P_c/B_c$ and $K_n = P_n/N$ where

P_c = phytoplankton production (g C m^{-2} d^{-1} calculated from 24 h, simulated *in situ* ^{14}C-uptake experiments as described by Malone et al. 1983a),

B_c = phytoplankton biomass (g C m^{-2} calculated from chlorophyll *a* as described by Chervin, Malone and Neale 1981),

P_n = DIN assimilation by phytoplankton (g N m^{-2} d^{-1} calculated from 0.18 x P_c in which 0.18 is N/C Redfield ratio by weight), and

N = DIN (g N m^{-2}).

Results shown in Fig. 6 are area weighted means based on 1376 sets of measurements made from 1973 to 1981 with minimum and maximum monthly totals of 16 (October) and 283 (March).

Variations in K_c and K_n follow similar annual cycles (Fig. 6) suggesting that nitrogen supply and assimilation are roughly in balance. K_c exceeds K_n as would be expected if the photosynthetic production of biomass occurs more rapidly than heterotrophic assimilation and nitrogen remineralization, as is probably the case (Riley 1972; Nixon 1981; Boynton et al. 1982). Seasonal variation from a December minimum to a May-June maximum reflects the fact that both nitrogen assimilation and carbon specific growth are light-limited on a seasonal scale (Malone 1976; Garside 1981). Variations in K_n relative to K_c are due to different patterns of change in the pool sizes of DIN and phytoplankton biomass, e.g., K_n decreases relative to K_c in May as a consequence of a rapid decline in biomass (Fig. 4) relative to DIN (Fig. 6). Such variations occur when rates of new nitrogen input and biomass export change with respect to each other.

The annual cycle of K_n/K_c indicates that the balance between new and regenerated production varies seasonally (Fig. 7). Regenerated production as a proportion of total production varies from a minimum of 10% during winter to a

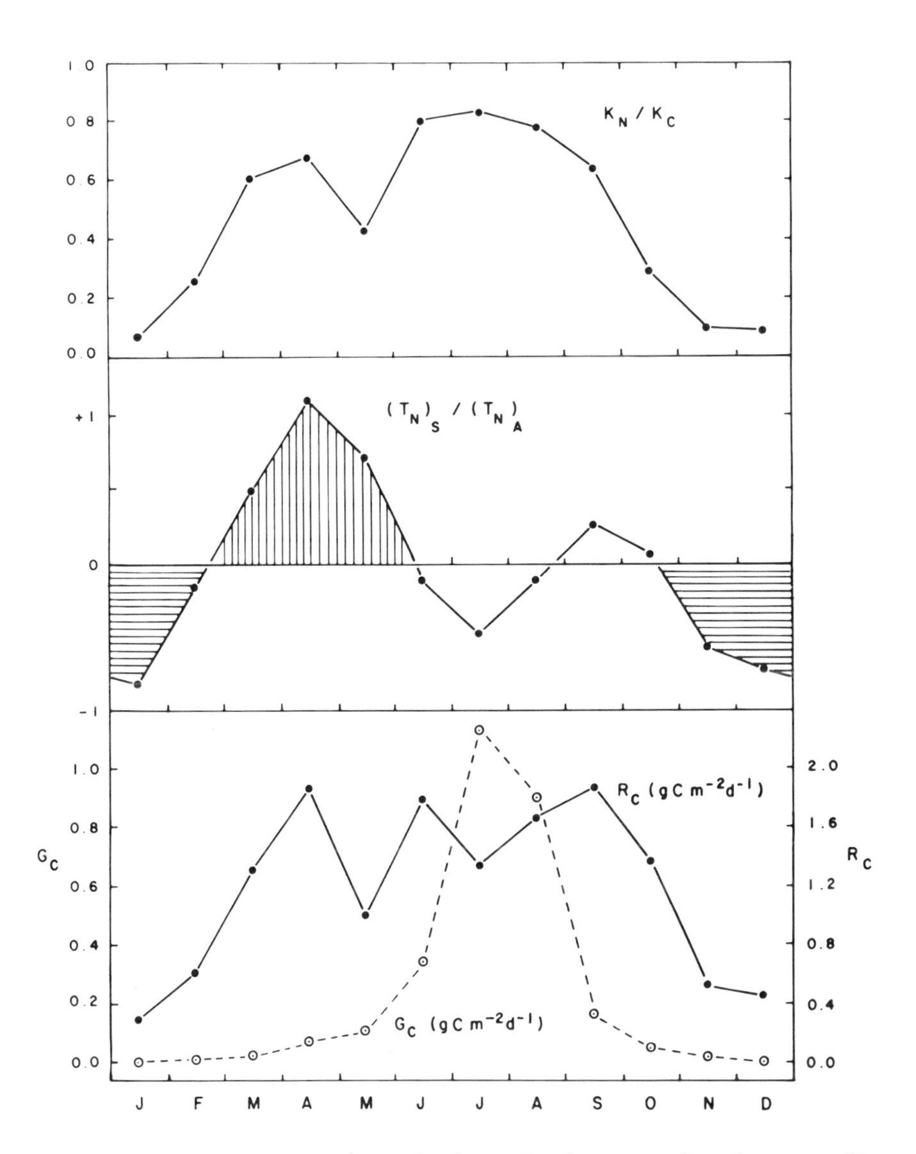

FIGURE 7. Proportion of phytoplankton production attributed to regenerated production (K_n/K_c), net exchange of DIN between plume and adjacent coastal water [$(T_N)_S$] relative to sewage-nitrogen input [$(T_N)_A$], phytoplankton mortality due to copepod grazing (G_c, from Chervin 1978; Chervin et al. 1981), and total plankton respiration (R_c, from Garside and Malone 1978 and described in Table 1) in the plume based on monthly means; units for dependent variables scaled on the ordinate are indicated on the respective graphs.

maximum of 80% during summer with a secondary peak of 65% during the spring bloom. Annually, regenerated production accounts for 60% of total production which is typical of coastal environments (Dugdale and Goering 1967; Eppley and Peterson 1979). Based on these percentages, nitrogen is recycled an average of 1.6 times annually, with an annual minimum of 0.1 during January and a maximum of 4.9 during July, before being lost from the plume.

With the notable exception of the spring peak, seasonal deviations from the annual mean of 60% are consistent with the temperature dependent cycle of nitrogen regeneration suggested by changes in the proportion of ammonium in the DIN pool (Malone 1976) and by variations in uptake rates of nitrate and ammonium (Garside 1981). Copepods, which dominate the macrozooplankton (Chervin 1978; Chervin et al. 1981), probably account for a large fraction of ammonium regeneration during summer (Fig. 7) when copepods graze up to 70% of phytoplankton production daily (July).

The spring peak in regenerated production occurs prior to the seasonal increase in copepod grazing (Fig. 7) and is probably unrelated to grazing by macrozooplankton. However, water column respiration exhibits a spring peak and accounts for 76% of the seasonal variation in K_n/K_c (Fig. 7). This suggests that most respiration during the spring bloom is due to microheterotrophs. Since peaks in respiration and regeneration also coincide with maximum phytoplankton biomass (Fig. 4) and production (Fig. 6), it seems likely that the spring increase in nitrogen regeneration is a response of microheterotrophs to high concentrations of organic substrates produced by phytoplankton.

NEW NITROGEN AND ASSIMILATION CAPACITY

Phytoplankton production in the plume is ultimately dependent on inputs of new nitrogen which are derived primarily from sewage and offshore sources. The input of sewage-nitrogen is known to a first approximation (Fig. 4), and the net transport of DIN between plume and shelf waters can be calculated over monthly intervals as follows:

$$(T_n)_S = (P_n)_{new} + \Delta N - (T_n)_A$$

where

$(T_n)_S$ = net transport of DIN between plume and shelf waters,

$(P_n)_{new}$ = assimilation of new DIN by phytoplankton within the plume estimated by $[1-(K_n/K_c)]\ P_n$,

ΔN = change in DIN content of the plume (from Fig. 6), and

$(T_n)_A$ = input of sewage-nitrogen (Fig. 4).

When $(T_n)_s$ is positive, the plume is a DIN sink and net transport is onshore into the plume. A negative sign indicates that the plume is a source, and net transport is offshore from the plume, i.e., in effect, the input of sewage-nitrogen is in excess of phytoplankton demand.

Net exchange of DIN with shelf water relative to the input of sewage nitrogen varies seasonally (Fig. 7). Maximum input of new nitrogen occurs during the spring bloom (March-May) when transport from offshore averages ca. 80% of the sewage-nitrogen input. Net exchange with shelf water is negligible during June-October when sewage-nitrogen satisfies the phytoplankton demand for new nitrogen. By November, the plume is exporting DIN to adjacent coastal water and an equivalent of 60% of the sewage-nitrogen input is exported during November-February.

These variations in net exchange are consistent with seasonal variations in hydrographic and climatic conditions which are likely to facilitate or inhibit exchange (Bowman and Wunderlich 1977; Malone et al. 1983 a,b). High transport of DIN from offshore coincides with a period when fresh water flow and storm frequency are high and the water column is weakly stratified. Net exchange of DIN with coastal water is minimal during summer when storm frequency is low and the plume is bounded by a strong thermocline which roughly coincides with the halocline. Net export of DIN from the plume is initiated with the fall overturn and continues during the period of low incident radiation and minimum vertical stability.

Seasonal variations in the balance between new and regenerated production should reflect the balance between phytoplankton production and heterotrophic metabolism within the plume since nitrogen regeneration is a by-product of heterotrophic metabolism, and new nitrogen sets an upper limit on the amount of organic matter that can be exported from the plume (cf. Eppley and Peterson 1979). A comparison of phytoplankton production and plankton respiration during the three periods defined by net DIN exchange with coastal water (Fig. 7) shows this to be the case (Table 1). The balance between phytoplankton production and plankton respiration within the plume shifts from a large excess during spring to a small deficit during winter, a pattern that is

TABLE 1. Balance between phytoplankton production and plankton respiration in the plume during periods when the plume is a sink for DIN (Mar.-May), a source (Nov.-Feb.), and when net exchange with shelf water is negligible (June-Oct.).

	10^6 *Kg C*		
	Mar.-May	*June-Oct.*	*Nov.-Feb.*
Phytoplankton Production[a]	363	327	59
Plankton Respiration[b]	159	305	72
Production-Respiration	204	22	- 13
New Production[c]	143	94	48

a This paper, Fig. 6
b Calculated from chlorophyll α specific respiration (R^B, gC[chl a]$^{-1}$) and chlorophyll α concentration of the plume (Fig 4); R^B was estimated from a model II, least square regression of R^B on surface temperature ($R^B = 6.36\ e^{0.062T}$, $r^2 = 0.86$) based on water column respiration rates (1978). given by Garside and Malone
c $[1-(K_n/K_c)]\ P_c$ from Figs. 6 and 7.

consistent with the seasonal progression of the plume from a system in which most particulate organic matter is packaged as phytoplankton during spring to one in which most particulate organic matter is packaged as detritus during fall (Malone and Chervin 1979; Chervin et al. 1981). New production follows a similar trend, decreasing from a maximum during the spring bloom to a minimum during winter when phytoplankton production is low (Table 1).

To the extent that organic matter is not accumulating within or below the plume, new production during the spring bloom period must be transported into the estuary or exported seaward. Phytoplankton biomass (Fig. 4) and particulate organic matter (Malone and Chervin 1979) accumulate within the plume on time scales of days-weeks but not seasonally. Accumulation on a seasonal scale also does not occur in the water column or sediments below the plume (Malone et al. 1983 a,b), and transport into the estuary is on the order of 9×10^4 kg C d^{-1} (Malone et al. 1980), or 6% of new production. These considerations indicate that most new production during the spring bloom period is exported seaward, a conclusion that is supported by cross-shelf distributions of chlorophyll α (Fig. 8).

Exports of sewage-nitrogen to adjacent coastal water, either as phytoplankton biomass or as DIN, are small in the context of the New York Bight as a whole. Inputs of new nitrogen from the Gulf of Maine and adjacent slope water are on the order of 6×10^6 kg N d^{-1} (Malone et al. 1983 a) compared to sewage inputs of $1.2 - 1.6 \times 10^5$ kg N d^{-1}. Stoddard (1983) estimates that new nitrogen input

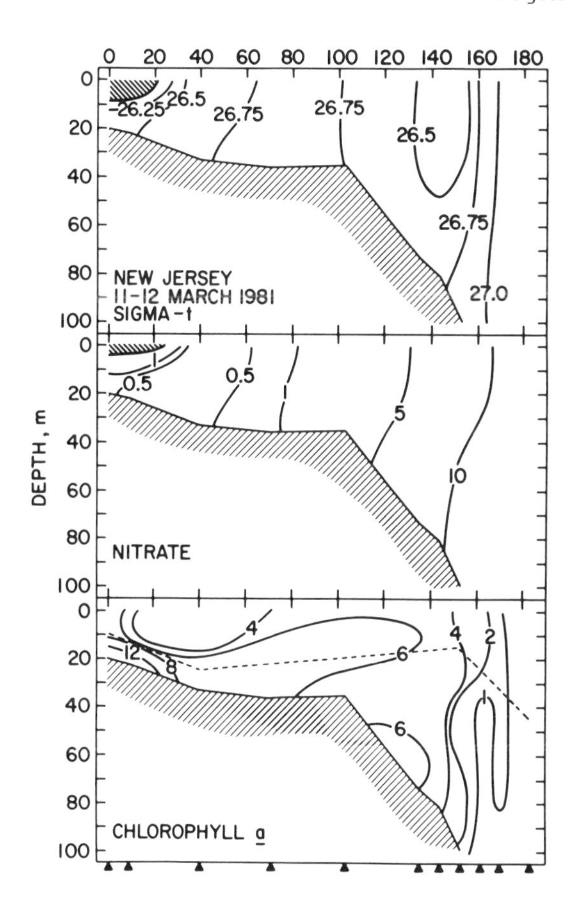

FIGURE 8. Vertical distributions of sigma-t, nitrate (μg-at liter^{-1}) and chlorophyll a (μg liter^{-1}) with distance (km on the abscissa) offshore from the plume to the shelf-break (----, 1% light depth) (▲, station locations).

would have to increase by an order of magnitude before the assimilation capacity of the Bight is exceeded. This would make sewage input roughly equivalent to other inputs of new nitrogen.

CONCLUSIONS

Sewage-nitrogen is a major source of new nitrogen throughout the year in the Hudson plume. Net export of DIN is negligible during March-October when sewage-nitrogen accounts for 54% (March-May) to 121% (June-October) of new production. During the remainder of the year, most new nitrogen is exported to adjacent coastal water and phytoplankton production within the plume is unaffected by the input of sewage-nitrogen. By this accounting, new production has increased in response to sewage-nitrogen loading by 77 x 10^6 kg C during March-May and by nearly 94 x 10^6 kg C during June-October. Thus, the discharge of sewage wastes into the lower Hudson estuary has resulted in a 30% increase (171 x 10^6 kg C/578 X 10^6 kg C from Table 1) in phytoplankton production over pre-discharge levels.

The fate of sewage-nitrogen (and new nitrogen from offshore) assimilated within the plume varies seasonally. During spring when the plume is a DIN sink for both sources of new nitrogen and when new production is highest, most sewage-nitrogen is exported into adjacent coastal water as phytoplankton biomass. Net export of DIN during the summer period of thermal stratification is small and most phytoplankton production is metabolized within the plume. The turnover rate of phytoplankton biomass is at its seasonal maximum under these conditions. This suggests that most new production is exported via the plankton food web during summer, either as biomass or as fecal material. Thus, new nitrogen inputs are balanced by physically forced lateral export of phytoplankton biomass during the spring bloom period and by biologically forced food web export during summer. During the remainder of the year (November-February) most of the input of new nitrogen is exported with little uptake within the plume. Clearly, the capacity of the plume to "filter" and metabolize sewage-nitrogen is greatest during summer when turnover rates of DIN and phytoplankton biomass are highest. This is probably a consequence of stable physical conditions and close coupling between phytoplankton production and heterotrophic consumption within the plume.

These variations in the response of plankton communities to new nitrogen inputs to the plume have important implications in terms of the influence of DIN of sewage origin on dissolved oxygen. Dissolved oxygen concentration varies seasonally and vertically with maximum concentrations (90-105% of saturation) during November-May and minimum concentrations (60-80% saturation) in bottom water during July-August of the stratified period (O'Connor, Thomann and Salas 1978; Malone 1982). Exceptions occur during summer in the region of the sludge and dredge-spoil dump sites where bottom oxygen depletion can reduce oxygen to less than 40% of saturation (Segar and Berberian 1976; Thomas et al. 1976). Lateral transport of phytoplankton biomass during the spring bloom period prevents accumulations of organic matter which would otherwise remain and provide organic substrates for oxygen below the plume as the seasonal thermocline forms. During summer when bottom water is most isolated from inputs of dissolved oxygen, over 80% of phytoplankton production is metabolized within the plume resulting in a proportionate decrease in the flux of organic matter into bottom water. Assuming that the remaining 20% of total production is metabolized below the plume during summer, this amounts to 9% of annual production. Thus, higher phytoplankton production in response to increases in sewage-nitrogen loading has had a small impact on dissolved oxygen compared to a potential impact on the order of 30%.

REFERENCES CITED

Bowman, M.J. and L.D. Wunderlich. 1977. Distribution of hydrographic properties in the New York Bight apex, pp. 58-68. *In:* M.G. Gross (ed.), *Middle Atlantic Shelf and the New York Bight.* Limnol. Oceanogr. Spec. Symp., 2.

Boynton, W.R., W.M. Kemp and C.W. Keefe. 1982. A comparative analysis of nutrients and other factors influencing estuarine phytoplankton production, pp. 69-90. *In:* V.S. Kennedy (ed.), *Estuarine Comparisons.* Academic Press, New York.

Chervin, M.B. 1978. Assimilation of particulate organic carbon by estuarine and coastal copepods. *Mar. Biol.* *49:*265-275.

Chervin, M.B., T.C. Malone and P.J. Neale. 1981. Interactions between suspended organic matter and copepod grazing in the plume of the Hudson River. *Estuar. Coast. Shelf Sci. 13:*169-184.

Deck, B.L. 1981. Nutrient-element distributions in the Hudson estuary. Ph.D. Dissertation, Columbia University, New York. 396 pp.

Dugdale, R.C. and J.J. Goering. 1967. Uptake of new and regenerated forms of nitrogen in primary productivity. *Limnol. Oceanogr. 12:*196-206.

Eppley, R.W., A.F. Carlucci, O. Holm-Hansen, D. Kiefer, J.J. McCarthy, E. Venrick and P.M. Williams. 1971. Phytoplankton growth and chemical composition in shipboard cultures supplied with nitrate, ammonium, or urea as the nitrogen source. *Limnol. Oceanogr. 16:*741-751.

Eppley, R.W. and B.J. Peterson. 1979. Particulate inorganic matter flux and planktonic new production in the deep ocean. *Nature 282:*677-680.

Eppley, R.W., E.H. Renger and W.G. Harrison. 1979. Nitrate and phytoplankton in southern California coastal waters. *Limnol. Oceanogr.. 24:*483-494.

Eppley, R.W., E.H. Renger, E.L. Venrick and M.M. Mullin. 1973. A study of plankton dynamics and nutrient cycling in the central gyre of the north Pacific ocean. *Limnol. Oceanogr. 18:*534-551.

Garside, C. 1981. Nitrate and ammonia uptake in the apex of the New York Bight. *Limnol. Oceanogr. 26:*731-739.

Garside, C. and T.C. Malone. 1978. Monthly oxygen and carbon budgets of the New York Bight apex. *Estuar. Coastal. Mar. Sci. 6:*93-104.

Garside, C., T.C. Malone, O.A. Roels and B.A. Sharfstein. 1976. An evaluation of sewage-derived nutrients and their influence on the Hudson estuary and New York Bight. *Estuar. Coast. Mar. Sci. 4:*281-289.

Goldman, J.C. 1976. Identification of nitrogen as a growth limiting nutrient in waste waters and coastal marine waters through continuous culture algal assays. *Water Res. 10:*97-104.

Harrison, W.G. 1978. Experimental measurements of nitrogen re-mineralization in coastal waters. *Limnol. Oceanogr. 23:* 684-694.

Harrison, W.G. 1980. Nutrient regeneration and primary production in the sea, pp. 433-460. *In:* P.G. Falkowski (ed.), *Primary Productivity in the Sea.* Plenum Press, New York.

Leith, H. and R.H. Whittaker. 1975. *Primary Productivity of the Biosphere.* Springer-Verlag, New York. 339 pp.

Malone, T.C. 1976. Phytoplankton productivity in the apex of the New York Bight: environmental regulation of productivity/chlorophyll *a*, pp. 260-272. *In:* M.G. Gross (ed.). *Middle Atlantic Shelf and the New York Bight.* Limnol. Oceanogr. Spec. Symp., 2.

Malone, T.C. 1982. Factors influencing the fate of sewage-derived nutrients in the lower Hudson estuary and New York Bight, pp. 301-320. *In:* G.F. Mayer (ed.), *Ecological Stress and the New York Bight: Science and Management.* Estuarine Research Foundation, Columbia, South Carolina.

Malone, T.C. and M.B. Chervin. 1979. The production and fate of phytoplankton size classes in the plume of the Hudson River, New York Bight. *Limnol. Oceanogr. 24:*683-696.

Malone, T.C., P.G. Falkowski, T.S. Hopkins, G.T. Rowe and T.E. Whitledge. 1983a. Mesoscale response of diatom populations to a wind event in the plume of the Hudson River. *Deep-Sea Res. 30:*149-170.

Malone, T.C., T.S. Hopkins, P.G. Falkowski and T.E. Whitledge. 1983b. Production and transport of phytoplankton biomass over the continental shelf of the New York Bight. *Continental Shelf Res. 1:*305-337.

Malone, T.C., P.J. Neale and D. Boardman. 1980. Influences of estuarine circulation on the distribution and biomass of phytoplankton size fractions, pp. 249-262. *In:* V.S. Kennedy (ed.) *Estuarine Perspectives.* Academic Press, New York.

Milliman, J.D. 1981. Transfer of river-borne particulate material to the oceans, pp. 5-12. *In:* *River Inputs to Ocean Systems.* United Nations, New York.

Nixon, S.W. 1981. Freshwater inputs and estuarine productivity, pp. 31-57. *In:* R. Cross and D. Williams (eds.), *Proceedings of the National Symposium on Freshwater Inflow to Estuaries.* U.S. Fish and Wildlife Service, Office of Biological Services, FWS/OBS-81/04.

O'Connor, D.J., R.V. Thomann and H.J. Salas. 1977. *Water Quality.* MESA New York Bight Atlas Monograph 27, New York Sea Grant Institute, Albany, New York. 104 pp.

Officer, C.B. and J.H. Ryther. 1977. Secondary sewage treatment versus ocean outfalls: An assessment. *Science 197:* 1056-1060.

Riley, G.A. 1972. Patterns of production in marine ecosystems, pp. 91-112. *In:* J.A. Viens (ed.), *Ecosystem Structure and Function.* U. Oregon Press, Corvallis, Oregon.

Ryther, J.H. and W.H. Dunstan. 1971. Nitrogen, phosphorus, and eutrophication in the coastal marine environment. *Science 171:*1008-1013.

Segar, D.A. and G.A. Berberian. 1976. Oxygen depletion in the New York Bight apex: causes and consequences, pp. 220-239. *In:* M.G. Gross (ed.), *The Middle Atlantic Shelf and the New York Bight.* Limnol. Oceanogr. Spec. Symp., 2.

Stoddard, A. 1983. Mathematical model of oxygen depletion in the New York Bight: an analysis of physical, biological, and chemical factors in 1975 and 1976. Ph.D. Dissertation, University of Washington, Seattle. 364 pp.

Thomas, J.P., W.C. Phoel, F.W. Steimle, J.E. O'Reilly and C.A. Evans. 1976. Seabed oxygen consumption - New York Bight apex, pp. 354-369. *In:* M.G. Gross (ed.) *Middle Atlantic Shelf and the New York Bight.* Limnol. Oceanogr. Spec. Symp., 2.

Thomas, W.H. 1970a. On nitrogen deficiency in tropical Pacific Ocean phytoplankton: photosynthetic parameters in poor and in rich water. *Limnol. Oceanogr. 15:*380-385.

Thomas, W.H. 1970b. Effect of ammonium and nitrate concentration on chlorophyll increases in natural tropical Pacific phytoplankton populations. *Limnol. Oceanogr. 15:*386-394.

Van Bennekom, A.J. and W. Salomons. 1981. Transfer of river-borne nutrients to the oceans, pp. 33-51. *In: River Inputs to Ocean Systems*. United Nations, New York.

Walsh, J.J. 1976. Herbivory as a factor in patterns of nutrient utilization in the sea. *Limnol. Oceanogr. 21*:1-13.

Walsh, J.J., G.T. Rowe, R.L. Iverson and C.P. McRoy. 1981. Biological export of shelf carbon is a sink of the global CO_2 cycle. *Nature 291*:196-201.

THE ESTUARY EXTENDED - A RECIPIENT-SYSTEM STUDY OF ESTUARINE OUTWELLING IN GEORGIA

Charles S. Hopkinson, Jr.
Frederick A. Hoffman

University of Georgia Marine Institute
Sapelo Island, Georgia

Abstract: Organic carbon budgets are presented for the marsh, estuarine water bodies and the inner portion of the nearshore region of Georgia. Budgets were determined by balancing external inputs and outputs from rivers and the mid-shelf against measurements of community production, metabolism and internal organic matter storage. On the marsh, primary production exceeds sedimentation and respiration by a factor of 2.6. The aquatic system is distinctly heterotrophic, with the primary production to community respiration ratio being 0.63:1. Our current best estimate is that the autotrophic marsh/estuarine system has 947 $g\ C \cdot m^{-2}$ potentially available for export. The nearshore system requires an input of 210 $g\ C \cdot m^{-2} \cdot yr^{-1}$ in addition to primary production to sustain its high rate of community production. Allochthonous organic inputs from rivers constitute 12% of total inputs to the marsh/estuarine/nearshore interface system. We calculate that roughly 50% of carbon inputs to the interface system are unaccounted for. About 10% of net primary production of the marsh must be exported to sustain heterotrophic levels in the estuarine aquatic system and the nearshore. We conclude that the nearshore region is a component of a coupled system and is strongly dependent on the adjacent marsh/estuary and river for a portion of its organic matter. As such, the nearshore can be considered as a functional extension of the estuary into the ocean. The estuary is a "filter" in the sense that materials received at the land-sea interface will be processed, transformed and exchanged within the marsh/estuary and the nearshore zones.

ISBN 0-12-405070-0

INTRODUCTION

By Pritchard's classical definition, estuaries are restricted to those bodies of water which are partially enclosed (Cameron and Pritchard 1963). There is no accommodation for coastal bodies of water which are mixtures of seawater and fresh water from land drainage but which are beyond the confines of emergent land masses. By this definition up to 100% of the brackish water associated with the Amazon and Mississippi Rivers is not estuarine. From a functional viewpoint, it may be desirable to extend the definition of estuarine boundaries to include more completely the interface system that couples continent to ocean. For the Georgia Bight this would include the nearshore region in addition to the classically defined estuary. The estuarine-nearshore complex would then be viewed as an interactive system whose internal dynamics were controlled largely by external forces.

Since the estuary is to be viewed as a receiving and "filtering" body in this symposium, we consider that the entire interface system should be examined. Here we present preliminary organic carbon budgets for the marsh and estuarine water bodies and for the inner portion of the nearshore region of the Georgia Bight. These budgets are then integrated in an attempt to examine the degree of coupling between the inner shelf and the classically defined estuary. The importance of the phenomenon of organic matter "outwelling" from southeastern estuaries to the adjacent continental shelf is assessed.

SITE DESCRIPTION

The Georgia coastal zone is centrally located in the most western portion of the Georgia Bight. The continental shelf is broad and gently sloping. Wave energy along the coast is quite low due to frictional damping across the shallow bottom, and tidal amplitude is the greatest in the southeastern U.S. Coastal barrier islands are located about 7 km offshore, with their long axes parallel to the coast. A band of marshes situated between the islands and the coastal plain exhibits a mixture of lagoonal and deltaic development (Frey and Basan 1978). These wetlands are approximately 80% intertidal and 20% aquatic in nature (Pomeroy and Wiegert 1981). The lagoonal marshes are dissected by several coastal plain and piedmont rivers of which the Altamaha and Savannah are the largest. Deltaic marshes extend upriver 20-30 km to the extent of tidal wave penetration. Inlets connecting marshes to the ocean are spaced roughly 15 km apart along the coast. The nearshore region of the coastal zone is the shallow-water (<15 m deep) portion of the continental shelf seaward of the barrier islands (Blanton and Atkinson 1978). Nearshore waters are highly turbid in a region of estuarine and riverine plumes which often extends 6-10 km offshore. Water further offshore is less turbid. A frontal zone 15-20 km offshore is a dynamic barrier to momentum and to the transport of dissolved and particulate material between the nearshore zone and the mid-shelf region (Blanton 1981).

The coastal zones discussed in this paper are identified as:

1. Marsh - intertidal wetland vegetated by emergent macrophytes.
2. Estuarine water bodies - sound, bays, and creeks. Items 1 and 2 are collectively called the marsh/ estuary.
3. Nearshore zone - 0 to 15-20 km offshore of barrier islands.
4. Estuarine plume region of the nearshore - turbid inner 6-10 km of the nearshore zone. Also called the inner nearshore.
5. Midshelf - region beyond the nearshore zone.
6. Coastal interface system - items 1-4 collectively.

Organic carbon budgets constructed in this paper are based on values from the literature for the Sapelo Island marsh/estuary and coastal plain rivers and from new, original measurements in the Altamaha River and the estuarine plume region of the nearshore. In the Altamaha River, samples were from the normal upstream extent of the tidal wave (~30 km). The river is approximately 200 m wide and average depth is 2 m at this point. The nearshore study site is 1.6 km off-shore from the center of Sapelo Island. Mean water depth is 4.6 m and mean tidal range is 2.1 m. More detailed descriptions of the site including sediment characteristics and faunal assemblages can be found in Hopkinson and Wetzel (1982), Leiper (1973), and Smith (1973).

METHODS

The conceptualization of organic carbon flow shown in Figure 1 was used as the framework for determining a preliminary carbon budget for the coastal interface system. The budget was determined by balancing external inputs and outputs from rivers and the mid-shelf against measurements of community production, metabolism and internal organic matter storage of the estuarine-nearshore complex. Only the inner portion of the estuarine plume region (0-3.2 km) was considered in this preliminary budget because metabolic measurements conducted 16 km offshore do not hold to the edge of the zone (Wetzel and Hopkinson, unpublished). The budget is conceptualized for an average 1 m wide band of the interface system extending perpendicularly from the coastal uplands to 3.2 km offshore.

Mid-Shelf

On the basis of Blanton's (1981) hydrodynamic analysis of the coastal current and frontal zone, exchange across the nearshore mid-shelf boundary is thought to be greatly inhibited. Although Turner, Woo and Jitts (1979) argued for

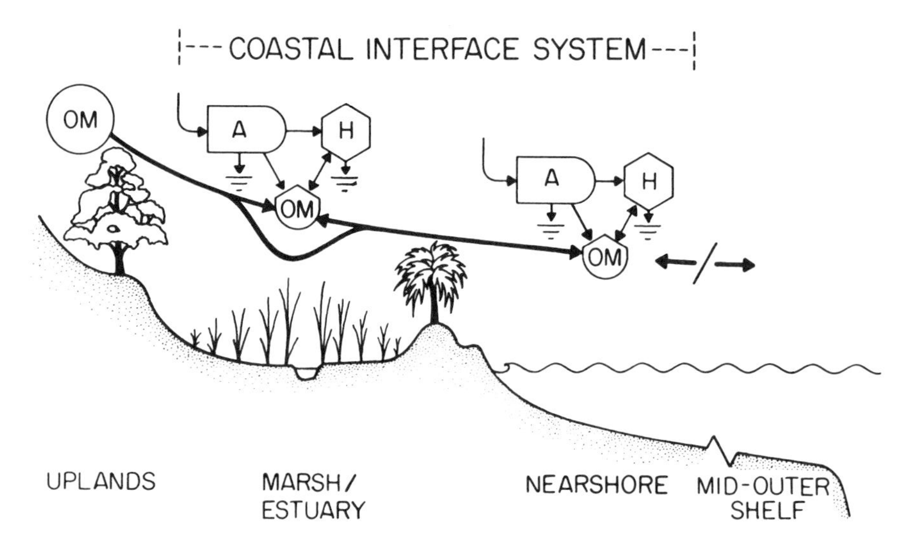

FIGURE 1. Conceptualization of organic carbon flux in the coastal interface system.

cross-shelf linkages, Bishop, Yoder and Paffenhofer (1980) observed no nearshore-midshelf coupling and we assume no significant transfer of organic material across this boundary.

Uplands

Terrigenous inputs of organic matter were calculated using the Altamaha River as a model piedmont stream and the data of Mulholland (1981) for coastal plain streams. Flow-weighted mean concentrations of particulate and dissolved organic carbon (POC and DOC, respectively) were multiplied by average discharge of piedmont and coastal plain rivers (J. L. Pearman, U.S.G.S., Doraville, GA, personal communication) in the Georgia Bight to calculate carbon loading. As a simplification we assumed an even distribution of river inputs along the coast. Loading was therefore divided by coastline length in the Georgia Bight to determine input per meter of coastline.

Carbon analyses were conducted on Altamaha River water sampled from mid-channel periodically from May 1982 until August 1983. Acid-washed glass bottles were filled at a depth of approximately 40 cm. The bottles were stored at 0°C pending processing within 3 hours. Fractionation of the sample was done by filtration through 1.1 μ (nominal retentivity) pre-combusted glass fiber filters (Gelman A/E). The filtrate was analyzed for DOC according to the method of Menzel and Vaccaro (1964) on an Oceanographic International Corp. apparatus. Particulate organic matter was determined by weight loss after ashing of filters at 450°C for 5 hours. Particulate organic carbon was considered to be 50% of the particulate organic matter.

Estuarine-Nearshore Interface

P/R ratios (primary production:community respiration) were calculated for wetland and water bodies of the marsh/estuary and for the inner nearshore region to evaluate the importance of organic matter export and allochthonous organic matter inputs relative to primary production. The autotrophic or heterotrophic nature of the marsh/estuary was calculated from literature reports. For the nearshore region it was determined by measuring and then balancing rates of community production with benthic and pelagic metabolism at a site 1.5 km offshore from Sapelo Island. The methodology used to measure the metabolic parameters is similar to that used by Hopkinson and Wetzel (1982). To summarize, short term measurements of benthic and water column uptake were made *in situ* in opaque acrylic hemispheres and *in vitro* in 20 l carboys, respectively, on a monthly basis over an annual period. Total carbon dioxide flux (Σ CO_2) was measured during three seasons to enable the determination of respiratory quotients (RQ) needed to convert oxygen-based respiration to carbon equivalents. The applicability of Thomas' (1966) measurements of primary production in an area adjacent to the 1.6 km site was determined with occasional measurements using standard ^{14}C techniques (Strickland and Parsons 1972).

RESULTS AND DISCUSSION

Carbon transported by rivers draining into the Georgia Bight is substantial (Table 1). Although carbon concentration is higher in coastal plain rivers, discharge is so much greater from piedmont streams that total input is predominated by the latter. Mulholland (1981) showed that there is little degradation or loss of carbon from the rivers during transit, hence most is discharged directly to coastal systems. We calculate that on average 1.8 metric tons of organic carbon are delivered annually to each m of coastline in the Georgia Bight. Mulholland (1981) found a high rate of POC formation when coastal plain river water and seawater were mixed, indicating that at least part of the coastal input could be considered as an external source of particulate organic carbon to the marsh/estuary. Lacking further measurements of the fate of carbon upon reaching coastal estuaries, it is uncertain whether carbon transported to the coast represents an input to the estuary (degradation, secondary production, or burial), a direct input to the nearshore region, or both.

The marsh/estuarine carbon budget (Table 2) indicates the overwhelmingly autotrophic nature of this portion of the interface system. On the marsh, primary production exceeds sedimentation and respiration by a factor of 2.6. However, creeks, rivers and sounds are distinctly heterotrophic (P/R = 0.63), indicating the great importance of allochthonous inputs. Incorporating hydrologic information from Imberger et al. (1983) into a salt marsh model, Wiegert, Christian and Wetzel (1981) estimated tidal export from the Duplin River marsh watershed to be 586 $g\ C \cdot m^{-2} \cdot yr^{-1}$. Our budget indicates an additional 361 $g\ C \cdot m^{-2} \cdot yr^{-1}$ must be accounted for which has not been observed to be deposited, degraded nor transported out of the estuary. It may be that metabolism has been underestimated for the marsh, as

TABLE 1. Riverine inputs of organic carbon to the Georgia Bight coastal zone.

Parameter	*Value*
Discharge[a]	
Coastal Plain	$1.0 \cdot 10^{10}$ m$^3 \cdot$ yr^{-3}
Piedmont	$4.8 \cdot 10^{10}$ m$^3 \cdot$ yr^{-3}
POC Concentration	
Coastal Plain[b]	0.7 g/m^3
Piedmont[c]	2.1 g/m^3
DOC Concentration	
Coastal Plain[d]	18.0 g/m^3
Piedmont[c]	9.3 g/m^3
Carbon flux to coast	$74.1 \cdot 10^{10}$ g $\cdot$ yr^{-1}
Shoreline length	400,000 m
Carbon flux per unit shoreline	1852 kg $\cdot$ m^{-1} $\cdot$ yr^{-1}

[a] *15-50 year averages (J. L. Pearman, U.S. Geological Survey, Doraville, GA, personal communication).*

[b] *Mulholland and Kuenzler (1979).*

[c] *This study.*

[d] *Mulholland (1981).*

R. Howarth (MBL, Woods Hole, MA, personal communication) has consistently observed rates of $SO_4^=$ reduction in Sapelo marshes which indicate that CO_2 fluxes used in Table 2 may be low by a factor of 2 (J. Hall and R. Christian, East Carolina Univ., Greenville, NC, personal communication). Thus our current best estimate is that the autotrophic marsh/estuarine system has between 586 and 947 g C $\cdot$ m^{-2} potentially available annually for export to the adjacent nearshore region.

In the nearshore environment, increased water clarity relative to the estuary, inputs of new nutrients and high rates of recycling contribute to high primary productivity (Table 3). However, the nearshore system requires an input of 210 g C $\cdot$ m^{-2} $\cdot$ yr^{-1} in addition to primary production to sustain its high rate of community respiration. During mid-summer and winter, P/R ratios are generally <0.6 and are <0.4 in December. It is only during late spring and mid-fall that the estuarine plume region of the nearshore approaches or becomes autotrophic (Hopkinson, personal observation). The annual P/R ratio of 0.72 for the nearshore region clearly indicates dependency on allochthonous carbon inputs from either terrigenous or marsh/estuarine sources, or both.

TABLE 2. Annual carbon budget for the Sapelo Island marsh/estuarine region.

Process	*Flux* ($g\ C \cdot m^{-2} \cdot yr^{-1}$)
Primary Production	
Aquatic* (gross)	326[a]
Marsh‡ (net)	2025[bc]
Community Respiration	
Aquatic*	520[d]
Marsh (excluding living macrophytes)‡	738[e]
Sedimentation or Storage†	29[f]
Export†	+586[f]
Export and Balance†	+947

[a] *E. Sherr, U. Georgia Marine Institute, Sapelo Island, GA.*

[b] *Includes phytoplankton production when marsh is flooded (E. Sherr, personal communication).*

[c] *Aerial production and relative area of creekbank versus high marsh (Gallagher et al. 1980); belowground production twice aerial production (Hopkinson and Schubauer 1984); factor of 0.45 to convert dry mass to carbon.*

[d] *Intertidal mudflat (Teal and Kanwisher 1961); planktonic respiration 0.64 g* $C \cdot m^{-2} \cdot d^{-1}$ *(Ragotzkie 1959).*

[e] CH_4 *(King and Wiebe 1978);* CO_2 *flux from bare mud (J. Hall and R. Christian, East Carolina Univ., Greenville, NC, personal communication); insects (Pomeroy and Wiegert 1981); standing dead Spartina alterniflora (Gallagher and Pfeiffer 1977); denitrification (Sherr 1977). Macrophyte respiration excluded because comparison is made with macrophyte net production which also excludes respiration.*

[f] *Pomeroy and Wiegert (1981).*

**Units -* $g\ C \cdot m^{-2}$ *aquatic area* $\cdot\ yr^{-1}$ *(21% of total).*

‡Units - $g\ C \cdot m^{-2}$ *marsh* $\cdot\ yr^{-1}$ *(79% of total).*

†Units - $g\ C \cdot m^{-2}$ *marsh and aquatic area* $\cdot\ yr^{-1}$*.*

TABLE 3. Carbon budget for the estuarine plume region of the nearshore Georgia bight.

Process	*Flux* $(g\ C \cdot m^{-2} \cdot yr^{-1})$
Primary Production[a]	
Pelagic (gross)	539
Community Respiration	
Pelagic	409
Benthic	340
Sedimentation	0
Balance	-210

[a]*High turbidity causes light penetration to be insufficient for support of benthic primary production.*

By integrating metabolic parameters of the coastal interface system for a zone extending from the landward edge of the marsh/estuarine region to 3.2 km offshore, we can evaluate the relative importance of each of the subsystems (Table 4). Total fixed carbon inputs for the interface system are 15254 kg $C \cdot m^{-1} \cdot yr^{-1}$. Allochthonous inputs from the river contribute 12% of this total. Of the locally fixed carbon, the marsh overwhelmingly dominates, supplying 85% of all organic matter produced. Also, we see that the nearshore aquatic community fixes 3.6 times more organic carbon than the comparable aquatic community in the bays, rivers and tidal creeks of the estuary.

The salt marsh is the site of the greatest respiratory activity. Over half of all the carbon degraded in the interface system is degraded on the marsh. Slightly more than three times as much organic matter is remineralized offshore than in the estuarine aquatic system.

Of the three interface subsystems (the marsh, estuarine water bodies, and estuarine plume nearshore region), only the marsh is autotrophic. Present best estimates suggest that there is about 2.6 times more C fixed in photosynthesis than there is consumed in respiration and lost in sedimentation. The budget suggests that the excess marsh production is exported not only to the adjacent estuarine water bodies but also to the inner nearshore zone where it, in addition to allochthonous riverine inputs, sustains a level of metabolism that is unattainable on the basis of pelagic primary productivity alone.

TABLE 4. Carbon budget for the coastal interface system in the Georgia Bight.

Region	*Production*	*Remineralization* ($kg\ C \cdot m^{-1} \cdot yr^{-1}$)	*Balance*
Terrigenous Input[a]	1,852		+1,852
Marsh/Estuary			
Marsh[b]	11,198	4,081	
Aquatic[c]	479	764	
Total[e]	11,677	5,048	+6,629
Nearshore[d]	1,725	2,397	- 672
Sum	15,254	7,445	+7,815

[a] *From Table 1.*

[b] *From Table 2; 7000* m^2 *marsh/estuarine zone (Pomeroy and Wiegert 1981); 79% of total zone is marsh.*

[c] *From Table 2 and footnote b, Table 4.*

[d] *From Table 3; 3200 m wide zone.*

[e] *Including burial in marsh/estuary.*

Organic carbon export to the ocean is even more demonstrably evident when the balance between production and remineralization is compared for the nearshore region, the marsh/estuary, and riverine organic inputs (Table 4). With this analysis, the marsh/estuary as a whole appears to be an autotrophic subsystem. In contrast, the estuarine plume region of the nearshore is markedly heterotrophic and is dependent on the marsh/estuary and river for 28% of its metabolic requirement for carbon. In balance, organic matter excess of the marsh/estuary exceeds the organic input from rivers by a factor of 3.6. The heterotrophic requirements of the nearshore are not sufficient to account for all of the excess carbon. As mentioned earlier, sediment respiration in the salt marsh may be underestimated but even after incorporating the higher rate of Howarth (personal communication), there is still an unaccounted for excess of 7000 kg $C \cdot m^{-1} \cdot yr^{-1}$. One possible scenario to explain the fate of the excess involves the export of the full 8481 kg $C \cdot m^{-1} \cdot yr^{-1}$ to the nearshore and beyond to the midshelf. The nearshore zone extends well beyond the first 3.2 km of the estuarine plume region. Wetzel and Hopkinson (personal observations) have measured an overall level of net heterotrophy extending to at least 10 km during summer. Certainly some of the excess will be required to sustain metabolic activity through 10 km and perhaps to the frontal boundary of the nearshore at about 20 km offshore. Sotille (1973) showed that the majority of DOC in estuarine water is resistant to decay. It is quite likely that labile DOC (about 10% of total) which is exported from the river and estuary is degraded in the nearshore; the remainder would diffuse across the frontal zone to be mixed into midshelf water. This hypothesis is supported by Gardner and Stephens' (1978) report of a stable distribution of terrestrially derived refractory organic compounds to 100 km off the Georgia coast.

The outwelling hypothesis (Odum 1968) is a deeply entrenched paradigm for the operation of marsh/estuarine systems along the U.S. Gulf and Atlantic coasts. However, persistent controversy surrounds the reality of the phenomenon (e.g., see Nixon 1980 and Haines 1979b). Although the present results do not unequivocably prove an export of *Spartina* carbon from marshes to the ocean, they, in conjunction with the results of marsh/estuarine metabolic studies (see Pomeroy and Wiegert 1981), strongly suggest it. We can make some estimates of the amount of carbon required to be exported from each m^2 of marsh based on the relative potential availability of organic carbon from the river and the marsh. From Table 4, for each m of coastline (7 km x 1 m band), 6957 and 1852 kg $C \cdot yr^{-1}$ are potentially available for export from the marsh and river, respectively. We then assume that the level of heterotrophy in excess of local production is sustained by the marsh and river at the ratio of 3.8:1 (6957:1852), respectively. Again from Table 4, 328 and 672 kg C $\cdot$ yr^{-1} are respired and sedimented in estuarine water bodies and the inner nearshore region, respectively, in excess of local production. To sustain this level of net heterotrophy, 181 g C, or about 9% of net primary production, must be exported from an average square meter of marsh surface annually. This relatively minor amount of marsh-derived carbon may partially explain the paradox of low $\delta^{13}C$ values for suspended carbon in the Duplin River (Haines 1977) in the face of marsh export. In terms of the export of marsh detritus to the nearshore, ≈5% of net marsh production is required to sustain observed levels of metabolism offshore.

The combination of independently conducted surveys of community production and metabolism in several adjacent coastal environments presented here reveals information about functional interaction between these subsystems, although one must exercise caution in interpreting these data because of possible random and systematic errors. Estimates of community metabolism on the marsh and in the estuarine water bodies are for the most part based on the summation of the respiration of individual populations or community components. There is some uncertainty as to the inclusiveness of the components. Needed are additional whole community measurements of metabolism similar to those conducted on the marsh in spring (Teal and Kanwisher 1961) or in the nearshore region. On the basis of the present carbon budget, we conclude that the estuarine plume region of the nearshore zone constitutes a coupled system strongly dependent on either the adjacent marsh/estuary or the river, or both, for a portion of its organic matter requirements. As such, the nearshore region can be considered as a functional extension of the estuary into the ocean.

The similarity between a number of descriptive parameters illustrates that whereas the inner nearshore is coupled to the estuary, it is strongly separated from the midshelf (Table 5). From the functional point of view, the midshelf is metabolically balanced and has a low rate of production. In contrast, the inner nearshore and estuarine water bodies are very active metabolically with a pronounced seasonality and net heterotrophy. Whereas nutrient concentrations, POC, and chlorophyll levels are very low on the midshelf, they are quite high inshore. The low salinity and high turbidity in the estuary and inner nearshore strongly contrast with the high salinity and clarity of the midshelf. Sediments along the coast also strongly contrast with those on the midshelf. Pilkey and Frankenberg (1964) showed that Recent sediments

TABLE 5. A comparison of parameters characterizing adjacent marine environments.

System Parameter	Value		
	Estuarine Water Body	*Inner Nearshore*	*Midshelf*
Primary production ($g\ C \cdot m^{-2} \cdot yr^{-1}$)[a]	350	550	125
Community metabolism ($g\ C \cdot m^{-2} \cdot yr^{-1}$)[b]	781	749	125
Nutrient concentrations (inorganic N-μM)[c]	<10	<2	<0.1
Water clarity (extinction coef)[d]	1-3	1-2	0.1
Salinity (ppt)[e]	8-35	15-35	>35
Chlorophyll *a* (mg/m^3)[f]	1-8	2-8	<1
Particulate organic carbon (mg/l)[g]	4-70	4-70	<2
Sediment characteristics (size)[h]	fines	fines	coarse
Seasonal patterns[i]	yes	yes	no

[a]*E. Sherr et al., U. Georgia Marine Institute (personal communication); Thomas (1966); Haines and Dunstan (1975).* [b]*Pomeroy and Wiegert (1981); this study; assuming balanced P-R ratio for midshelf.* [c]*Haines (1979a,b); Bishop et al. (1980).* [d]*Haines (1979b); Oertel and Dunstan (1981).* [e]*Haines (1979a); Blanton (1981); Bishop et al. (1980).* [f]*Sherr et al. (personal communication); Bishop et al. (1980); Haines (1979b).* [g]*Oertel and Dunstan (1981); Manheim, Meade and Bond (1970); Haines and Dunstan (1975).* [h]*Pilkey and Frankenberg (1964).* [i]*Pomeroy and Wiegert (1981); Bishop et al. (1980); Thomas (1966).*

which are predominantly fine-grained sands with some silts and clays are found in the marsh/estuary and out to a depth of about 11 m (≃10 km offshore). Beyond the sharp boundary at 11 m, medium to coarse Pleistocene sands are found.

CONCLUSIONS

Several lines of evidence ranging from abiotic sediment characteristics to rates and patterns of community metabolism and production indicate functional and structural similarities between the marsh/estuary, especially the aquatic portion, and the inner nearshore region of the Georgia Bight. It is only by definition that the nearshore zone is separate from the estuary. Clearly the nearshore is a functional component of the interface system that couples the terrestrial and oceanic environments.

With regard to the concept of the estuary as a filter, we can conclude that signals received at the land-sea interface will be processed, transformed, and exchanged between the marsh/estuary and the nearshore zone. From the overall carbon budget, it is evident that materials coming in to the marsh and taken up by marsh macrophytes will be incorporated into organic matter and transported to the adjacent bays and nearshore region by the ebb and flood of tides. There the material may be incorporated into pelagic food webs. The high degree of coupling between the marsh/estuary and nearshore region must be considered in management decisions concerning the coastal zone.

ACKNOWLEDGMENTS

We sincerely thank Lorene Gassert, Mary Robertson and Sarita Marland for expeditious help with drafting and manuscript preparation. We acknowledge the helpful manuscript reviews of Steve Newell and Alice Chalmers. We appreciate the help of several individuals who assisted under water with the benthic work in the ocean, especially Mary Robertson and Bob Fallon.

This is Contribution No. 503 from the University of Georgia Marine Institute. This work was supported by the Georgia Sea Grant College Program, part of the National Sea Grant College Program maintained by the National Oceanic and Atmospheric Administration, U.S. Department of Commerce.

REFERENCES CITED

Bishop, S. S., J. A. Yoder and G. A. Paffenhofer. 1980. Phytoplankton and nutrient variability along a cross-shelf transect off Savannah, Georgia, U.S.A. *Est. Coast. Mar. Sci. 11:*359-368.

Blanton, J. O. 1981. Ocean currents along a nearshore frontal zone on the continental shelf of the southeastern United States. *J. Phys. Oceanogr. 11:*1627-1637.

Blanton, J. O. and L. Atkinson. 1978. Physical transfer processes between Georgia tidal inlets and nearshore waters, pp. 515-532. *In:* M. L. Wiley (ed.), *Estuarine Interactions*. Academic Press, New York.

Cameron, W. M. and D. W. Pritchard. 1963. Estuaries, pp. 306-324. *In:* M. N. Hill (ed.), *The Sea. Volume 2.* John Wiley and Sons, New York.

Frey, R. W. and P. Basan. 1978. Coastal salt marshes, pp. 101-169. *In:* R. A. Davis (ed.), *Coastal Sedimentary Environments*. Springer-Verlag, New York.

Gallagher, J. L. and W. J. Pfeiffer. 1977. Aquatic metabolism of the standing dead plant communities in salt and brackish water marshes. *Limnol. Oceanogr. 22:*562-564.

Gallagher, J. L., R. Reimold, R. Linthurst and W. Pfeiffer. 1980. Aerial production, mortality, and mineral accumulation-export dynamics in *Spartina alterniflora* and *Juncus roemerianus* plant stands in a Georgia salt marsh. *Ecology 61:*303-312.

Gardner, W. S. and J. Stephens. 1978. Stability and composition of terrestrially derived dissolved organic nitrogen in continental shelf surface waters. *Marine Chem. 6:*335-342.

Haines, E. B. 1977. The origins of detritus in Georgia salt marsh estuaries. *Oikos 29:*254-260.

Haines, E. B. 1979a. Nitrogen pools in Georgia coastal waters. *Estuaries 2:*34-39.

Haines, E. B. 1979b. Interactions between Georgia salt marshes and coastal waters: a changing paradigm, pp. 35-47. *In:* R. J. Livingston (ed.), *Ecological Processes in Coastal and Marine Systems*. Plenum Press, New York.

Haines, E. B. and W. N. Dunstan. 1975. The distribution and relation of particulate organic material and primary productivity in the Georgia Bight, 1973-1974. *Est. Coast. Mar. Sci. 3:*431-441.

Hopkinson, C. S. and J. P. Schubauer. 1984. Static and dynamic aspects of nitrogen cycling in the salt marsh graminoid, *Spartina alterniflora*. *Ecology*. (in press).

Hopkinson, C. S. and R. L. Wetzel. 1982. *In situ* measurements of nutrient and oxygen fluxes in a coastal marine benthic community. *Mar. Ecol. Prog. Ser. 10:*29-35.

Imberger, J., T. Berman, R. Christian, E. Sherr, D. Whitney, L. Pomeroy, R. Wiegert and W. Wiebe. 1983. The influence of water motion on the distribution and transport of materials in a salt marsh estuary. *Limnol. Oceanogr. 28:*201-214.

King, G. M. and W. J. Wiebe. 1978. Methane release from soils of a Georgia salt marsh. *Geochim. Cosmochim. Acta 42:* 343-348.

Leiper, A. S. 1973. Seasonal change in the structure of three sublittoral marine benthic communities off Sapelo Island, Georgia. Ph.D. Thesis. Univ. Georgia, Athens.

Manheim, F. T., R. H. Meade and G. Bond. 1970. Suspended matter in surface waters of the Atlantic continental margin from Cape Cod to Florida Keys. *Science 167:*371-376.

Menzel, D. W. and R. F. Vaccaro. 1964. The measurement of dissolved organic and particulate organic carbon in seawater. *Limnol. Oceanogr. 9:*138-142.

Mulholland, P. J. 1981. Formation of particulate organic carbon in water from a southeastern swamp-stream. *Limnol. Oceanogr. 26:*790-795.

Mulholland, P. J. and E. Kuenzler. 1979. Organic carbon export from upland and forested wetland watersheds. *Limnol. Oceanogr. 24:*960-966.

Nixon, S. W. 1980. Between coastal marshes and coastal waters -- a review of twenty years of speculation and research on the role of salt marshes in estuarine productivity and water chemistry, pp. 437-525. *In:* P. Hamilton and H. MacDonald (eds.), *Estuarine and Wetland Processes*. Plenum Press, New York.

Odum, E. P. 1968. A research challenge: evaluating the productivity of coastal and estuarine water, pp. 63-64. *In:* Proc. 2nd Sea Grant Conf., Grad. School of Oceanography., Univ. Rhode Island, Kingston.

Oertel, G. F. and W. N. Dunstan. 1981. Suspended sediment distribution and certain aspects of phytoplankton production off Georgia, U.S.A. *Marine Geol. 40*:171-197.

Pilkey, O. H. and D. Frankenberg. 1964. The relict-recent sediment boundary on the Georgia continental shelf. *Bull. Ga. Acad. Sci. 22*:37-40.

Pomeroy, L. R. and R. G. Wiegert (eds.). 1981. *The Ecology of a Salt Marsh*. Springer-Verlag, New York. 271 p.

Pomeroy, L. R., L. Shenton, R. Jones and R. Reimold. 1972. Nutrient flux in estuaries, pp. 274-291. *In:* G. E. Liken (ed.), *Nutrients and Eutrophication*. Amer. Soc. Limnol. Oceanogr. Spec. Symp.

Ragotzkie, R. A. 1959. Plankton productivity in estuarine waters of Georgia. *Publ. Inst. Mar. Sci. Univ. Texas 6*:146-158.

Sherr, B. F. 1977. The ecology of denitrifying bacteria in salt marsh soils -- an experimental approach. Ph.D. Thesis, Univ. Georgia, Athens. 154 p.

Smith, K. L. 1973. Respiration of a sublittoral community. *Ecology 54*:1065-1075.

Sotille, W. S. 1973. Studies of microbial production and utilization of dissolved organic carbon in a Georgia salt marsh-estuarine ecosystem. Ph.D. Thesis, Univ. Georgia, Athens. 153 p.

Strickland, J. and T. Parsons. 1972. A practical handbook of seawater analysis. *Bull. Fish. Res. Bd. Canada 167 (3rd edition)* Ottawa. 311 p.

Teal, J. M. and J. Kanwisher. 1961. Gas exchange in a Georgia salt marsh. *Limnol. Oceanogr. 6:*388-399.

Thomas, J. P. 1966. The influence of the Altamaha River on primary production beyond the mouth of the river. M.S. Thesis, Univ. Georgia, Athens. 88 p.

Turner, R. E., S. Woo and H. Jitts. 1979. Estuarine influences on a continental shelf plankton community. *Science 206:*218-220.

Wiegert, R. G., R. R. Christian and R. L. Wetzel. 1981. A model view of the marsh, pp. 183-218. *In:* L. Pomeroy and R. Wiegert (eds.), *The Ecology of a Salt Marsh.* Springer-Verlag, New York.

SURFACE FOAM CHEMISTRY AND PRODUCTIVITY IN THE DUCKABUSH RIVER ESTUARY, PUGET SOUND, WASHINGTON

Robert C. Wissmar
Charles A. Simenstad

Fisheries Research Institute
College of Ocean and Fishery Sciences
University of Washington
Seattle, Washington

Abstract: Chemical and biotic constituents of surface foam were compared in estuarine and neritic waters of Puget Sound, WA. The role of surface foam formation as a major "filtering" medium in an estuary was indicated by entrainment and concentration of constituents in foam several times higher than that of subsurface and neritic waters. Stable carbon isotopic ratios ($\delta^{13}C$) and organic solute fractions (hydrophobic and hydrophilic) indicated that organic matter in foams originated from dissolved organic carbon compounds released from seagrasses and benthic algae. Concentrations of particulate organic carbon and nitrogen, chlorophyll *a*, adenosine triphosphate, and bacteria counts showed enrichment in foams due to entrained sediment particles, algae, and bacteria. Inorganic nutrient concentrations, primary production and respiration rates of estuarine foam were also several-fold greater than neritic waters. The high levels of organic and inorganic chemical consituents, and metabolic rates of the microbial community indicate that surface foams are important interfaces for cycling of nutrients, formation of detrital matter, and production of microbial forage for consumers.

ISBN 0-12-405070-0

INTRODUCTION

Surface foam can be seen floating in shallow waters of eelgrass beds and littoral flats in Puget Sound during the growing season. Surface foams form at the leading edge of the early flood tide intrusion where turbid flow conditions can resuspend and entrain inorganic and organic materials (Fig. 1). Important components of surface foams may be dissolved (i.e., extracellular) and particulate organic carbon (DOC and POC) originating from autotrophes in estuaries (Percival and McDowell 1967; Sellner 1981; Mann 1982).

Extracellular release of dissolved organic compounds by metabolically active estuarine primary producers can be naturally high (Turner 1974; Penhale and Smith 1977; Faust and Chrost 1981; Sellner 1981; Pregnall 1983) and intensified during stress due to desiccation and physical disruption (Velimirov 1980; Pregnall 1983). Potential sources of POC include particle formation due to flocculation of DOC at freshwater/seawater interfaces (Sholkovitz 1976; Morris et al. 1978; Mulholland 1981; Jensen and Søndergaard 1982), interactions of surface-active agents and air bubbles (MacIntyre 1974; Pellenbarg and Church 1979; Wallace and Duce 1978; Avnimelech, Taylor and Reed 1982; Durako, Medlyn and Moffler 1982; Hunter

FIGURE 1. Surface foam at leading edge of flood tide intrusion (July) in Duckabush River estuary, Puget Sound, WA.

and Liss 1982), and fragmentation of eelgrass, algae, and marsh plants (Mann 1982).

This paper reports on short-term pilot experiments that compare chemical and biological constituents of surface foam in estuarine and neritic waters of Puget Sound, WA. The purpose was to obtain an initial understanding of the composition, metabolic activity, and sources of matter and biota in surface foam.

METHODS

Surface foam and subsurface water were studied during July 1982 in eelgrass (*Zostera marina*) and macroalgal habitats in the Duckabush River estuary and adjacent neritic waters of Hood Canal, a fjord in Puget Sound, WA. The eelgrass habitat was located at the outer margin of the littoral flat and the macroalgal habitat was located higher on the flat, approximately 125 m from the river channel. The principal macroalgae were *Ulva lactuca*, *U. expansa* and *Enteromorpha linza*. Surface and subsurface neritic waters immediately adjacent to the estuary were sampled during ebb slack tide, prior to sampling of the eelgrass and macroalgae habitats as the foam front flooded across the littoral flat. Foam was sampled by suction on the water surface and subsampled by decanting following the collapse of foam to liquid form. Subsurface waters, 0.2 m in the eelgrass and macroalgal habitats and 2 m in the neritic, were sampled with a battery powered pump and a 3 L Van Dorn bottle. Temperatures (°C), salinities (°/oo), and dissolved oxygen (mg L^{-1} O_2) were measured with a Yellow Springs Instrument (YSI) meter. NH_4^+-N, NO_2^--N, NO_3^--N, and PO_4^{-2}-P were measured on an Autoanalyzer II and total nitrogen (ΣN) and phosphorus (ΣP) were estimated after Valderrama (1981). Dissolved inorganic nitrogen, DIN = [NH_4^+-N] + [NO_2^--N] + [NO_3^--N]. POC and particulate organic nitrogen (PON) were measured as described by Wissmar et al. (1981), and DOC determined after Menzel and Vaccaro (1964). To characterize dissolved organic materials further, DOC was fractionated into hydrophobic and hydrophilic solutes (Huffman Laboratories, Denver, CO; Leenheer and Huffman 1976).

The natural stable carbon isotopic composition ($\delta^{13}C$) of organic carbon in foam was used to assess the autotrophic origins of carbon. Assessments were made by comparisons with a $\delta^{13}C$ data base from Simenstad and Wissmar (unpubl. data). POC $\delta^{13}C$ samples were filtered and acid-washed (5% HCl) on pre-combusted glass-fiber filters (<1.0 μm; Strickland and Parsons 1972). Filters were wrapped in pre-combusted aluminum foil and placed in polypropylene vials with desiccant.

DOC $\delta^{13}C$ samples (the filtrates of POC samples) were placed in acid-washed, pre-soaked polypropylene bottles, fixed with mercuric chloride, and stored under refrigeration prior to analysis. POC and DOC $\delta^{13}C$ analysis was performed by Coastal Science Laboratories, Inc., Austin, TX.

Adenosine triphosphate (ATP) and chlorophyll *a* measurements were after Holm-Hansen and Booth (1966) and Strickland and Parsons (1972), respectively. Bacterial counts were made using the epifluorescent procedures of Hobbie, Daley and Jaspers (1977). Phytoplankton were identified by Dennis Kunkel, University of Washington, Seattle, WA, using a Lietz inverted microscope.

Primary production and total respiration of the plankton community were estimated using changes in dissolved oxygen concentrations (Strickland and Parsons 1972) in a shaking water bath at 20°C under controlled light (130 μEinst. $s^{-1}m^{-2}$). Time series samples were taken at 2 to 6 hour intervals during 15 hour incubation periods. Prior to filling light-dark 300 ml BOD bottles, homogeneous foam concentrations (collapsed foam) and undersaturated O_2 levels were obtained by mixing and passing N_2 through a 20 L carboy. Oxygen concentrations were determined colorimetrically after Broenkow and Cline (1969). Oxygen was converted to carbon equivalents by assuming 1 ml of O_2 equals 546 μg C and photosynthetic and respiratory quotients equal to 1.0 (Johnson, Burney and Sieburth 1981).

Possible surface foam enrichments in chemical and biological matter were examined using enrichment ratios, i.e., the concentration of constituents in foam relative to concentrations in subsurface waters.

RESULTS AND DISCUSSION

Chemical Characteristics of Neritic and Estuarine Waters

Surface and subsurface samples had relatively high concentrations of O_2 (9.2 to 13.5 mg L^{-1}) in all habitats (Table 1). Salinities (31.0°/oo) and temperatures (28.1°C) were highest over the eelgrass beds, reflecting influences of air exposure and solar heating during ebb tide, while lower salinities (13.5°/oo) and temperatures (20.5°C) were evident in the more riverine macroalgal habitat.

Ammonium comprised >94% of the total DIN (1.6 to 14.5 μM; Table 1) in all waters. The principal differences among the neritic and estuarine habitats were the higher concentrations (2X to 66X) of PON, ΣN, PO_4^{-2}-P, and ΣP in both estuarine habitats. The eelgrass foam had the highest concentrations

TABLE 1. Eelgrass and macroalgal samples from Duckabush River estuary and neritic samples from Hood Canal, Puget Sound, waters. Subsurface depths were -0.2 m for eelgrass and macroalgal sites and -2 m for neritic. Enrichments calculated as ratio of surface foam to subsurface concentration where >1.0 indicates enrichment and <1.0 is depletion.

Waters	NH_4^+-N	NO_2^--N	NO_3^--N	DIN[a]	PON[b]	ΣN	PO_4^{-2}-P	ΣP	O_2	Sal	Temp
	(μM)								(mg L^{-1})	(‰)	(°C)
NERITIC											
Foam	7.3	0.0	0.1	7.4	4.8	16.0	0.4	2.2	9.2	26.2	17.5
Subsurface	8.2	0.0	0.1	8.3	5.6	18.5	0.6	3.5	10.3	26.0	16.5
Enrichment:	0.9	1.0	1.0	0.9	0.9	0.9	0.7	0.6	-	-	-
EELGRASS											
Foam	11.9	0.0	0.1	12.0	315.0	323.4	2.0	34.5	13.5	31.0	27.8
Subsurface	5.1	0.0	0.1	5.2	40.7	37.1	1.1	4.3	13.5	30.5	28.1
Enrichment:	2.3	1.0	1.0	2.3	7.7	8.7	1.8	8.0	-	-	-
MACROALGAE											
Foam	13.7	0.0	0.8	14.5	177.9	124.4	1.8	21.0	9.5	13.5	20.5
Subsurface	1.5	0.0	0.1	1.6	12.1	12.9	1.5	1.5	9.6	13.5	20.4
Enrichment:	9.1	1.0	8.0	9.1	14.7	9.6	1.2	14.0	-	-	-

[a] *DIN = total dissolved inorganic nitrogen (NH_4^+-N) + NO_2^--N) + (NO_3^--N).*
[b] *Means of two to three samples; standard errors range from 4% (surface) to 36% (foam).*

of PON (315.0 μM), ΣN (323.4 μM), ΣP (34.5 μM) and PO_4^{-2}-P (2.0 μM). High enrichment ratios for NO_3^--N (8.0), NH_4^+-N (2.3 to 9.1), PON (7.7 to 14.7), ΣN (8.7 to 9.6), and ΣP (8.0 to 14.0) in estuarine foams indicate possible influences of turbulent mixing, chemical and biological reactions. In contrast, enrichment ratios for neritic samples were low (0.6 to 1.0), indicating depletion for most nutrient and particulate constituents.

The highest concentrations of DOC (3867 μM) and POC (2808 μM) were found in the eelgrass foam and were 4 and 30X higher than in neritic samples, respectively (Table 2). Maximum enrichment of POC was 9.7 in the macroalgae foam and 3.6 for DOC in the eelgrass foam. There was no enrichment in neritic waters.

Enrichment ratios observed in Tables 1 and 2 are similar to values reported for organic (Williams 1967; Nishizawa 1971; Sieburth et al. 1976; Carlson 1982, 1983) and inorganic (Williams 1967; Nishizawa 1971) enrichments in surface microlayers of estuarine, coastal, and oceanic waters.

Stable carbon isotope compositions of DOC and POC showed the highest $\delta^{13}C$-depletion (-29.6°/oo) in the subsurface DOC of the macroalgae habitat (Table 2), similar to $\delta^{13}C$ values for associated salt marsh vegetation (-24.2 to -31.0°/oo; Simenstad and Wissmar, unpubl. data). Subsurface and foam POC in the same habitat had $\delta^{13}C$ values -20.4°/oo and -19.5°/oo, respectively, corresponding with $\delta^{13}C$ of macroalgae carbon sources. In contrast, isotopic values for eelgrass foam and subsurface POC and DOC were enriched (-12.5°/oo and -16.1°/oo), indicating both eelgrass and macroalgae carbon sources (-9.1°/oo to -21.9°/oo; Simenstad and Wissmar, unpubl. data). Neritic POC and DOC were more depleted (-20.5°/oo to -23.7°/oo), similar to phytoplankton carbon sources (-20.2°/oo to -25.8°/oo; Simenstad and Wissmar, unpubl. data). An exception, the enriched DOC in neritic surface waters (-14.6°/oo), suggested import from eelgrass beds.

Separation of DOC into hydrophilic and hydrophobic solutes (Table 2) indicated that water soluble, hydrophilic, organic carbon compounds predominated (52% to 94%). Such hydrophilic solutes may include various low molecular weight amino acids, polysaccharides, and organic acids (i.e., hydroxy and fatty acids) which could be readily utilized by heterotrophic microorganisms (McKnight et al. 1982). Similar dissolved organics such as proteinaceous and lipid residues from exudates of seagrasses, macroalgae, and kelp and associated surface active agents have been implicated in the formation of marine foam (Percival and McDowell 1967; Velimirov 1980). The importance of hydrophilic compounds in foam entrainment of particles may involve their capacities for hydrogen

TABLE 2. Concentrations of dissolved (DOC) and particulate organic carbon (POC), respectively; characterization of DOC and POC by stable carbon isotope values ($\delta^{13}C$) and DOC by fractionation into hydrophilic and hydrophobic solutes (%). Waters and enrichment calculations as in Table 1.

Waters	*DOC*	*POC*	*DOC*	*POC*	*Hydrophobics (%)*				*Hydrophilic (%)*
	(μM)		*($\delta^{13}C$)*		*Acids*	*Neutrals*	*Bases*	*Total*	*Total*
NERITIC									
Foam	1041[a]	99[b]	-14.6	-20.5	0	18	0	18	82
Subsurface	1010	106	-23.7	-21.2	5	43	0	48	52
Enrichment:	1.0	0.9							
EELGRASS									
Foam	3867	2808	-15.9	-15.1	2	36	0	38	62
Subsurface	1085	364	-16.1	-12.5	9	29	0	38	62
Enrichment:	3.6	7.2							
MACROALGAE									
Foam	1839	1041	M[c]	-19.5	10	37	0	47	53
Subsurface	1210	107	-29.6	-20.4	2	4	0	6	94
Enrichment:	1.5	9.7							

[a]*Means of two to three samples; standard errors range from 8% (neritic and subsurface) to 27% (foams).*

[b]*Means of two to three samples; standard errors range from 1% (neritic and subsurface) to 22% (foams).*

[c]*M denotes missing data due to analytical difficulties.*

bonding and ion-exchange reactions with sediments (Leenheer and Huffman 1976).

Hydrophobic organic solutes comprised between 6% and 48% of the DOC in our samples (Table 2). The highest values occurred in neritic subsurface (48%), macroalgae foam (47%), and in foam and subsurface waters in the eelgrass habitat (38%). The neutral group, presumably various hydrocarbons (Leenheer and Huffman 1976), was most prevalent. The remaining fractions were in the acid group, which contains high molecular weight organic acids similar to fulvic acids (Stuber and Leenheer 1978).

Biological Characteristics of Neritic and Estuarine Waters

The highest chlorophyll *a*, bacterial cell, and ATP concentrations were in the eelgrass and macroalgae foams (Table 3). Concentrations of chlorophyll *a* (80.0 and 216.6 μg L^{-1}), bacteria cells (16.92 and 21.63 X 10^6 cells ml^{-1}), and ATP (202 μg L^{-1}) were respectively 127X, 5X, and 6X greater in the estuarine than in neritic habitats. Enrichment ratios for the estuary samples were also high. Substantial ATP concentrations

TABLE 3. Concentrations of chlorophyll *a*, bacterial cells, and adenosine triphosphate. Waters and enrichment calculations as in Table 1. M = missing data.

Waters	*Chlorophyll a (ug L^{-1})*	*Bacterial cells (10^6·cell ml^{-1})*	*ATP (μg L^{-1})*
NERITIC			
Foam	1.7	4.30	35
Subsurface	1.8	4.31	22
Enrichment:	0.9	1.0	1.6
EELGRASS			
Foam	216.6	21.63	202
Subsurface	16.1	3.89	54
Enrichment:	13.5	5.6	3.7
MACROALGAE			
Foam	80.0	16.92	M
Subsurface	4.7	1.85	47
Enrichment:	17.0	9.2	-

in all waters reflected high biomasses of phytoplankton, bacteria, and microzooplankton (<100 μm size fraction).

Concentrations and enrichment ratios for biological characteristics (Table 3) compare with ATP and chlorophyll *a* values from surface microlayers of the North Atlantic Ocean and Maine coastal waters (0.0 to 21.5 μg L^{-1} ATP, 0.02 to 7.3 enrichment; 0.0 to 1020 μg L^{-1} chlorophyll *a*, 0.1 to 192.4 enrichment; Sieburth et al. 1976; Carlson 1982) and for bacterial cell counts in coastal surface foams (1.3 X 10^5 to 1.2 X 10^6 cells ml^{-1}; Velimirov 1980).

The high chlorophyll *a* concentrations and microalgae taxa composition in the samples suggest that algae were an important carbon component of the surface foams (Tables 3 and 4). Eelgrass and macroalgae foams and subsurface waters had the greatest variety of species (Table 4). The most common genus in surface foams at the eelgrass site was the filamentous blue green *Cylindrospermum* sp. and at the macroalgae site, the diatom *Melosira monoliformis*. Taxa occurring exclusively in the eelgrass foam were *Asteromorphalus* sp., *Coscinodiscus* sp., and *Cylindrospermum* sp. and in the eelgrass subsurface water were *Amphiprora* sp., *Coscinodiscus cosinus*, and *Rhozoselenia* sp. Similarly, *Ankistrodesmus* sp. was unique to the macroalgae foam and *Coscinodiscus excentricus* to the macroalgae subsurface waters. The diatom assemblages of the estuarine foams resembled epiphytic microalgae on eelgrass (Kentula 1982) and benthic genera (Pamatmat 1968). Common epiphyte genera included *Cocconeis*, *Navicula*, and *Nitzschia* (Kentula 1982).

Species unique to the neritic habitat were *Chaetocerus curvisatus* in the surface and *Melosira sulcata*, *Protoperidinium* sp., and thecate dinoflagellates in subsurface waters.

Foam and Subsurface Incubation Experiments

Light-dark bottle incubations were conducted for both foam and subsurface waters on July 21-22 and July 23-24 (Table 5). The average photosynthetic rates of the eelgrass foam (68.1 and 72.1 mg C $m^{-3}hr^{-1}$) and subsurface waters (95.9 mg C $m^{-3}hr^{-1}$) were respectively 5 to 7X greater than for neritic waters (13.0 mg C $m^{-3}hr^{-1}$). The neritic production rate was comparable to the lowest rates for primary production during July and August in Hood Canal (13 to 79 mg C $m^{-3}hr^{-1}$; Shuman 1978; Welschmeyer 1982).

Production rates in the eelgrass surface foam were 2 to 7X higher than rates observed in surface films in a Georgia saltmarsh (Gallagher 1975) and in an estuarine mesocosm (Johnson et al. 1981), but 4 to 5X less than in subsurface waters of a temperate saltmarsh (Johnson et al. 1981) and

TABLE 4. Phytoplankton species list for neritic surface film, eelgrass and macroalgae samples. S indicates subsurface waters and + the presence of phytoplankton species; waters as in Table 1.

Phytoplankton species	*Waters*					
	Neritic		*Eelgrass*		*Macroalgal*	
	Foam	*S*	*Foam*	*S*	*Foam*	*S*
Achnanthes sp.			+	+	+	
Ankistrodesmes sp.					+	
Amphiprora sp.				+		
Asteromophalus sp.			+			
Chaetocerus radicans	+	+				
C. didymus	+	+				
C. curvisatus	+					
Cocconeis sp.			+	+	+	+
Coscinodiscus sp.			+			
C. excentricus						+
C. cosinus				+		
Cylindrospermum sp.			+			
Grammatophora marina			+	+	+	
Leptocylindrus sp.				+	+	
Licomophora sp.			+	+		
Melosira sp.			+		+	
M. monoliforimis				+	+	+
M. sulcata		+				
Navicula sp.			+	+	+	
Nitzschia closterium			+			+
N. seriata			+	+	+	+
Pleurosigma sp.			+	+	+	
Protoperidinium sp.		+				
Rhizoselenia sp.				+		
Thalassionema nitzschioides		+			+	
Tropidoneis antarctica var. *Polyplasta*			+	+	+	

estuary (Williams 1966) (Table 5). The most striking similarities between production rates of eelgrass surface foam in this study and that of the Georgia saltmarsh were the high production to respiration ratios ranging from 1.1 to 9.3. These P/R ratios were high compared to ratios in the neritic samples of this study and water columns (0.5 to 1.7; Williams 1966; Gallagher 1975; Johnson et al. 1981), indicating that microbiota were more autotrophically active.

Changes in production rates and DOC concentrations during eelgrass foam incubations suggested extracellular

TABLE 5. Comparison of microalgal primary production and P/R ratios in our study with others. P and R rates for this study represent 3.5 hour incubations. P = light bottle O_2 production, and R = O_2 consumption by the total community (algae, bacteria, and microzooplankton).

Environment	*Primary Production (P) (mg C m^{-3} hr^{-1})*	*P/R*	*References*
Neritic subsurface	13.0	0.7	This study
Eelgrass subsurface	95.9	5.8	This study
Eelgrass foam	68.1-72.1	1.1-5.0	This study
Salt marsh surface film (Georgia)[a,b]	10.1-21.2	1.8-9.3	Gallagher 1975
Salt marsh water column (Georgia)[a,b]	16.4-79.9	0.7	Gallagher 1975
Salt marsh water column (Rhode Island)[c]	260.6-353.8	0.5-1.0	Johnson et al. 1981
Simulated estuary water column (MERL)[c]	34.8	1.3	Johnson et al. 1981
Estuary water column (North Carolina)[d]	315.0	1.7	Williams 1966

[a] *O_2 flux converted to carbon equivalents after Johnson et al. (1981).*
[b] *P and R measured using light-dark bottle changes in O_2 (Strickland and Parsons 1972).*
[c] *P and R calculated from diel changes in O_2 concentrations.*
[d] *Weighted mean of converted O_2 values.*

release of dissolved organic matter by algae. This was especially evident during the July 23-24 experiment when production rates were highest (520 and 337 mg C m^{-3} hr^{-1}) and DOC/POC ratios increased to 12.6 (Fig. 2a and b). Also, the possible use of DOC exudates by bacteria was implied by DOC/POC ratios of <0.5 in the respiration bottles. DOC mean percent of time series concentrations relative to initial concentrations showed the same trends (Fig. 3).

During the July 23-24 experiment, mean percentages for DIN (46% and 42%, respectively) in both production and respiration bottles implied cellular uptake (Fig. 3). During the incubation, NH_4^+-N concentrations comprised 2% to 93% of the DIN in the production bottles and 2% to 34% in respiration bottles. In contrast, possible release of PO_4^{2-}-P (120%) was evident in the respiration bottles. Essentially no change occurred for PON, ΣN, ΣP, and POC.

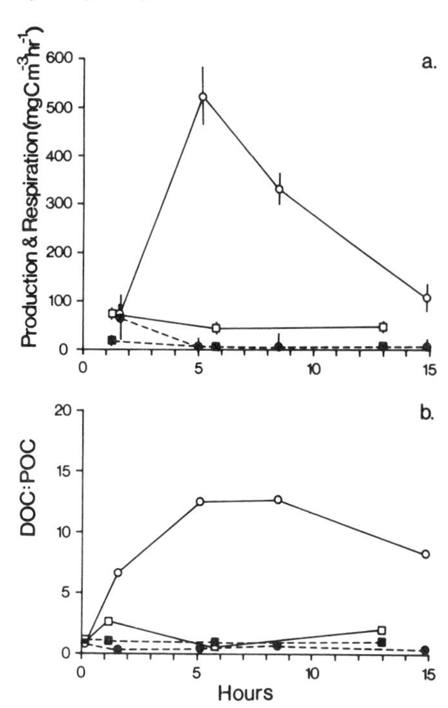

FIGURE 2(a). Primary production and total community respiration rates of eelgrass surface foams in Duckabush River estuary. (b). Changes in DOC/POC in eelgrass surface foams from the estuary. Square symbols represent the 21-22 July incubation time series experiment and circles the 23-24 July experiment. Open symbols indicate production bottles; shaded symbols indicate respiration bottles.

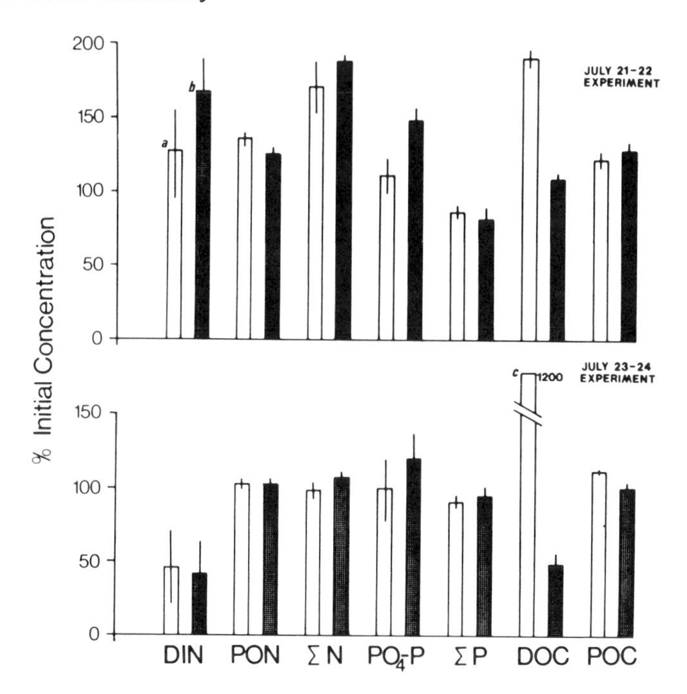

FIGURE 3. Percent changes in concentration (μM) of nitrogen, phosphorus, and carbon constituents in primary production (P) and total community respiration (R) incubation bottles during the July 21-22 and July 23-24 time series experiments. Symbols a *and* b *represent P (light bars) and R bottles (dark bars), respectively. Percentages are mean of time series concentrations. Vertical lines indicate percent standard errors. Symbol* c *denotes mean value of 1200% ± 202%.*

In contrast to the July 23-24 experiment, the July 21-22 experiment exhibited lower production rates (47 to 72 mg C m^{-3} hr^{-1}) and all nitrogen, phosphorus (except ΣP), and carbon concentrations exceeded (>100%) the initial concentrations (Figs. 2 and 3). These high percentages suggest the release of DIN and PO_4^{2-}-P and accumulation of PON, ΣN, and POC. Higher percentages for most constituents in both production and respiration bottles and low metabolic activities indicates possible influences of variable compositions of foams and cellular stresses caused by incubation conditions.

The variable nature of surface foams was also suggested by our simultaneous incubations of killed dark bottles (buffered formalin) which showed O_2 consumption. The results imply that dissolved and particulate aggregates in the surface foams exhibited some abiotic O_2 losses and behaved like sediment slurries. For example, if aggregates contain anaerobic-reducing microenvironments, reduced inorganic constituents such as Fe^{+3} and HS^- might occur with the high concentrations of organics (Moore et al. 1979) causing some chemical oxidation (O_2 consumption) which yields no CO_2 (Pamatmat 1975). Apparent killed dark bottle chemical oxidation rates were 28% higher than the respiration rates in the more productive 23-24 July experiment and 50% lower in the less metabolically active 21-22 July experiment.

CONCLUSIONS

This study of the biological and chemical characteristics of surface foam in a small Puget Sound estuary indicates that foams during July contained productive microalgal communities and were enriched in dissolved and particulate carbon, nitrogen, and phosphorus. Although estimates of production and respiration rates require better definition, our measurements of chlorophyll *a*, ATP, and bacterial counts in conjunction with organic solute fractions and dissolved nutrients indicate that the foam represents an extremely important environment for microbial communities and the cycling of cellular exudates and nutrients. $\delta^{13}C$ values indicated that organics in estuarine foams also contained carbon from eelgrass and macroalgae. In summary, the biotic composition and high concentrations of organics and nutrients indicate that the surface foams concentrate important food resources for possible use by higher consumers.

ACKNOWLEDGMENTS

This research was supported by Grant R/F-37 from Washington Sea Grant. Invaluable logistic and facilities support was provided by the Washington Department of Fisheries; we are particularly indebted to Kurt Fresh, Mark Carr, Gene Sanborn, and the WDF personnel at the Point Whitney Shellfish Laboratory for their assistance and advice. Contribution No. 645, School of Fisheries, University of Washington, Seattle, WA.

REFERENCES CITED

Avnimelech, Y., B. W. Taylor and L. W. Reed. 1982. Mutual flocculation of algae and clay: Evidence and implications. *Science 216*:63-65.

Broenkow, W. W. and J. D. Cline. 1969. Colorimetric determination of dissolved oxygen at low concentration. *Limnol. Oceanogr.* 14:450-454.

Carlson, D. J. 1982. Phytoplankton in marine microlayers. *Can. J. Microbiol. 28*:1226-1234.

Carlson, D. J. 1983. Dissolved organic materials in surface microlayers: Temporal and spatial variability and relation to sea state. *Limnol. Oceanogr. 28*:415-431.

Durako, M. S., R. A. Medlyn and M. D. Moffler. 1982. Particulate matter resuspension via metabolically produced gas bubbles from benthic estuarine microalgae communities. *Limnol. Oceanogr. 27*:752-756.

Faust, M.A. and R. J. Chrost. 1981. Photosynthesis, extracellular release and heterotrophy of dissolved organic matter in Rhode River estuarine plankton, pp. 447-464. *In*: B. J. Neilson and L. E. Cronin (eds.). *Estuaries and Nutrients*, The Humana Press, Inc., Clifton, N.J.

Gallagher, J. L. 1975. The significance of the surface film in saltmarsh plankton metabolism. *Limnol. Oceanogr. 20*:120-123.

Hobbie, J. E., R. T. Daley and S. Jaspers. 1977. Use of nucleopore filters for counting bacteria by fluorescent microscopy. *Appl. Environ. Microbiol. 33*:1225-1228.

Holm-Hansen, O. and R. C. Booth. 1966. The measurement of adenosine triphosphate in the ocean and its ecological significance. *Limnol. Oceanogr. 11*:510-519.

Hunter, K. A. and P. S. Liss. 1982. Organic matter and the surface charge of suspended particles in estuarine waters. *Limnol. Oceanogr. 27*:322-335.

Jensen, L. M. and M. Søndergaard. 1982. Abiotic formation of particles from extracellular organic carbon released by phytoplankton. *Microb. Ecol. 8*:47-54.

Johnson, K. M., C. M. Burney and J. McN. Sieburth. 1981. Enigmatic marine ecosystem metabolism measured by direct diel total CO_2 and O_2 flux in conjunction with DOC release and uptake. *Mar. Biol. 65*:49-60.

Kentula, M. E. 1982. Production dynamics of a *Zostera marina* bed in Netarts Bay, Oregon. Ph.D. Dissertation. Oregon State Univ., Corvallis, OR. 158 pp.

Leenheer, J. A. and E. W. D. Huffman, Jr. 1976. Classification of organic solutes in water by using macroreticular resins. *J. Res. U.S. Geol. Surv. 4*:737-751.

MacIntyre, F. G. 1974. Chemical fractionation and sea surface microlayer processes, pp. 245-299. *In*: E. D. Goldberg (ed.), *The Sea, Volume 5*. John Wiley and Sons, Inc., New York.

Mann, K. H. 1982. *Ecology of Coastal Waters: A Systems Approach*. Blackwell Sci. Publ., Oxford. 322 pp.

McKnight, D. M., W. E. Pereira, M. L. Ceazan and R. C. Wissmar. 1982. Characterization of dissolved organic materials in surface waters within the blast zone of Mount St. Helens, Washington. *Org. Geochem. 4*:85-92.

Menzel, D. W. and R. F. Vaccaro. 1964. The measurement of dissolved and particulate carbon in seawater. *Limnol. Oceanogr. 9*:138-142.

Moore, R. M., J. D. Burton, P. J. LeB. Williams and M. L. Young. 1979. The behavior of dissolved organic material, iron and manganese in estuarine mixing. *Geochim. Cosmochim. Acta. 43*:919-926.

Morris, A. W., R. F. C. Mantoura, A. J. Bale and R. J. M. Howland. 1978. Very low salinity regions of estuaries: Important sites for chemical and biological reactions. *Nature 274*:678-680.

Mulholland, P. J. 1981. Formation of particulate organic carbon in water from a southeastern swamp-stream. *Limnol. Oceanogr. 26*:790-795.

Nishizawa, S. 1971. Concentration of organic and inorganic materials in the surface skin at the equator, 155°W. *Bull. Plankton Soc. Jap. 18*:42-44.

Pamatmat, M. M. 1968. Ecology and metabolism of a benthic community on an intertidal sandflat. *Internat. Revue ges. Hydrobiol. 53*:211-298.

Pamatmat, M. M. 1975. *In situ* metabolism of benthic communities. *Cah. Biol. Mar. 16*:613-633.

Pellenbarg, R. E. and T. M. Church. 1979. The estuarine surface microlayer and trace metal cycling in a salt marsh. *Science 203*:1010-1012.

Penhale, P. A. and W. O. Smith, Jr. 1977. Excretion of dissolved organic carbon by eelgrass (*Zostera marina*) and its epiphytes. *Limnol. Oceanogr. 22*:400-407.

Percival, E. and R. McDowell. 1967. *Chemistry and Enzymology of Marine Polysaccharides*. Academic Press, London. 219 pp.

Pregnall, A. M. 1983. Release of dissolved organic carbon from the estuarine intertidal macroalgae, *Enteromorpha prolifera*. *Mar. Biol. 73*:37-42.

Sellner, K. G. 1981. Primary productivity and the flux of dissolved organic matter in several marine environments. *Mar. Biol. 65*:101-112.

Sholkovitz, E. R. 1976. Flocculation of dissolved organic and inorganic matter during mixing of river water and seawater. *Geochim. Cosmochim. Acta 40*:831-845.

Shuman, F. R. 1978. The fate of phytoplankton chlorophyll in the euphotic zone - Washington coastal waters. Ph.D. Dissertation. Univ. Washington, Seattle, WA. 243 pp.

Sieburth, J. McN., P.-J. Willis, K. M. Johnson, C. M. Burney, D. M. Lavoie, K. Hinga, D. A. Caron, F. W. French III, P. W. Johnson and P. G. Davis. 1976. Dissolved organic matter and heterotrophic microneuston in the surface microlayers of the North Atlantic. *Science 194*:1415-1418.

Strickland, J. D. H. and T. R. Parsons. 1972. *A Practical Handbook of Seawater Analysis*. Bulletin 167, Fish. Res. Board Can., Ottawa. 311 pp.

Stuber, H. A. and J. A. Leenheer. 1978. Assessment of a resin based on fractionation procedure for monitoring organic solutes from oil shale retorting wastes, pp. 266-272. *In*: Everett, L. G. and K. D. Schmidt (eds.), *Establishment of Water Quality Monitoring Programs*. American Water Resources Association, Denver, CO.

Turner, R. E. 1974. Community plankton respiration in a salt marsh tidal creek and estuary and in the continental shelf waters of Georgia. Ph.D. Dissertation. Univ. Georgia, Athens, GA. 110 pp.

Valderrama, J. C. 1981. The simultaneous analysis of total nitrogen and phosphorus in natural waters. *Mar. Chem.* *10*:109-122.

Velimirov, B. 1980. Formation and potential trophic significance of marine foam near kelp beds in the Benguela upwelling system. *Mar. Biol.* *58*:311-318.

Wallace, G. T. Jr. and R. A. Duce. 1978. Transport of particulate organic matter by bubbles in marine waters. *Limnol. Oceanogr.* *23*:1155-1167.

Welschmeyer, W. A. 1982. The dynamics of phytoplankton pigments: Implications for zooplankton grazing and phytoplankton growth. Ph.D. Dissertation, Univ. Washington, Seattle, WA. 176 pp.

Williams, P. M. 1967. Sea surface chemistry: Organic carbon and organic and inorganic nitrogen and phosphorus in surface films and subsurface waters. *Deep-Sea Res.* *14*:791-800.

Williams, R. B. 1966. Annual phytoplanktonic production in a system of shallow temperate estuaries, pp. 699-716. *In*: H. Barnes (ed.), *Some Contemporary Studies in Marine Science*. George Allen and Unwin, Ltd., London.

Wissmar, R. C., J. E. Richey, R. F. Stallard and J. M. Edmond. 1981. Plankton metabolism and carbon processes in the Amazon River, its tributaries, and floodplain waters, Peru-Brazil, May-June 1977. *Ecology* *62*:1622-1633.

MICROBIOLOGICAL CHANGES OCCURRING AT THE FRESHWATER-SEAWATER INTERFACE OF THE NEUSE RIVER ESTUARY, NORTH CAROLINA

Robert R. Christian

Biology Department
East Carolina University
Greenville, North Carolina

Donald W. Stanley
Deborah A. Daniel

Institute for Coastal and Marine Resources
East Carolina University
Greenville, North Carolina

Abstract: Previously, other researchers have observed rapid biogeochemical changes at the freshwater-seawater interface (FSI) of the Tamar Estuary in England. The postulated cause of these changes is a sequence of processes triggered by mass mortality of freshwater phytoplankton. Such mortality can be viewed as an example of selective "filtration" operating at the FSI. We examined a number of physical, chemical, and microbiological variables in the Neuse River Estuary, North Carolina, within the context of this hypothesis. Nitrate nitrogen, orthophosphate, and number of phytoplankton taxa all decreased significantly ($p<0.05$) going downriver across the FSI, while bacterial density and productivity each rose sharply ($p<0.05$). Other parameters that were monitored (phytoplankton density, chlorophyll *a*, ammonia nitrogen, and light absorption coefficient), changed in the vicinity of the FSI, but none of the changes were statistically significant at the 0.05 level. In the laboratory, we tested the effects of low salinities on the riverine microbial community. Of seven variables, only bacterial density and bacterial productivity showed any significant treatment effects. Overall, we could not confirm the hypothesis that exposure to

ISBN 0-12-405070-0

salt in the FSI affects the microbial community. Alternate hypotheses involving nutrient availability, increased residence time of water, light availability, and population interactions are discussed.

INTRODUCTION

In 1978, Morris et al. published an intriguing paper on the importance of the freshwater-seawater interface (FSI) as a site of rapid changes in biogeochemical interactions in river-estuarine systems. Those changes in such variables as dissolved organic carbon, chlorophyll *a* and several inorganic chemical species were found to occur at oligohaline salinities in the Tamar Estuary, England. Morris et al. (1978) hypothesized that the cause of these changes was "a sequence of processes starting with mass mortality of freshwater halophobic phytoplankton incapable of withstanding the sharp sudden osmotic and compositional changes of the FSI." Thus, the FSI was seen as a "filter" within an estuary, capable of removing living biomass so that the "filtrate" possessed new biogeochemical properties.

During the summer of 1982, we examined a number of physical, chemical and microbiological variables along a stretch of the Neuse River Estuary emcompassing the FSI. We also tested, by means of laboratory experiments, the effects of low salinities on the riverine microbial community. These studies are described herein, and the results are considered in the context of the hypothesis of Morris et al. (1978) as well as the work of others.

THE STUDY AREA

The Neuse River drains 5700 km^2 of eastern North Carolina (Fig. 1) and is an important water supply source to cities in the moderately populated (800,000) basin. There are over thirty municipal and industrial wastewater discharges, even though two-thirds of the land is either agricultural or forested. The largest single discharge is from a paper pulp mill above New Bern near the river's mouth. Excess nitrogen and phosphorus nutrients have resulted in blue-green algal blooms in the lower Neuse River in some recent years (North Carolina Department of Natural Resources and Community Development 1983), but not during the period of this study in the summer of 1982 (Stanley 1983).

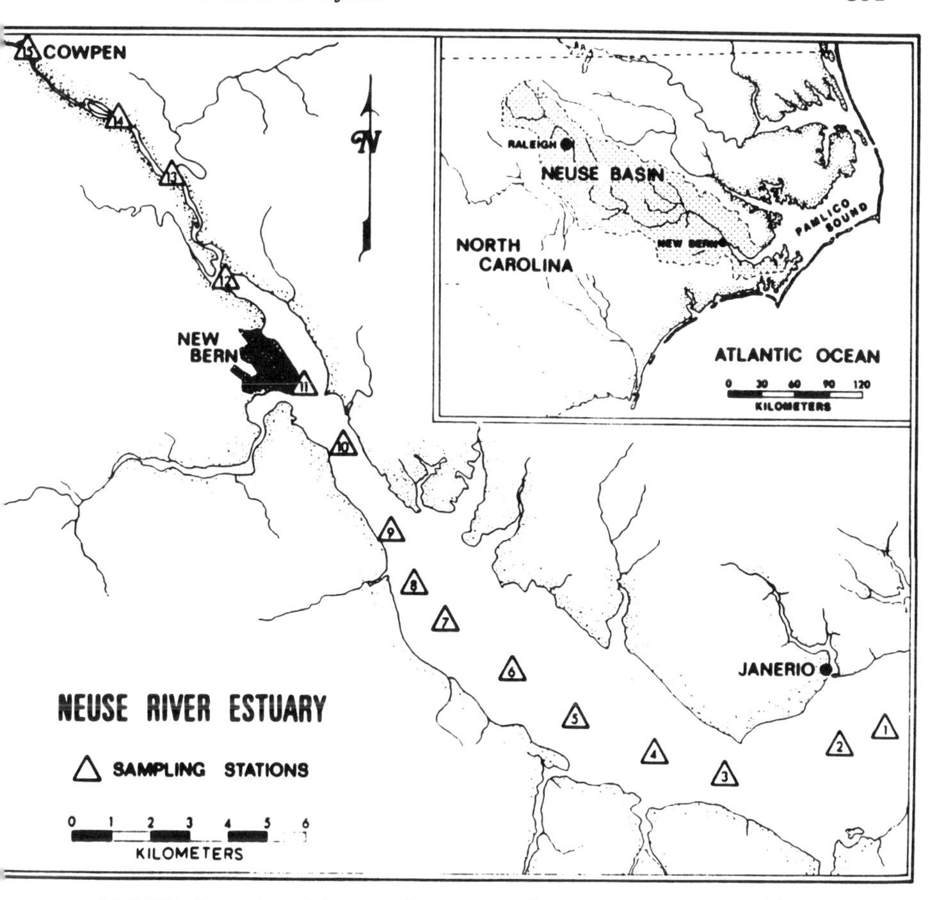

FIGURE 1. Locations of Neuse River Estuary sampling stations.

Below New Bern the Neuse River Estuary extends eastward for 60 km before emptying into Pamlico Sound behind the Outer Banks (barrier islands) which isolate much of North Carolina's large estuarine system from the Atlantic Ocean. The Neuse is a shallow (4.6 m mean depth), moderately eutrophic estuary valued for both commercial and recreational uses. Salinity ranges up to 20 ppt at the mouth, depending on freshwater runoff and wind direction. The Outer Banks dampen the lunar tides to less than 20 cm, but wind-driven tides can exceed 1 m (Hobbie and Smith 1975; Matson et al. 1983). Salt wedges are common in the upper estuary during intermediate and low flow. The freshwater-seawater interface migrates from below New Bern at high flow (usually in late winter) to about 6 km above New Bern in late summer when river flow slackens.

METHODS AND MATERIALS

During the summer of 1982 we made four transects of a 49-km stretch of the Neuse River Estuary between Cowpen Landing and Janerio, N.C. (Fig. 1). On each date at 15 stations, salinity, temperature, and dissolved oxygen measurements were made (YSI meters - Models 33 and 57) just below the surface. Water samples were collected and chilled on ice for processing later. In addition, the penetration of photosynthetically available radiation (PAR) was measured at 0.5-m depth intervals with an underwater quantum sensor (Lambda Instruments Company Model LI-192S). Within a few hours, the water samples were brought to the laboratory, and subsamples were taken for chemical and biological measurements. Water filtered through precombusted Whatman GF/C glass-fiber filters was analyzed for NO_3-N plus NO_2-N (Strickland and Parsons 1968), NH_4-N (Scheiner 1976), and PO_4-P (Environmental Protection Agency 1979). Phytoplankton samples were preserved in the field with Lugol's acetic acid solution and later concentrated on membrane filters (Millipore HA) for density determinations (American Public Health Association 1975). At least 25 microscope fields or 400 cells were counted per sample, and cells were identified to species when possible and reported as numbers of individuals per liter. Biomass was estimated by the chlorophyll *a* method (Environmental Protection Agency 1979). Acridine orange direct counts (AODC) of bacteria were made by epifluorescence microscopy on samples preserved in the field with formalin (Hobbie, Daley and Jasper 1977; Newell and Christian 1981). These samples were also used for determination of bacterial productivity by the frequency of dividing cells (FDC) technique of Hagström et al. (1979). Details of this method, along with the calibration method used, can be found in Newell and Christian (1981) and Christian, Hanson and Newell (1982).

Two experiments to test effects of salt on the Neuse microbial community were conducted on freshwater collected from the most upriver station (15) (Fig. 1). In the laboratory, 12-L subsamples were dispensed into four glass carboys. Two of the carboys also received enough artificial seawater mix (Instant Ocean) to raise the salinity to 1 ppt; the other two carboys served as controls. Carboys were incubated under natural light at close to *in situ* temperatures (approximately 28°C). Twenty-four hours later (T_o + 1 day), subsamples were taken from each carboy for the chemical and biological analyses described above and for nitrate uptake, ammonium uptake, and photosynthesis rate measurements. Then, to simulate the increase in salinity that the freshwater community would encounter in the estuary, another salt addition was made. This time the salinity was raised to 4 ppt in the two experimental carboys. The next day (T_o + 2 days) and again 3 days later (T_o + 5 days), subsamples for chemical and biological analyses, and for N uptake and photosynthesis measurements were removed from both the experimental and control carboys.

Uptake of NH_4-N and NO_3-N was measured using the stable isotope ^{15}N as a tracer (Dugdale and Goering 1967). First, $Na^{15}NO_3$ or $^{15}NH_4Cl$ (99 atom-%) was added to 150-ml water samples in screw-capped glass bottles. After incubation the samples were filtered onto Whatman GF/C glass-fiber filters which were then dried and stored in a desiccator. Later, particulate material on the filters was converted by Dumas combustion to N_2 for emission spectrometry. Particulate nitrogen was measured in a Coleman nitrogen analyzer. Phytoplankton photosynthesis was estimated by the ^{14}C method (modified from Steemann Nielsen 1952). Details of the ^{14}C and ^{15}N methodologies can be found in Stanley (1983).

RESULTS

Field Studies

Our concern here is whether or not the microbial community in the Neuse River Estuary is affected by increasing salinity in the vicinity of the FSI. During the study period, variations in river flow and wind tides caused the penetration of fresh water to range from 11.2 to 17.0 km below Station 15 at Cowpen Landing. This movement of the FSI makes station-by-station data comparisons over time misleading and inappropriate for our purposes. Instead, for each sampling date, we have grouped the stations according to salinity: freshwater (0 ppt), oligohaline (>0 to 5 ppt), and mesohaline (>5 ppt) (Table 1). In fact, some stations in the "mesohaline" region had salinities slightly above those normally used for this category. For each date and salinity regime we then computed the arithmetic means of the nine variables under consideration (Fig. 2). Overall differences in the variable means among salinity regimes were analyzed by the nonparametric Friedman's method for randomized blocks (Sokal and Rohlf 1969). The significance values are given in Fig. 2.

TABLE 1. Segment characteristics along a transect of the Neuse River Estuary on each of four sampling dates in 1982.

Date		Segment		
		Fresh	*Oligohaline*	*Mesohaline*
6/24	Surface salinity (ppt)	0	0.4-4.0	6.2-7.4
	Segment length (km)	17	10	22
	Stations included	10-15	6-9	1-5
7/27	Surface salinity (ppt)	0	0.5-4.3	5.8-14.0
	Segment length (km)	11	12	24
	Stations included	12-15	8-11	2-7
8/20	Surface salinity (ppt)	0	0.5-4.5	6.0-12.5
	Segment length (km)	17	20	12
	Stations included	10-15	4-9	1-3
9/3	Surface salinity (ppt)	0	1.2-4.8	5.2-9.6
	Segment length (km)	11	9	29
	Stations included	12-15	9-11	1-8

Four non-microbiological variables were examined. The absorption of light was greatest in freshwater, while in oligohaline and mesohaline waters light penetration increased (absorption coefficients decreased) (Fig. 2A). Unfortunately, light penetration was measured on only three of the four transects, which is too few to allow statistical testing by the method used. Nitrate plus nitrite concentrations (Fig. 2B) were highest in fresh water and decreased rapidly in the oligohaline region. Concentrations fell from as high as 56 µM at one (freshwater) station to <1 µM over a distance of <20 km. In contrast, ammonium concentrations were

FIGURE 2. Changes in nine variables as a function of salinity near the FSI in the Neuse River Estuary during the summer of 1982. Sampling dates and stations are listed in Table 1. F = fresh; O = oligohaline; M = mesohaline. Significance values are provided above each graph.
● *6/24* ○ *7/27* ■ *8/20* □ *9/3*

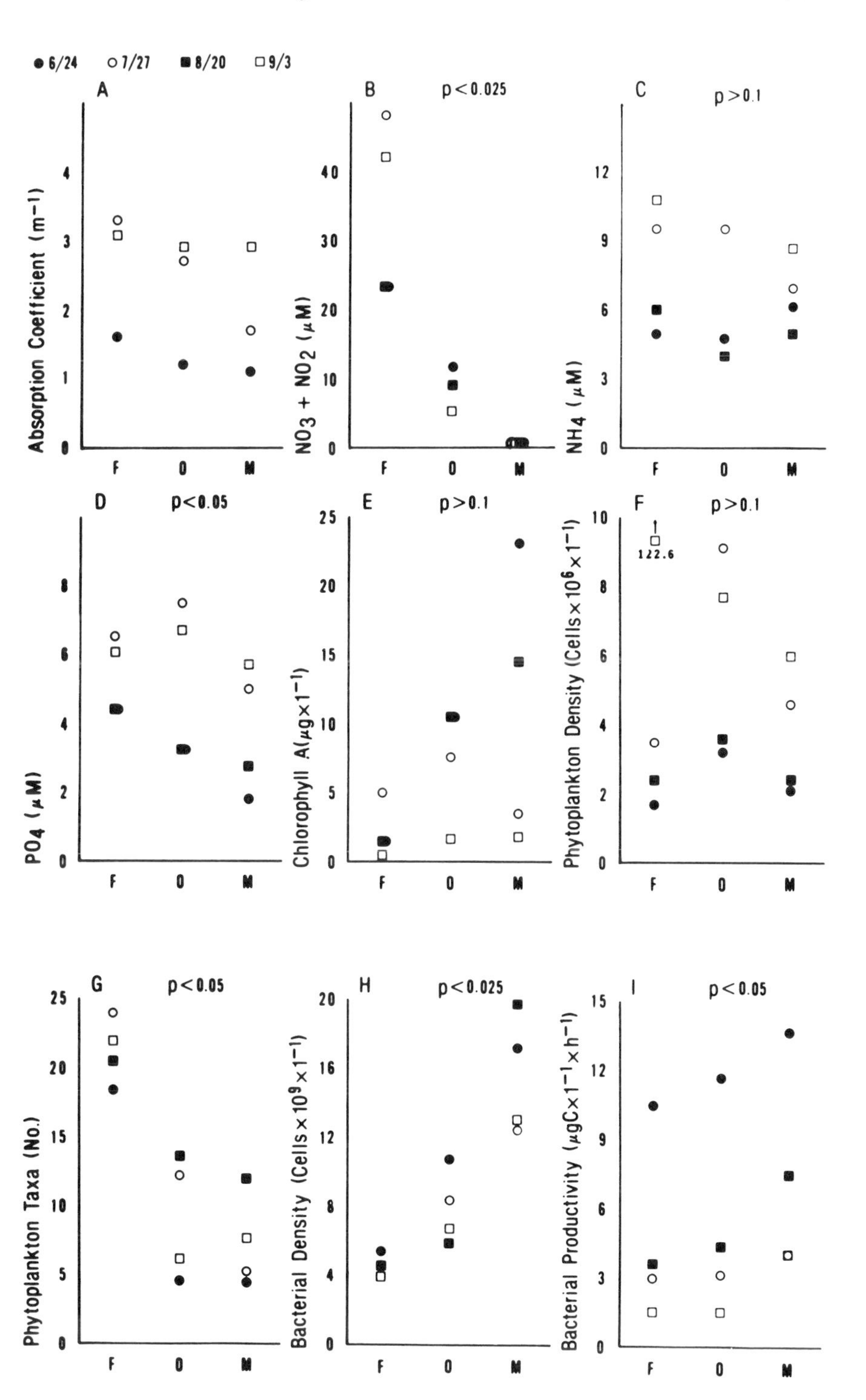
● 6/24 ○ 7/27 ■ 8/20 □ 9/3
A
Absorption Coefficient (m⁻¹)
B p<0.025
NO3 + NO2 (μM)
C p>0.1
NH4 (μM)
D p<0.05
PO4 (μM)
E p>0.1
Chlorophyll A(μg×1⁻¹)
F p>0.1
Phytoplankton Density (Cells×10⁶×1⁻¹)
122.6
G p<0.05
Phytoplankton Taxa (No.)
H p<0.025
Bacterial Density (Cells×10⁹×1⁻¹)
I p<0.05
Bacterial Productivity (μgC×1⁻¹×h⁻¹)
F O M

generally 4-10 μM and showed no consistent pattern of increase or decrease with either location or salinity (Fig. 2C). Soluble reactive phosphorus (PO_4-P) declined slightly with increasing salinity (Fig. 2D).

Three phytoplankton-related variables were examined, two of which were most often highest in the oligohaline section of the estuary. Except on 27 July, chlorophyll *a* concentrations (Fig. 2E) were lowest in freshwater, rose in oligahaline waters, then either continued to increase, decrease, or stay the same in mesohaline waters. Phytoplankton densities (Fig. 2F) were lowest in freshwater on all dates except 3 September, when a bloom of an unidentified nannoplankter was found there. On all other dates highest densities occurred in the oligohaline region. Because of these inconsistencies in both chlorophyll *a* concentration and cell density patterns, statistically significant differences were not found.

Phytoplankton taxa numbers, however, exhibited a dramatic, statistically significant drop at the FSI, but little further change was seen between the oligohaline and mesohaline regions (Fig. 2G). Phytoplankton taxa were distinguished on the basis of size and morphology, and most larger phytoplankters were identified at least to genus. Nannoplankton, because of their small size, were often not identifiable taxonomically; instead distinct morphologies were considered as taxa. Thus, there could be some ambiguity in our results; however, the trends seen would probably not be altered qualitatively by a more thorough taxonomic identification.

Both bacterial density (Fig. 2H) and bacterial productivity (Fig. 2I) were assessed. Each demonstrated statistically significant increases from fresh to mesohaline waters. However, the increase in density appeared more pronounced than the rise in productivity.

Thus, for a number of variables monitored in the four transect studies, significant differences were found among the three salinity regions of the Neuse River Estuary. It cannot, however, be assumed that these differences were directly linked to changes in salinity.

Experimental Studies

The results of two experiments testing the effects of salinity on the freshwater microbial community are summarized in Tables 2 and 3. Values in the tables are means from replicate control and experimental carboys. Also listed are the overall significance levels of differences between

TABLE 2. Summary of data from first salt-effects experiment on 4-9 August, 1982.

Parameter	Day 1		Day 2		Day 5	
Salinity	0 ppt	1 ppt	0 ppt	4 ppt	0 ppt	4 ppt
Primary productivity ($\mu g\ C \cdot l^{-1} \cdot h^{-1}$)	96	104	127	73	241	331[a]
Chlorophyll concentration (μg chl $a \cdot l^{-1}$)	2.4	4.4	5.6	5.2	16.0	18.4[a]
Phytoplankton taxa (No.)	28	30	24	23	30	24[a]
NH_4-N uptake ($\mu M \cdot h^{-1}$)	0.7	0.8	1.4	1.4	1.4	1.4[a]
NO_3-N uptake ($\mu M \cdot h^{-1}$)	0.6	0.6	1.8	1.1	2.6	3.0[a]
Bacterial density (No. x $10^9 \cdot l^{-1}$)	6.1	7.7	5.3	7.2	3.4	2.6[b]
Bacterial productivity ($\mu g\ C \cdot l^{-1} \cdot h^{-1}$)	3.0	4.7	5.2	7.9	5.1	6.3[c]

[a] *No significant difference between treatments with $p>0.05$.*
[b] *Significant treatment effect at $p<0.05$ and significant interaction between day and treatment at $p<0.05$.*
[c] *Significant treatment effect at $p<0.05$.*

TABLE 3. Summary of data from second salt-effects experiment on 8-13 September, 1982.

Parameter	*Day*					
	1		*2*		*5*	
Salinity	0 ppt	1 ppt	0 ppt	4 ppt	0 ppt	4 ppt
Primary productivity ($\mu g\ C \cdot l^{-1} \cdot h^{-1}$)	372	434	330	234	217	375[a]
Chlorophyll concentration (μg chl $a \cdot l^{-1}$)	20	18	38	20	16	21
Phytoplankton taxa (No.)	9	12	10	8	13	11
NH_4-N uptake ($\mu M \cdot h^{-1}$)	2.3	1.6	0.9	1.2	1.1	1.8
NO_3-N uptake ($\mu M \cdot h^{-1}$)	0.6	0.4	0.6	0.5	0.5	0.5
Bacterial density (No. x $10^9 \cdot l^{-1}$)	5.9	4.5	6.5	10.9	13.8	5.3

[a]*No significant differences between treatments with p>0.05 for all variables measured.*

control and experimental values. The significance levels are based on two-way analyses of variance with replication (Sokal and Rohlf 1969).

Of the seven variables examined during the first experiment in August, only bacterial density and bacterial productivity showed significant treatment effects (Table 3). (Note that bacterial productivity was not estimated in the second experiment). Thus, within the framework of our experimental design, we could not confirm the hypothesis that oligohaline salinities affect the microbial community.

DISCUSSION

Caution must be exercised when one hypothesizes about changes in the concentration-salinity gradients of substances in an estuary. One simple explanation may be that the substance behaves conservatively, so that its concentration at a given point is a linear function of the concentrations in the river and in the ocean. This occurs typically for materials that are not active in biogeochemical cycles within the estuary (Biggs and Cronin 1981). Usually, the turnover rate of such a substance is very slow relative to water turnover within a segment of the system (Imberger et al. 1983). Nonconservative parameters, on the other hand, are those for which there are likely to be net gains or losses in the estuary. There is ample evidence that the biogeochemical parameters that we have considered are subject to short-term gains and losses in the Neuse (Stanley 1983; Stanley and Christian, unpublished data) relative to the time required for water movement through the system (Woods 1969). Thus, they would not act conservatively and should be expected to exhibit longitudinal concentration patterns that deviate from linearity and that cannot be explained by simple dilution.

In the Neuse, most of the parameters that we followed in the field studies did indeed exhibit sharp, nonlinear declines or increases in the vicinity of the FSI. One of the most dramatic changes was the decline in NO_3-N plus NO_2-N concentration, which is consistent with the patterns observed for other river-estuaries in North Carolina (Stanley 1983), for the Chesapeake and Delaware Bays (T. Fisher, Horn Point Environmental Laboratories, personal communication), and for the British estuary investigated by Morris et al. (1978). Other notable changes were the decline in phytoplankton taxa number and the increases in phytoplankton density and chlorophyll *a*. Also, as might be predicted by Morris et al.'s (1978) hypothesis, bacterial density and productivity increased. Overall, then, our field observations tend to confirm those of Morris and his co-workers (1978).

Even though our field data suggest that salinity may be a causal agent for rapid changes within the FSI, they are not adequate as a test of the Morris et al. (1978) hypothesis. In 1982 we did not observe in the Neuse a "mass mortality of freshwater halophobic phytoplankton." Most often, such an event would not be very noticeable in the Neuse anyway, because the riverine algal biomass is usually low relative to that in the estuary (Stanley 1983; Paerl 1984). High turbidity at most times is the probable cause for the paucity of algae in the river. Nevertheless, in some years (e.g., 1983) the summertime flow becomes quite low so that turbidity lessens and, as mentioned above, there is substantial growth of blue-green algae in the lower Neuse. Preliminary analysis of the 1983 Neuse data shows that indeed there was a chlorophyll minimum at the FSI, surrounded by very high concentrations of blue-green algae upriver (>100 μg chlorophyll *a* l^{-1}) and relatively high concentrations (40 μg chlorophyll *a* l^{-1}) of non-blue-green (species not yet determined) algae below the FSI (Stanley and Christian, unpublished data). Perhaps this was a case of replacement of halophobic freshwater species killed at the FSI. If so, then we should be able to produce the same result in a controlled experiment such as those described above.

However, when we examined the direct, short-term effects of low salinity, no consistent results were obtained, even though we tested Neuse River water collected during the same time period as the field studies and monitored the same parameters as in the field studies. Thus, we could not demonstrate in microcosms the changes noted in the field and cannot confirm the hypothesis of direct salinity effects.

Much has been written on the influences of salinity on microorganisms, including reviews by Brown (1964, 1976), Gessner and Shramm (1971), MacLeod (1971), and Remane and Schlieper (1971). It is clear that many organisms which are found in freshwater are detrimentally affected by salt. It is not our intent here to review this literature thoroughly; however, two points are worth noting. First, most studies have not dealt with the low salinities found in the FSI region. Second, a major mechanism of effect addressed has been water activity (i.e., water potential or osmotic pressure). Thus, it would seem worthwhile to calculate the expected influence of low FSI salinity levels (0-5 ppt) on the water activity experienced by microorganisms. From data relating water potential to chlorinity (Riley and Chester 1971), we have calculated that 1 ppt salinity is equivalent to approximately -0.7 bars at 25°C. Thus, oligohaline, FSI waters would be characterized by water potentials more positive than -4 bars and with water activities >0.997. Such degrees of water activity rarely have been demonstrated to inhibit the growth of microorganisms or cause their death (Brown 1976). In fact, many microbial growth media used in the laboratory produce lower water availability than this. Although little evidence has been found for direct salt effects at such high water activities, compositional changes in the water at oligohaline salinities may have effects on microorganisms. These compositional effects may involve ionic interactions with cell membranes (Brown 1964; MacLeod 1971).

What alternate hypotheses might explain our field observations? Albright and co-workers (1983a; 1983b; Bell and Albright 1981; Valdés and Albright 1981) found stimulation of microbial activity in the plume of the Fraser River as it mixed with the Strait of Georgia, British Columbia. It was postulated that the plume waters provided more inorganic nutrients to the microbes than the two parent waters. This is not likely in the Neuse River Estuary, however, given that N and P concentrations are generally much higher upriver than in the estuary. Nevertheless, during intermittent periods of stratification the bottom waters do become enriched in ammonium and phosphate (Matson et al. 1983). The rates of exchange between these bottom waters and surface waters, which we sampled, is unknown. Also unknown are the rates of recycling of nutrients in the various portions of the system.

Wright and Coffin (1984) found a "mid estuary" peak in bacterial density and productivity in New England estuaries. They ascribed this to prolonged residence of water in the estuary and contact with neighboring salt marshes which release dissolved organic matter. However, in our case, although the residence time of water in the upper Neuse River Estuary is of the order of several days (Woods 1969), there are no major marshes to provide a source of organic matter.

One last alternate hypothesis is under consideration. As described above, light penetration in the Neuse increases as river water moves into the estuary. Also, during stratification, the depth of the mixing layer is reduced (Christian and Stanley, unpublished data), exposing microbes to light for a longer portion of the day. Enhanced light availability may promote greater primary productivity and hence more bacterial productivity (Durand 1979). This hypothesis requires confirmation. Also, it does not directly explain the loss of taxa in the oligohaline region. Changes resulting from competition or predator-prey interactions may occur, but these have not been studied.

In summary, some of the parameters that we followed in the 1982 field studies did exhibit changes that would be expected as a consequence of removal of halophobic algae at the FSI. And, in 1983, when the freshwater algal biomass was dominated by blue-greens, there was some evidence that the salt-filtration mechanism was operating. However, we could not demonstrate experimentally that the observed changes were the direct result of salinity increases. Alternate hypotheses involving nutrient availability, increased residence time of water, light availability, and population interactions remain to be tested.

ACKNOWLEDGMENTS

For assistance with field sampling and laboratory analyses, we are grateful to Judy Heath, Loede Harper, Will Sanderson, and Roger Barnaby. Support for this work was provided by the National Oceanic and Atmospheric Administration Office of Sea Grant under grant NA83AA-D-0012, and the State of North Carolina, Department of Administration. The grant is administered by the University of North Carolina Sea Grant College Program.

REFERENCES CITED

Albright, L.J. 1983a. Heterotophic bacterial biomasses, activities, and productivities within the Fraser River plume. *Can. J. Fish. Aquatic Sci. 40:*216-220.

Albright, L.J. 1983b. Influence of river-ocean plumes upon bacterioplankton production of the Strait of Georgia, British Columbia. *Mar. Ecol. Prog. Ser. 12:*107-113.

American Public Health Association. 1975. *Standard Methods for the Examination of Water and Wastewater.* American Public Health Association, New York. 1193 p.

Bell, C.R. and L.J. Albright. 1981. Attached and free-floating bacteria in the Fraser River estuary, British Columbia, Canada. *Mar. Ecol. Prog. Ser. 6:*317-327.

Biggs, R.B. and L.E. Cronin. 1981. Special characteristics of estuaries, pp. 3-24. *In:* B.J. Neilson and L.E. Cronin (eds.), *Estuaries and Nutrients.* Humana Press, Clifton, New Jersey.

Brown, A.D. 1964. Aspects of bacterial response to the ionic environment. *Bacteriol. Rev. 28:*296-329.

Brown, A.D. 1976. Microbial water stress. *Bacteriol. Rev. 40:*803-846.

Christian, R.R., R.B. Hanson and S.Y. Newell. 1982. Comparison of methods for measurement of bacterial growth rates in mixed batch cultures. *App. Environ. Microbiol.* *43:*1160-1165.

Dugdale, R.C. and J.J. Goering. 1967. Uptake of new and regenerated forms of nitrogen in primary productivity. *Limnol. Oceanogr. 12:*196-206.

Durand, J.B. 1979. Nutrient and hydrological effects of the pine barrens on neighboring estuaries, pp. 195-211. *In:* R.T.T. Forman (ed.), *Pine Barrens*. Academic Press, New York.

Environmental Protection Agency. 1979. Methods for chemical analyses of water and wastes. EPA-600/4-79-020. Cincinnati, Ohio. 430 p.

Gessner, F. and W. Schramm. 1971. Salinity: plants, pp. 705-820. *In:* O. Kinne (ed.), *Marine Ecology,* Vol. 1, part 2. Wiley-Interscience, London.

Hagström, A., U. Larsson, P. Hörstedt and S. Normark. 1979. Frequency of dividing cells, a new approach to the determination of bacterial growth rates in aquatic environments. *Appl. Environ. Microbiol. 37:*805-812.

Hobbie, J.E., R.J. Daley and S. Jasper. 1977. Use of Nuclepore filters for counting bacteria by fluorescence microscopy. *Appl. Environ. Microbiol. 33:*1225-1228.

Hobbie, J.E. and N.W. Smith. 1975. Nutrients in the Neuse River estuary. University of North Carolina Sea Grant Program. Report UNC-SG-75-21. Raleigh, N.C. 183 p.

Imberger, J., T. Berman, R.R. Christian, E.B. Sherr, D.E. Whitney, L.R. Pomeroy, R.G. Wiegert and W.J. Wiebe. 1983. The influence of water motion on the distribution and transport of materials in a salt marsh estuary. *Limnol. Oceanogr.* *28:*201-214.

MacLeod, R.A. 1971. Salinity: bacteria, fungi and blue-green algae, pp. 689-703. *In:* O. Kinne (ed.), *Marine Ecology,* Vol. 1, part 2. Wiley-Interscience, London.

Matson, E.A., M.M. Brinson, D.D. Cahoon and G.J. Davis. 1983. Biogeochemistry of the sediments of the Pamlico and Neuse River Estuaries, North Carolina. University of North Carolina Water Resources Research Institute. Report 191. Raleigh, N.C. 103 p.

Morris, A.W., R.F. Mantoura, A.J. Bale and R.J.M. Howland. 1978. Very low salinity regions of estuaries: important sites for chemical and biological reactions. *Nature 274:* 678-680.

Newell, S.Y. and R.R. Christian. 1981. Frequency of dividing cells as an estimator of bacterial productivity. *Appl. Environ. Microbiol. 42:*23-31.

North Carolina Department of Natural Resources and Community Development. 1983. Nutrient Management Strategy for the Neuse River Basin. Report 83-050. NCDNRCD, Division of Environmental Management. Raleigh, N.C. 29 p.

Paerl, H. 1984. The effects of salinity on blue-green algal bloom potential in the Neuse River estuary. University of North Carolina Sea Grant Program Report. (in press).

Remane, A. and C. Shlieper. 1971. *Biology of Brackish Water*, 2nd edition. Wiley Interscience, New York. 372 p.

Riley, J.P. and R. Chester. 1971. *Introduction to Marine Chemistry*. Academic Press, London. 465 p.

Scheiner, D. 1976. Determination of ammonia and Kjeldahl nitrogen by indophenol method. *Wat. Res. 10:*31-36.

Sokal, R.R. and F.J. Rohlf. 1969. *Biometry*. W.H. Freeman and Co., San Francisco. 776 p.

Stanley, D.W. 1983. Nitrogen cycling and phytoplankton growth in the Neuse River, North Carolina. University of North Carolina Water Resources Research Institute. Report 204. Raleigh, N.C. 85 p.

Steemann Nielsen, E. 1952. The use of radioactive carbon (^{14}C) for measuring organic carbon production in the sea. *J. Cons. Int. Explor. Mer. 18:*117-140.

Strickland, J.D. and T.R. Parsons. 1968. A practical handbook of seawater analysis. *Fish. Res. Bd. Can. Bull. 167.* 310 p.

Valdés, M. and L.J. Albright. 1981. Survival and heterotrophic activities of Fraser River and Strait of Georgia bacterioplankton within the Fraser River plum. *Mar. Biol. 64:*231-241.

Woods, W.J. 1969. Current study in the Neuse River and estuary of North Carolina. University of North Carolina Water Resources Research Institute. Report 13. Raleigh, N.C. 35 p.

Wright, R.T. and R.B. Coffin. 1984. Factors affecting bacterioplankton density and productivity in salt marsh estuaries. Proceedings of Third International Symposium of Microbial Ecology, August 7-12, 1983. (in press).

INFLUENCES OF SUBMERSED VASCULAR PLANTS ON ECOLOGICAL PROCESSES IN UPPER CHESAPEAKE BAY

W. Michael Kemp

University of Maryland
Horn Point Environmental Laboratories
Cambridge, Maryland

Walter R. Boynton

University of Maryland
Chesapeake Biological Laboratory
Solomons, Maryland

Robert R. Twilley
J. Court Stevenson
Larry G. Ward

University of Maryland
Horn Point Environmental Laboratories
Cambridge, Maryland

Abstract: Physical, chemical and biological influences of submersed vascular plants (dominated by *Potamogeton perfoliatus* and *Ruppia maritima*) on their surrounding environment are summarized for portions of upper Chesapeake Bay. Rates of accretion of organic matter in these ecosystems were high owing to the combined effects of vascular plant and associated algal production and the trapping of particulate organics of phytoplanktonic origin. Time-series observations of seston along transects traversing vegetated bottoms indicated significantly less turbid water over the plant beds, due both to increased deposition and to decreased resuspension of fine-grain sediments. Submersed plants provided a preferred habitat for many animal populations,

ISBN 0-12-405070-0

and abundance of fishes (predominantly juveniles) was significantly greater in these plant beds than in adjacent unvegetated areas. Recent declines in several species of migrating waterfowl which feed directly on plant material were highly correlated with contemporaneous decreases in plant distribution. Rapid uptake of dissolved inorganic nitrogen (N) and phosphorus (P) was demonstrated for these communities, with subsequent incorporation into plant material via both growth and facultative increases in percent N and P composition. Upon senescence and death, submersed vascular plants decayed at moderate rates, with relatively slow releases of nutrients and low dissolved oxygen (O_2) demand compared to algae (micro and macro) and to marsh grass. Thus, organic carbon (C) from these submersed plants is transferred to microbial food-chains, with minimal secondary effects of O_2 depletion and nutrient enrichment. Part of the influence of these plant communities on the upper Bay is summarized in terms of three materials budgets for 1960, where these plants contributed 33% to the organic C budget, while acting as a seasonal sink for 210 and 7% of the total sediment and nitrogen inputs (respectively) to the estuary.

INTRODUCTION

The wide range of physical, chemical and biological interactions between submersed vascular plants and their surrounding environments has been cited often to emphasize the importance of these plants (e.g. Wood, Odum and Zieman 1969; Thayer, Adams and La Croix 1975; Kemp 1983). Submersed plant communities are considered to be among the most productive natural ecosystems (McRoy and McMillan 1977). Although direct grazing on this organic production is limited, these plants are utilized indirectly in myriad detrital food-chains (Thayer et al. 1975). Populations of small fish and invertebrates, which seek food supplies and refuge from predation, tend to be more abundant and diverse in areas dominated by these vascular plants than in adjacent unvegetated regions (Adams 1976a,b; Brook 1978; Orth and Heck 1980). The physical structure of the plants, while functioning as shelter and substrate, also tends to damp wave and tidal energies, thus increasing sedimentation (Harlin, Thorne-Miller and Boothroyd 1982), reducing resuspension (Ginsburg and Lowenstam 1958) and shore erosion (Christiansen et al. 1981). Finally, the ability of these plant communities to absorb pulses of nutrient inputs

(Howard-Williams 1981) and to modify their sediment chemistry (Iizumi, Hattori and McRoy 1980; Kenworthy, Zieman and Thayer 1982) represents a major factor in some aquatic nutrient cycles.

While submersed vascular plants are abundant in the shallow waters of lakes, estuaries and seas, most of the research documenting their influence on adjacent environments has been confined to lacustrine and marine systems. In estuaries and brackish waters throughout the world, a mixture of halo-tolerant freshwater species are found commonly in waters up to 20°/oo salinity, (Bourn 1932; Mathiesen and Nielsen 1956; Haller, Sutton and Barlowe 1974; Stevenson and Confer 1978; Spence, Barclay and Bodkin 1979), with marine species (seagrasses) dominating the higher salinity regimes. The ubiquitous genus, *Ruppia*, often occurs throughout the full salinity range (e.g., Verhoeven 1979). Despite their widespread distribution in brackish waters, submersed plants have been studied only to a limited extent in these habitats.

Dramatic declines in the distribution and abundance of submersed vegetation in the brackish waters of upper Chesapeake Bay during the last two decades (Bayley et al. 1978; Stevenson and Confer 1978; Orth and Moore 1983) have served to focus attention on the potential ecological importance of these macrophytes. In the present paper we summarize results from various studies, which have addressed questions as to the mechanisms and extent of influence exerted by these submersed plants on ecological processes in surrounding environments. Here, we consider the effects of these plants specifically in terms of: primary production and associated plant and animal abundances; trapping and binding of sediments; and immobilization and recycling of nutrients. Finally, to place these results into a broad regional perspective, we provide first-order estimates of the former role of submersed plants in Bay-wide budgets for three kinds of materials: organic carbon; inorganic sediments; and total fixed nitrogen.

METHODS AND STUDY AREAS

Study Areas and Designs

Most of the field research reported here was conducted during the 1980 growing season (May-Oct.) at a site (Todds Cove) located about 5.2 km upstream of the mouth of the Choptank River, a tributary of Chesapeake Bay (38°36'28"

N, 76°14'35" W). This small cove is typical of the dendritic shoreline of eastern Chesapeake Bay. The specific vegetated site covered an area of about 0.03 km^2 and was dominated by *Potamogeton perfoliatus*, with *P. pectinatus*, *Ruppia maritima* and *Zannichellia palustris* of lesser importance. An unvegetated reference area with similar gross sediment features (sandy-silts) was also located in this cove about 300 m west of the plant community. Mean depths at both locations were about 0.7 m, and the average tidal amplitude was 0.4 m. Annual ranges of temperature and salinity here were about 0-33°C and 8-15°/oo, respectively.

Additional studies to examine effects of nutrient enrichment were conducted during 1981 in eight experimental ponds located at Horn Point Environmental Laboratories near Todds Cove. These ponds, which were each about 27 m by 13 m at the water surface with a mean water depth of about 1.0 m, were flushed weekly with water from the nearby Choptank River estuary just prior to treatment. The pond ecosystems, which had been inoculated with sediments and vegetation from the estuary in the spring of 1980, had a plant species composition similar to the Todds Cove site. Duplicate ponds were treated weekly at three levels (plus controls) of nitrogen (N) plus phosphorus (P) enrichment (10:1, atomic) to achieve initial nominal concentrations of N at 30, 60 and 120 μM. Concentrations of NH_4^+, NO_2^-, NO_3^-, $PO_4^{\equiv}$, and total N and P were measured at 0, 1, 3 and 7 d following each treatment to investigate the fate of added nutrients. Carbon (C), N, and P contents of plant tissue were sampled prior to and at 26 and 60 d following initial treatment. Denitrification was measured in vegetated and unvegetated sediments of low and high treatment ponds using an acetylene blockage technique (Chan and Knowles 1979).

To measure the relative rates of nutrient recycling and attendant oxygen (O_2) consumption after death of these plants, gross decomposition processes were studied in microcosm experiments for three submersed vascular plants (*P. perfoliatus*, *R. maritima* and *Myriophyllum spicatum*), a microalgal species (*Chlorella sp.*), a macroalgal species (*Ulva lactuca*), and an emergent marsh grass (*Spartina alterniflora*). Approximately 120 g fresh weight of recently collected plant material was placed in nylon bags (1 mm mesh) which were, in turn, placed into glass aquaria with filtered (1 μm) estuarine water (25 L) and a 1.0 g inoculum of sediment. Concentrates of *Chlorella sp.* were obtained by centrifugation of mass culture stocks. Over the course of a 93 d experiment, the aquaria were maintained in darkness at constant temperature (23 ± 3°C), with continuous air bubbling between samplings. The following variables were measured over the experimental period: weight of material in bags and that collected by filtration (GF/C, 1.2 μm) of external water; respiration (as dissolved O_2 consumption) of the aquarium communities; C, N and P content of plant tissue; and dissolved inorganic and organic nutrients in the water of these batch systems.

Plant Biomass and Productivity

Standing stocks of submersed vascular plant shoots (including stems, leaves, and flowers) were obtained by collecting the material within a 0.25 m^2 quadrat frame randomly placed in the plant bed. At each sampling, six replicates were placed in individual plastic bags, returned to the laboratory, washed and sorted by species, and dried at 60°C to constant weight. Root biomass (including rhizomes) was collected using brass coring tubes which produced cores 50 cm^2 area and 25 cm deep. Six replicate cores, obtained in sediments where shoots had been clipped, were combined to constitute a single root/rhizome sample, with six total replicates corresponding to respective shoot samples. Cored sediments were sieved to 0.5 mm, and all living (non-photosynthetic) plant material was washed and dried at 60°C to constant weight.

Primary productivity of both vegetated and unvegetated communities was estimated in terms of apparent O_2 production (i.e., in excess of respiration) in closed, circulated chambers (190 L) and in open waters. Measurements of O_2 change were made with polarographic electrodes (Orbisphere Model 2603) which were calibrated daily. Short-term incubations (1-4 h) were used for chamber experiments, while diel O_2 monitoring was employed for open-waters (Odum and Hoskin 1958); both methods yielded similar rates. Potential problems with O_2 storage in vascular plant lacunae (e.g. Zieman and Wetzel 1980) were not evident, since waters surrounding plant leaves were well circulated (Westlake 1978), and because the dominant species have been demonstrated to exhibit more storage and transport of O_2 in and through their vascular systems (Lewis 1980; Sand-Jensen, Prahl and Stokholm 1982).

Fish Abundance

Finfish samples were obtained using 31 m haul seines (1.8 m deep, 0.6 cm stretch mesh) with a collection bag (purse) attached to one end (1.8 m x 1.8 m). With the purse-end fixed to a pole, the net was deployed by hand from a small boat in a circle enclosing 80 m^2. The net was drawn into a progressively smaller circle and finally pursed with a bottom chain in a manner similar to that described in Kjelson and Colby (1977). Collected fish were preserved in 10% buffered formalin for subsequent sorting, identification, and measurement for standard length and weight. All seining reported here was done at dusk to maximize capture efficiency (e.g. Kjelson and Colby 1977).

Physical and Chemical Analyses

Suspended particulate materials were sampled at 2-3 h intervals at vegetated and unvegetated sites over several tidal cycles during the summers of 1980 and 1981. Water samples were collected in van Dorn bottles, and 100-500 ml were passed through 2.5 cm GF/C filters (1.2 μm) for analyses of chlorophyll *a* and total particulate mass. Filters for chlorophyll *a* analysis were ground and pigments extracted in 90% acetone for 24 h in darkness. Fluorescence was measured by a Turner (Model III) fluorometer. The pre-weighed filters for particulate mass were dried to constant weight at 90°C (Strickland and Parsons 1972). Attenuation of diffuse down-welling photosynthetically active radiation (PAR) was estimated from vertical profiles of PAR measured at 10 cm intervals with a LICOR (1925B) quantum sensor. Dissolved inorganic nutrients (NH_4^+, NO_2^-, NO_3^-, $PO_4^{\equiv}$) were assayed using a Technicon AutoAnalyzer II with standard colorimetric techniques (Environmental Protection Agency 1979). Dissolved organic nitrogen and phosphorus were estimated using the persulfate digestion method of Valderrama (1981) followed by analysis for NO_3^- or $PO_4^{\equiv}$ as above. Total carbon and nitrogen of plant and sediment samples (upper 3 cm) were assayed with a Perkin Elmer 240B elemental analyzer. Sediment (upper 3 cm) pigments were extracted for 24 h in the dark as above. Stable carbon isotope ratios of plants, sediments, and seston ($\delta^{13}C$) were measured by mass-spectrometry in the laboratory of Patrick Parker, University of Texas, Port Aransas.

RESULTS AND DISCUSSION

Primary and Secondary Producers

Populations of submersed vascular plants have occurred in upper Chesapeake Bay in dense stands, with peak above-ground (shoots) standing stocks from 80-500 g dry weight (d.w.)/m^2 (Boynton 1982). These values are comparable to those reported for temperate seagrasses (McRoy and McMillan 1977); however, seasonal maxima at most of our study sites were at the middle to lower end of this range. The highest shoot biomass values observed in the upper Bay were for stands of *Myriophyllum spicatum* in an isolated cove (530 g d.w./m^2) and for a mixed assemblage dominated by *P. perfoliatus* and *R. maritima* (270 g d.w./m^2) in experimental ponds (Boynton 1982; Twilley et al. 1983). At the Todds Cove site, peak biomass was only about 80 g d.w./m^2, occurring in July (Fig. 1a). Biomass of below-ground plant parts here ranged from about 10-40% that of shoots, with highest proportions at the beginning and end of the growing season for these perennial plants (Fig. 1a). Such values for root biomass fraction are lower than those reported elsewhere for this species (e.g. Ozimek, Prejs and Prejs 1976) and for various seagrasses (Ziemen and Wetzel 1980).

At Todds Cove, apparent primary production (O_2) of the entire plant community, including associated micro-algae, was 2-5 times that at the nearby reference site without vascular plants (Fig. 1b). Peak production occurred when net accretion of macrophytic biomass was negligible, suggesting that a major portion of the O_2 generation was attributable to epiphytic (and other) micro-algae (c.f., Penhale 1977; Staver 1984). Maximum rates of net plant growth observed in (May-July) were equivalent to an O_2 evolution rate of about 1.3 g O_2 m^{-2} d^{-1} (assuming photosynthetic quotient of 1.0 and C at 50% dry weight, Zieman and Wetzel 1980). This value is about 30-50% of apparent O_2 production rates for the community (Fig. 1b), but the two measurements are not strictly comparable because the former considers only vascular plants, and the latter does not account for night respiration. These rates of community O_2 production are similar to those reported previously for various seagrasses (e.g. Zieman and Wetzel 1980). Thus, these submersed plant communities constitute a major source of organic matter derived both from macrophytic and from associated microphytic production.

During the study period (1979-1982), these plant populations exhibited distinct annual patterns, with growing

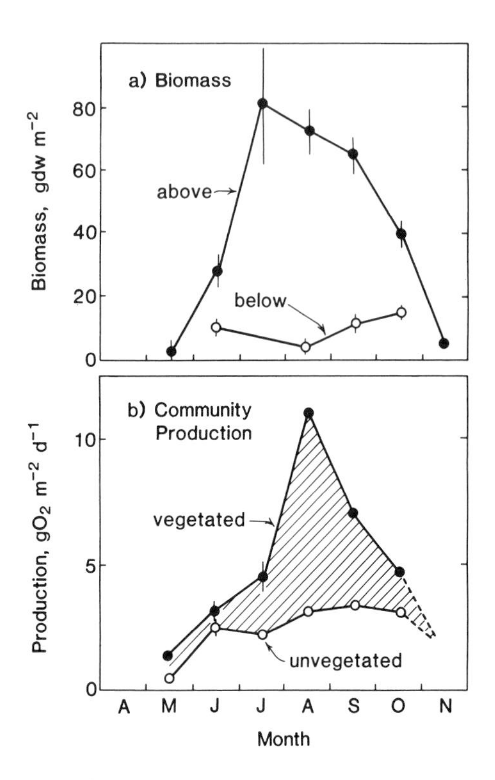

FIGURE 1. Vascular plant biomass (a) and apparent O_2 production (means ± S.E.) for a community dominated by Potomogeton perfoliatus at Todds Cove (Choptank River, Chesapeake Bay). Both above-ground (stems, leaves and flowers) and below-ground (roots, rhizomes, and stems) biomass are presented.

seasons limited to 4-6 mo between May and October. Previous reports (e.g., Purdy 1916) suggest that prior to 1965, above-ground biomass may have persisted in some areas of the Bay well past October. At one site in Eastern Bay which we visited for three successive years (1979-1981), we found a continually decreasing growing season from 6 to 4 to 2 mo (Boynton 1982). It appears that the general decline in abundance occurring in Chesapeake Bay has been accompanied by progressive shortening of the period of above-ground plant presence (Kemp et al. 1983b).

Direct herbivory on submersed vascular plants in upper Chesapeake Bay generally is limited to that of diving and dabbling ducks. The trend of prematurely senescing plants may have a particularly detrimental effect on these migrating waterfowl which return to the upper Bay in early autumn. Annual waterfowl census data summarized by Perry, Munro and Haramis (1981) indicated that several important duck populations are presently declining in Chesapeake Bay relative to their North American breeding stocks. These include canvasback (*Aythya valisneria*), redhead (*A. americana*), and common goldeneye (*Bucephala clangula*). Historical data for food habits of these ducks in the Bay indicate that submersed vegetation was the predominant food source around the early 1900's for two of these species, canvasback and redhead, with vegetation comprising 70% and 100% of their respective diets (Perry et al. 1981).

While we have not conducted direct experiments with waterfowl, we were able to establish strong correlations between relative plant abundance in the uppermost estuary (Bayley et al. 1978) and the ratio of Chesapeake Bay to North American duck populations for these two species from 1958-1975 (Fig. 2). The coincident declines in submersed plant and waterfowl abundances emphasize the importance of plant production as a source of duck food. With decreasing availability of these plants since the 1960's, canvasback shifted more to a carnivorous diet while redhead did not (Perry et al. 1981). However, most of the remaining redhead population in Chesapeake Bay is now centered in the lower Bay (Perry et al. 1981) where moderately abundant beds of *Zostera marina* still persist (Orth and Moore 1983).

Other animal populations are equally dependent on these plants although not as a direct source of food. Abundant epifaunal communities, including amphipods, isopods, gastropods, insect larvae and grass shrimp, occur in vegetated areas, feeding on epiflora, plankton and algal detritus associated with the vascular plant physical structure (e.g. Marsh 1973). Epifaunal and infaunal invertebrates are, in turn, consumed by large numbers of small fish. At the Todds Cove site we observed striking patterns of fish abundance and diversity in relation to vegetation (Fig. 3a). Fish densities were 1-10 times greater in the plant community, with maximum abundance (and difference between sites) occurring at times of peak plant biomass (July-September). Numerically dominant species in the vegetated area included white perch (*Morone americana*), menhaden (*Brevoortia tyrannus*), killifish (*Fundulus spp.*, *Lucania parva*) and sticklebacks (*Apeltes quadracus*), while spot (*Leiostomus xanthurus*) was by far the dominant fish in the nearby reference area. Species diversity (richness) of fish assemblages was significantly greater for the vegetated community over the 5 mo sampling period.

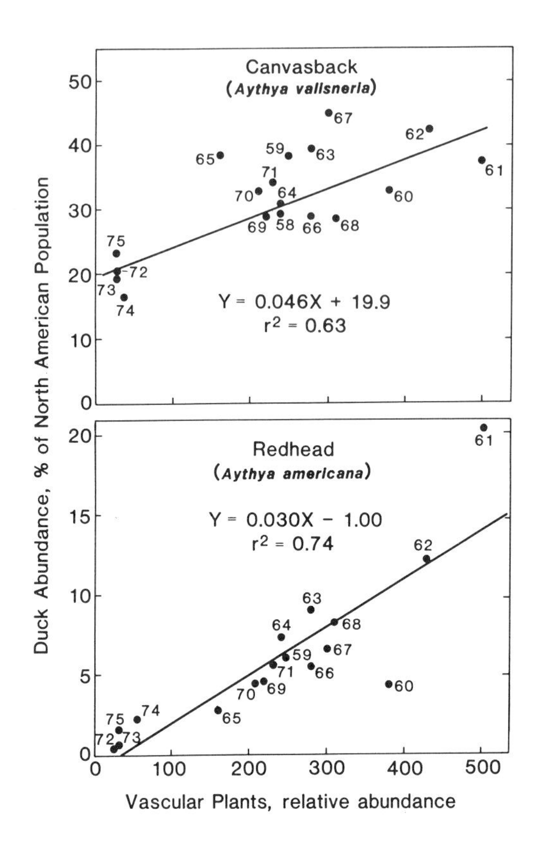

FIGURE 2. Correlation of annual abundance of two diving ducks in Chesapeake Bay as a percentage of the total count for North American breeding grounds (Perry et al. 1981) versus relative abundance of submersed vascular plants at upper Bay site (Bayley et al. 1978). Numbers adjacent to points indicate year (1958-1975).

Fishes captured in the vascular plant bed were also significantly smaller (and younger), on the average, than those sampled at the unvegetated site during the growing season (Fig. 3b), suggesting the importance of these beds as nursery and refuge areas for juvenile and small fish. While this concept has been considered previously in terms of absolute abundance and food habits of small fish (e.g. Carr and Adams 1973; Adams 1976a,b; Livingston 1982), few comparative studies have been reported. Weinstein and Brooks (1983) have recently shown that both seagrass beds and tidal marsh creeks represent productive habitats for juvenile nekton. Total

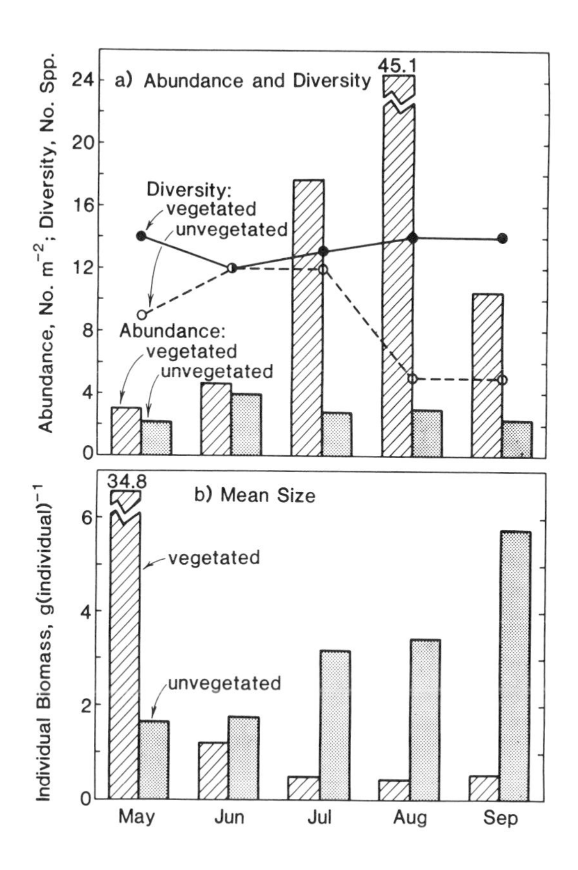

FIGURE 3. Numerical abundance (a) and species diversity (i.e., richness) for finfish at two sites in Todds Cove (Choptank River, Chesapeake Bay), one covered with Potamogeton perfoliatus, the other unvegetated. Mean size (b) for fish at these two sites.

fish biomass was also greater at sites with submersed plants in Todds Cove despite the predominance of juvenile forms, with a mean standing stock in the plant bed of 6.4 g d.w./m^2 (compared to 2.1 g d.w./m^2 in unvegetated area) over the study period. This value is considerably (2-4 times) higher than those previously reported for other systems with submersed plants, (e.g. Hoese and Jones 1963; Adams 1976a; Brook 1977; Robertson 1980).

Trapping Sediments and Organic Matter

The role of submersed vascular plants as "sedimentological filters" trapping fine-grain sediments in upper Chesapeake Bay, has been recently demonstrated by Ward, Kemp and Boynton (1984), who reported data from both the Todds Cove site and a nearby bed dominated by *R. maritima*. Patterns similar to that provided in Fig. 4 were observed over a 6 d time-series of hourly measurements, with 3-20 fold differences in seston concentrations between vegetated and unvegetated stations. These grass beds were shown to damp

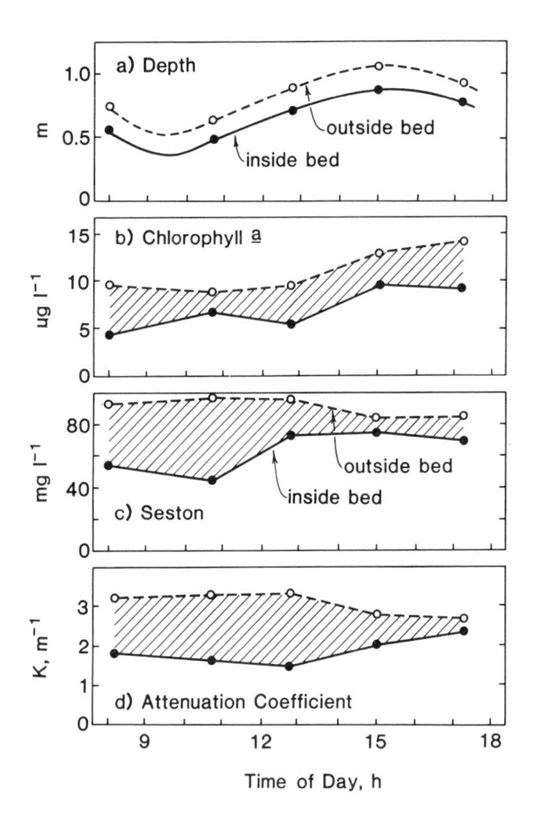

FIGURE 4. Example of time-series observations of (a) depth, (b) planktonic chlorophyll a, (c) seston, and (d) PAR attenuation coefficient for a typical tidal cycle in September 1980 in a submersed vascular plant bed dominated by Potamogeton perfoliatus and in an adjacent unvegetated site.

significantly wind-waves traversing the estuarine shoals and thereby reduce sediment resuspension and shoreline erosion. The reduction in resuspension was found to be a direct function of plant biomass, consistent with the observations of Fonseca et al. (1982) for *Z. marina*. Using several indirect methods, Ward et al. (1984) estimated that sediment accumulated in these beds at 2-3 mm/mo (over the 6 mo growing season) which, while less than the estimates of Harlin et al. (1982) for a flood-tide delta, is about 10-fold higher than for unvegetated areas in the upper Bay (e.g. Schubel and Carter 1977). However, the fate of this trapped material during the winter after plant senescence is uncertain, especially in view of the intensity of winter storms and the current trend of shortened growing seasons.

The organic content (combustibles) of seston in these shallow waters ranged from about 10-40% (Ward et al. 1984), and judging from the parallel patterns of seston and planktonic chlorophyll *a* (e.g. Fig. 4), much of this organic fraction was probably of algal origin. Based on the estimated rates of dry matter deposition in these beds and on the percent organics, we calculate the rate of organic carbon accumulation by this physical mechanism to be on the order of 1-5 g C $m^{-2}d^{-1}$ which is similar to *in situ* rates of summertime production. Thus, it appears that the physical structure of these plants contributes to trapping large quantities of allochthonous organic matter which might serve as an indirect source of food for secondary production.

This concept is further supported by comparing the sediments in vegetated and reference areas (Table 1). Sediments in the plant bed were significantly ($p < 0.05$) richer in both organic carbon and chlorophyll *a*. While vascular plant detritus could constitute the major source of this sediment chlorophyll *a* and organic C, it is unlikely that the intact pigment would be associated with *P. perfoliatus* because there was little recognizable plant material in these sediments, and because chlorophyll *a* decays much more rapidly than does the support tissue for these vascular plants (e.g. Knauer and Ayers 1977).

Data on stable carbon isotope in sediment organic matter also corroborate this idea, where $\delta^{13}C$ values for the reference area sediment were substantially lower than those for the vegetated sediments. Various submersed vascular plants examined in this area had a mean isotope ratio of $-12.6^{o}/oo$ which is intermediate between values reported for seagrasses (McMillan, Parker and Fry 1980) and freshwater macrophytes (LaZerte and Szalados 1982),

TABLE 1. Comparisons of selected sediment characteristics in shallow estuarine areas (Todds Cove) with and without submersed vascular plant communities.

Site	Organic C[a] (% dry mass)	Chlorophyll a[a] ($\mu g/m^2$)	$\delta^{13}C$[a] Value (°/oo)	$\delta^{13}C$[a] Planktonic origin (%)[b]
Vegetated	3.25 ± 0.83	47.8 ± 8.8	-16.4 ± 0.7	40
Unvegetated	1.76 ± 0.37	18.0 ± 5.3	-19.2 ± 2.2	70

[a] *Samples (mean ± S.E.) obtained in July-September 1980 from upper 3 cm of sediments.*

[b] *This estimate uses linear proportioning and assumes that submersed vascular plants and plankton were the only primary sources of organic matter at these sites, using mean values for $\delta^{13}C$ of plankton = -22.0°/oo and submersed vascular plants = -12.6°/oo.*

while seston here had a mean ratio of -22.0°/oo which is comparable to values reported for plankton (Fry, Scanlan and Parker 1977). Assuming that plankton and submersed vascular plants were the only important sources of organic C for these two systems, it would appear by linear proportioning that some 40% of the vegetated sediment C was of planktonic origin (Table 1). This ignores the possible role of benthic diatoms (Haines 1976); however, benthic production in these grass beds tends to be relatively low, probably limited by insufficient light (Murray 1983).

Thayer et al. (1978) reported relatively high $\delta^{13}C$ values for deposit-feeding invertebrates in a *Z. marina* bed (-15.0°/oo compared to -12.2°/oo for the seagrass). They interpreted these data to indicate that this seagrass bed was a comparatively young system whose sediments were too new to reflect the isotopic ratio of the vascular plants. An alternative explanation, which invokes the trapping of planktonic debris, would be equally plausible. This process of collecting seston may also help to explain how, in a steady-state context, some seagrasses might derive most of their nutrient demands from sediments (e.g., Patriquin 1972), where the decomposition of deposited planktonic detritus would provide the necessary nutrients.

Nutrient Uptake, Cycling and Decomposition

In addition to filtering the water column for organic particulates and fine-grain sediments, these submersed macrophytic communities are also capable of rapidly removing dissolved inorganic nutrients from surrounding waters. In experimental ponds treated with additions of inorganic N

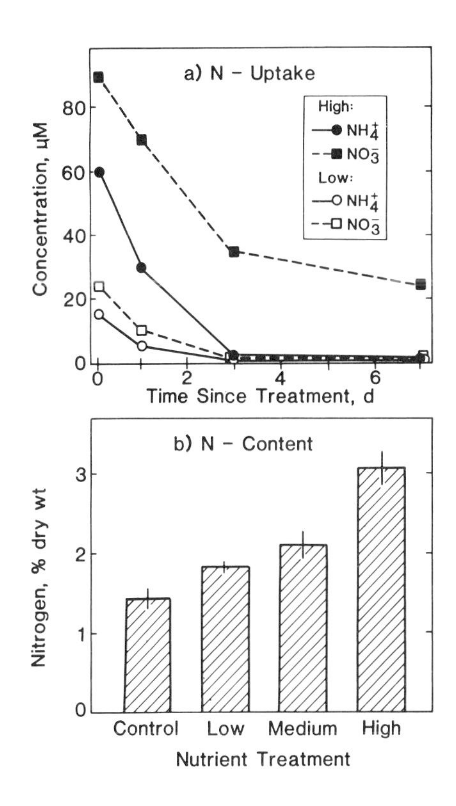

FIGURE 5. Removal of NH_4^+ and NO_3^- from water column (a) and incorporation of nitrogen into plant tissue (b), for experimental pond ecosystems containing submersed vascular plants and treated with 3 levels of nutrient enrichment (plus controls).

and P, we observed rapid uptake of these nutrients after weekly doses (Twilley et al. 1983). Presented in Fig. 5a are time-course trends for mean NH_4^+ and NO_3^- concentrations in high and low treatment ponds over a 1 wk period, where values of both N species were reduced below 1 μM within 3 d at low dosage. At high dosage, NH_4^+ again disappeared rapidly by the third day; however, NO_3^- decreased more slowly, and about 25% of initial concentrations remained after 7 d. Howard-Williams (1981) also reported rapid depletion of dissolved inorganic N and P in enclosures containing dense stands of *P. pectinatus*. The apparent preference for NH_4^+ uptake is consistent with results of kinetic experiments for *M. spicatum* (Nichols and Keeney 1976), *Z. marina* (Iizumi and Hattori 1982), and other species, and contrary to the field studies of Mickle and Wetzel (1978).

The fate of these nutrients removed from the water column was varied. About 10% of the NO_3^- was lost through denitrification; however, differences between rates in vegetated and unvegetated sediments were non-significant (Twilley et al. 1983). Denitrification rates observed here (30-80 μmol N $m^{-2}h^{-1}$) were similar to those reported for other vascular plant beds (Chan and Knowles 1979; Iizumi et al. 1980). While a portion of the nutrient amendments to these experimental ponds was also bound in sediments, uptake by vascular plants and associated algae accounted for the most important mechanism of nutrient immobilization (Twilley et al. 1983). In fact, *P. perfoliatus* exhibited a facultative increase in tissue nitrogen fraction with elevated N loading (Fig. 5b). Throughout the experiment, in low treatment ponds the entire nutrient addition could be accounted for as plant tissue. Although this percentage decreased at heavy fertilization rates, there was clear evidence that plant uptake can represent an important seasonal sink for nutrient inputs to these estuarine waters.

The sequence of plant mortality, decomposition, and associated O_2 consumption represents a pathway for recycling these nutrients. Some of the results of experiments comparing decomposition processes for three species of submersed angiosperms, along with a unicellular and a macrophytic algal species and an emergent marsh grass, are presented in Fig. 6 arranged in terms of the initial C:N ratio for each plant. Here we use C:N ratio as an index of plant structural integrity and relative availability of organic matter for bacterial metabolism. Degradation rates (as loss of mass) were similar to those reported previously for these or related species (e.g., Jewell 1971; Marinucci and Bartha 1982; Rogers and Breen 1982), and were inversely proportional to initial C:N (Godshalk and Wetzel 1978; Naiman and Melillo 1984). Oxygen consumption rates during decomposition would be expected to follow the same pattern, since decay and O_2 demand are both related to microbial metabolism. This was generally the case except for respiration in the *S. alterniflora* microcosms where the highest rates were exhibited. Large releases of dissolved organic matter during the early stages of *S. alterniflora* decomposition apparently

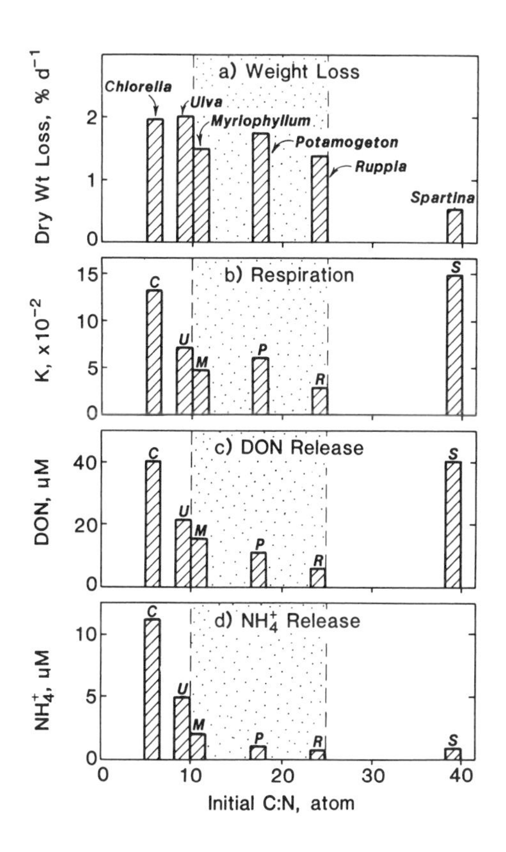

FIGURE 6. Comparison of decomposition processes for microcosms containing various aquatic autotrophs with differing ratios of tissue carbon to nitrogen: a) rate of loss of dry matter; b) first-order rate coefficient for oxygen consumption; c) release of dissolved organic nitrogen (DON); d) release of NH_4^+. Nitrogen release (c and d) indicated as mean concentration (as N) over the last 20 d of experiment minus control concentrations.

resulted from passive leaching (Marinucci and Bartha 1982), and this excretion of labile organic substrate (as DON in Fig. 6c) was presumably sufficient to sustain elevated bacterial metabolism. In these batch experimental microcosms there was little change in detrital C:N during decomposition so that the observed inverse relationship between NH_4^+ release and initial C:N (Fig. 6d) is reasonable.

Submersed vascular plants are able to remain upright through the buoyancy of their vascular system (Sculthorpe 1967). Therefore, they have less need (compared to emergent plants) for lignin-cellulose based structural material to maintain vertical orientation. Their intermediate C:N ratios appear to enable these submersed plants to be an effective storage for assimilated nutrients. During plant decomposition these nutrients were released at relatively slow rates with lower rates of O_2 demand compared to other aquatic producers. By assimilating N and P throughout the spring and summer, these plants reduce the availability of such nutrients for growth of phytoplankton and macrophytic algae, which would exert larger O_2 demands in decomposition especially during the warm summer months. The fact that most of the mortality of submersed vascular plants occurs in the cooler autumn months further decreases their contribution to O_2 depletion of bottom waters in partially stratified coastal plain estuaries such as Chesapeake Bay. Moreover, the considerable O_2 demand associated with release of large quantities of dissolved organics in the decay of *S. alterniflora* was not evident with these submersed plants.

Synthesis and Budgets

The data presented here can be grouped into three general categories concerning the influence of submersed vascular plants on associated ecosystems: 1) production and accumulation of organic matter and creation of habitat structure, all of which may enhance animal growth and recruitment; 2) reduction of wave and tidal energies, which contributes to the trapping and binding of sediments; and 3) temporal buffering of nutrient cycles, thereby reducing the potential for erratic phytoplankton blooms. In this section, we estimate a quantitative aspect of these three effects by constructing budgets for inputs of organic carbon, fine-grain sediments, and total fixed nitrogen to upper Chesapeake Bay.

The magnitude of impact of these plants on the three budgets (Table 2) ranges from moderate (carbon and nitrogen) to very large (sediments). The budgets are calculated on the basis of estimated former abundance and distribution of

TABLE 2. Estimated influence of submersed vascular plant communities on selected materials budgets for upper Chesapeake Bay in 1960.

• *Sources or (Sinks)*[a]	*Material Fluxes*[a]	
	Mass Flux (mT/y)	*% of Inputs*
	Organic Carbon ($\times 10^5$)[b]	
• Rivers	0.8	11
• Phytoplankton	3.8	56
• Macrophytes	2.2	33
	Sediments ($\times 10^6$)[c]	
• Rivers	2.2	65
• Shore Erosion	0.6	35
• (Macrophytes)	(3.6)	(210)
	Nitrogen ($\times 10^3$)[d]	
• Rivers	50	92
• Sewage	4.2	8
• (Macrophytes)	(3.6)	(7)

[a]*Estimates modified from Boynton (1982), including entire area of upper Bay plus tributaries above Potomac River mouth (1.5 x 10^9 m^2). Values in parentheses indicate that submersed vascular plant (macrophyte) communities act as material sinks.*

[b]*River inputs from Biggs and Flemer (1972); mean phytoplankton production (250 g C m^{-2} y^{-1}) taken from Flemer (1970); mean macrophyte production taken to be 360 g C m^{-2} y^{-1}, with areal distribution at 6 x 10^8 m^2 (Stevenson and Confer 1978).*

[c]*Sediment inputs from Schubel and Carter (1977); deposition in macrophyte beds estimated at 0.2 cm/mo (Ward et al. 1984) with a 3 mo season of significant accumulation, a dry bulk density of 1 g/cm^3 and 6 x 10^8 m^2 macrophyte coverage (Stevenson and Confer 1978).*

[d]*River inputs from Guide and Villa (1972) and sewage loading from Smullen, Taft and Macknis (1982); macrophyte uptake assumes 3% N (Twilley et al. 1983) and 200 gd.w./m^2 biomass (Boynton 1982).*

submersed plants for the upper Bay in 1960. The aggregated budget for inputs of organic carbon to the Bay illustrates the importance of autochthonous production which accounted for almost 90% of the total. Although the estuary is phytoplankton dominated, submersed vascular plants contributed one-third of the total input. This is similar to the estimated seagrass contribution in the much shallower Beaufort estuary (Penhale and Smith 1977) but lower than the calculations of Thayer et al. (1975) for North Carolina sounds.

In an average hydrologic and meterological year, about 2×10^6 mT of fine-grain sediments are delivered to Chesapeake Bay from its watershed and shores (Schubel and Carter 1977). According to our estimates, about twice this mass of material would have been deposited during the growing season in the formerly extensive grass beds inhabiting the Bay's shoals (Table 2). This calculation illustrates the potential sedimentological importance of submersed plant communities in estuaries such as Chesapeake Bay. As stressed earlier, the ultimate fate of these sediments following annual senescence of plant shoots is uncertain. However, these trapped sediments were probably held in place for at least 6-8 mo, thus increasing water clarity during the most productive part of the year. The quantitative significance of submersed vascular plants as sinks in nutrient budgets for upper Chesapeake Bay appears to have been less dramatic. Some 7% of the total N inputs could have been assimilated into plant structure, an amount similar to the total sewage loading (Table 2). However, the rapid removal of inputs of N and P in experimental ponds (Fig. 5) illustrates the ability of these plant communities to damp nutrient input pulses from runoff events (e.g. Mickle and Wetzel 1978).

With the recent decline in abundance of submersed plant populations in upper Chesapeake Bay, most of these interactions have been lost, and the few remaining communities tend to be relatively small and isolated in comparision with historical conditions. Thus, the observations from recent years provided here may represent conservative estimates of the previous importance of these plant populations. Furthermore, we have not considered all of the potentially significant effects of submersed plants on ecological processes in Chesapeake Bay. For example, we have not directly demonstrated the role of this vegetated habitat on growth and recruitment of estuarine fish and invertebrates, nor have we addressed the possible impact of these plants on the geochemistry of sulfur, iron, manganese and various other trace elements. Nevertheless, the information provided here documents some of the ways by which submersed plants can influence estuarine dynamics.

ACKNOWLEDGMENTS

We are indebted to our students and research associates for their efforts in helping collect these data, including: L. Lubbers, K. Kaumeyer, S. Bunker, K. Staver, M. Lewis, J. Cunningham, F. Lipschultz, W. Goldsborough, M. Shenton, A. Hermann, S. Bollinger, D. Marbury, G. Baptist, M. Meteyer, L. Winchell. This work was supported by grants from the U.S. EPA (No. R805932010 and X003248010) and the Maryland Dept. of Natural Resources (No. C18-80-430(82)). University of Maryland Center for Environmental and Estuarine Studies Contribution Number HPEL 84-1517.

REFERENCES CITED

Adams, S.M. 1976a. The ecology of eelgrass, *Zostera marina* (L.), fish communities. I. Structural analysis. *J. exp. mar. Biol. Ecol. 22:*269-291.

Adams, S.M. 1976b. Feeding ecology of eelgrass fish communities. *Trans. Amer. Fish. Soc. 105:*514-519.

Bayley, S., V.D. Stotts, P.F. Springer and J. Steenis. 1978. Changes in submerged aquatic macrophyte populations at the head of the Chesapeake Bay, 1958-1975. *Estuaries 1:*74-85.

Biggs, R.B. and D.A. Flemer. 1972. The flux of particulate carbon in an estuary. *Mar. Biol. 12:*11-17.

Bourn, W.S. 1932. Ecological and physiological studies on certain aquatic angiosperms. *Contr. Boyce Thompson Inst. Plant Res. 4:*425-496.

Boynton, W.R. 1982. Ecological role and value of submerged macrophyte communities: A scientific summary, pp. 428-502 *In:* E.G. Macalaster, D.A. Barker and M. Kasper (eds.) *Chesapeake Bay Program Technical Studies; A Synthesis.* U.S. Environmental Protection Agency. NTIS, Springfield, VA.

Brook, I.M. 1977. Trophic relationships in a seagrass community (*Thalassia testudinum)* in Card Sound, Florida. Fish diets in relation to macrobenthic and cryptic faunal abundance. *Trans. Amer. Fish. Soc. 106:*219-229.

Brook, I.M. 1978. Comparative macrofaunal abundance in turtlegrass (*Thalassia testudinum*) community in south Florida characterized by high blade density. *Bull. Mar. Sci. 28*:212-220.

Carr, W.S. and C.A. Adams. 1973. Food habits of juvenile marine fishes occupying seagrass beds in the estuarine zone near Crystal River, Florida. *Trans. Amer. Fish. Soc. 102*:511-540.

Chan, Y.-K. and R. Knowles. 1979. Measurement of denitrification in two freshwater sediments by an *in situ* acetylene inhibition method. *Appl. Environ. Microbiol. 37*:1067-1072.

Christiansen, C., H. Christoffersen, J. Dalsgaard and P. Nørnberg. 1981. Coastal and near-shore changes correlated with die-back in eelgrass (*Zostera marina L*). *Sedim. Geol. 28*:163-173.

Environmental Protection Agency. 1979. *Methods for Chemical Analysis of Water and Wastes*. USEPA-600/4-79-020. Environmental Monitoring and Support Laboratory, Cincinnati, OH.

Flemer, D.A. 1970. Primary production in the Chesapeake Bay. *Chesapeake Sci. 11*:117-129.

Fonseca, M.S., J.S. Fisher, J.C. Zieman and G.W. Thayer. Influence of the seagrass, *Zostera marina L.*, on current flow. *Estuar. Coast. Shelf Sci. 15*:351-364.

Fry, B., R.S. Scalan and P.L. Parker. 1977. Stable carbon isotope evidence for two sources of organic matter in coastal sediments: seagrass and plankton. *Geochim. Cosmochim. Acta.41*:1875-1877.

Ginsburg, R.N. and H.A. Lowenstam. 1958. The influence of marine bottom communities on the depositional environment of sediments. *J. Geol. 66*:310-318.

Godshalk, G.L. and R.G. Wetzel. 1978. Decomposition of aquatic angiosperms. III. *Zostera marina L.* and a conceptual model of decomposition. *Aquat. Bot. 5*:329-354.

Guide, V. and O. Villa. 1972. Chesapeake Bay nutrient input study. Techn. Rep. 47. Region III, EPA, Annapolis Field Office, Annapolis, MD.

Haines, E.B. 1976. Relation between the stable carbon isotope composition of fiddler crabs, plants, and soils in a salt marsh. *Limnol. Oceanogr. 21:*880-883.

Harlin, M.M., B. Thorne-Miller and J.C. Boothroyd. 1982. Seagrass-sediment dynamics of a flood-tidal delta in Rhode Island (USA). *Aquat. Bot. 14:*127-138.

Haller, W.T., D.T. Sutton and W.C. Barlowe. 1974. Effects of salinity on growth of several aquatic macrophytes. *Ecology 55:*891-894.

Hoese, H.D. and R.S. Jones. 1963. Seasonality of larger animals in a Texas turtle grass community. *Publ. Inst. Mar. Sci. 9:*37-47.

Howard-Williams, C. 1981. Studies on the ability of a *Potamogeton pectinatus* community to remove dissolved nitrogen and phosphorus compounds from lake water. *J. Appl. Ecol. 18:* 619-637.

Iizumi, H., A. Hattori and C.P. McRoy. 1980. Nitrate and nitrite in interstitial waters of eelgrass beds in relation to the rhizosphere. *J. exp. mar. Biol. Ecol. 47:*181-201.

Iizumi, H. and A. Hattori. 1982. Growth and organic production of eelgrass *(Zostera marina L.)* in temperate waters of the Pacific coast of Japan. III. Kinetics of nitrogen uptake. *Aquat. Bot. 12:*245-256.

Jewell, W.J. 1971. Aquatic weed decay: Dissolved oxygen utilization and nitrogen and phosphorus regeneration. *J. Water Pollut. Contr. Fed. 43:*1457-1467.

Kemp, W.M. 1983. Seagrass communities as a coastal resource: A preface. *Mar. Techn. Soc. J. 17:*3-5.

Kemp, W.M., W.R. Boynton, J.C. Stevenson, J.C. Means, R.R. Twilley and T.W. Jones (eds.). 1983a. Submerged aquatic vegetation in upper Chesapeake Bay: Studies related to possible causes of the recent decline in abundance. U.S. Environmental Protection Agency. NTIS, Springfield, VA. 298 pp.

Kemp, W.M., W.R. Boynton, R.R. Twilley, J.C. Stevenson and J.C. Means. 1983b. The decline of submerged vascular plants in upper Chesapeake Bay: Summary of results concerning possible causes. *Mar. Techn. Soc. J. 17:*78-89.

Kenworthy, W.J., J.C. Zieman and G.W. Thayer. 1982. Evidence for the influence of seagrasses on the benthic nitrogen cycle in a coastal plain estuary near Beaufort, North Carolina (USA). *Oecologia 54:*152-158.

Kjelson, M.A. and D.R. Colby. 1977. The evaluation and use of gear efficiencies in estimation of estuarine fish abundance, pp. 416-424. *In:* M. Wiley (ed.) *Estuarine Processes, Vol. 2.* Academic Press, New York.

Kanuer, G.A. and A.V. Ayers. 1977. Changes in carbon, nitrogen, adenosine triphosphate, and chlorophyll *a* in decomposing *Thalassia testudinum* leaves. *Limnol. Oceanogr. 22:*408-414.

LaZerte, B.D. and J.E. Szalados. 1982. Stable carbon isotope ratio of submerged freshwater macrophytes. *Limnol. Oceanogr. 27:*13-418.

Lewis, M.R. 1980. An investigation of some homeostatic properties of model ecosystems in terms of community metabolism and component interactions. MS Thesis, University Maryland, College Park. 150 pp.

Livingston, R.R. 1982. Trophic organization of fishes in a coastal seagrass system. *Mar. Ecol. Progr. Ser. 7:*1-12.

Marinucci, A.C. and R. Bartha. 1982. A component model of decomposition of *Spartina alterniflora* in a New Jersey salt marsh. *Can. J. Bot. 60:*1618-1624.

Marsh, A.G. 1973. The *Zostera* epifaunal community in the York River, Virginia. *Chesapeake Sci. 14:*87-97.

Mathiesen, H. and J. Nielsen. 1956. Botaniske undersogelser i Randers Fjord og Grund Fjord. *Bot. Tidssk. 53:*1-34.

McMillan, C., P.L. Parker and B. Fry. 1980. $^{13}C/^{12}C$ ratios in seagrasses. *Aquat. Bot. 9:*237-249.

McRoy, C.P. and C. McMillan. 1977. Production ecology and physiology of seagrasses, pp. 53-88 *In:* C.P. McRoy and C. Helfferich (eds.) *Seagrass Ecosystems: A Scientific Perspective.* Marcel Dekker, New York.

Mickle, A.M. and R.G. Wetzel. 1978. Effectiveness of submersed angiospermepiphyte complexes on exchange of nutrients and organic carbon in littoral systems. I. Inorganic nutrients. *Aquat. Bot. 4*:303-316.

Murray, L. 1983. Metabolic and structural studies of several temperate seagrass communities, with emphasis on microalgal components. Ph.D. Thesis, College of William and Mary, Williamsburg, VA. 90 pp.

Naiman, R. and J. Melillo. 1984. Factors controlling decay rates and detrital nutrient dynamics in northern streams: Implications for ecological progress. *Bull. Mar. Sci.* (In press).

Nichols, D.H. and D.R. Keeney. 1976. Nitrogen nutrition of *Myriophyllum spicatum:* Uptake and translocation of ^{15}N by shoots and roots. *Freshw. Biol. 6*:145-154.

Orth, R.J. and K.L. Heck. 1980. Structural components of eelgrass *(Zostera marina)* meadows in the lower Chesapeake Bay--fishes. *Estuaries 3*:278-288.

Orth, R.J. and K.A. Moore. 1983. Chesapeake Bay: An unprecedented decline in submerged aquatic vegetation. *Science 222*:51-53.

Odum, H.T. and C.M. Hoskin. 1958. Comparative studies on the metabolism of marine waters. *Publ. Inst. Mar. Sci. 8*:23-55.

Ozimek, T., A. Prejs and K. Prejs. 1976. Biomass and distribution of underground parts of *Potamogeton perfoliatus* L. and *P. Lucens* L. in Mikolajskie Lake, Poland, *Aquat. Bot. 2*: 309-316.

Patriquin, D.G. 1972. The origin of nitrogen and phosphorus for growth of the marine angiosperm *Thalassia testudinum. Mar. Biol. 15*:35-46.

Penhale, P.A. 1977. Macrophyte-epiphyte biomass and productivity in an eelgrass *(Zostera marina* L.*)* community. *J. exp. mar. Biol. Ecol. 26*:211-224.

Penhale, P.A. and W.O. Smith. 1977. Excretion of dissolved organic carbon by eelgrass *(Zostera marina)* and its epiphytes. *Limnol. Oceanogr. 22*:400-407.

Perry, M.C., R.E. Munro and G.M. Haramis. 1981. Twenty-five year trends in diving duck populations in Chesapeake Bay. *Trans. N. Am. Wildl. Nat. Res. Conf. 46*:299-310.

Purdy, W.C. 1916. Investigation of the pollution and sanitary conditions of the Potomac watershed -- the plankton, pp. 130-135. U.S. Treasury Dept. Hygienic Laboratory Bull. No. 104, U.S. Govt. Print. Off., Washington, D.C.

Robertson, A.I. 1980. The structure and organization of an eelgrass fish fauna. *Oecologia 47*:76-82.

Rogers, K.H. and C.M. Breen. 1982. Decomposition of *Potamogeton crispus L.:* The effects of drying on the pattern of mass and nutrient loss. *Aquat. Bot. 12*:1-12.

Sand-Jensen, K., C. Prahl and H. Stokholm. 1982. Oxygen release from roots of submerged aquatic macrophytes. *Oikos 38*:349-354.

Schubel, J.R. and H.H. Carter. 1977. Suspended sediment budget for Chesapeake Bay, pp. 48-62. *In:* M. Wiley (ed.) *Estuarine Processes Vol. 2.* Academic Press, New York.

Schulthorpe, C.D. 1967. *The Biology of Aquatic Vascular Plants.* Edward Arnold Ltd., London. 610 pp.

Smullen, J., J.L. Taft and J. Macknis. 1982. Nutrient and sediment loads to the tidal Chesapeake Bay system, pp. 147-262. *In:* E.G. Macalaster, D.A. Barker and M. Kasper (eds.) *Chesapeake Bay Program Technical Studies: A Synthesis.* U.S. Environmental Protection Agency. NTIS, Springfield, VA.

Spence, D.H.N., A.M. Barclay and P.C. Bodkin. 1979. Limnology and macrophytic vegetation of Loch Obisary, a deep brackish lake in the Outer Hebrides. *Proc. Royal Soc. Edinburgh 78B:* 123-138.

Staver, K.W. 1984. Responses of epiphytic algae to nitrogen and phosphorus enrichment and effects on productivity of the host plant, *Potamogeton perfoliatus L.,* in estuarine waters. MS Thesis, Univ. Maryland, College Park. 68 pp.

Stevenson, J.C. and N.M. Confer (eds.). 1978. *Summary of Available Information on Chesapeake Bay Submerged Vegetation.* FWS/OBS-78-66, Fish and Wildlife Service, U.S. Dept. of Interior, Washington, D.C.

Strickland, J.D.H. and T.R. Parsons. 1972. *A practical Handbook of Seawater Analysis.* Fish. Res. Bd., Canada Bull. 167. Ottawa. 310 pp.

Thayer, G.W., S.M. Adams and M.W. LaCroix. 1975. Structural and functional aspects of a recently established *Zostera marina* community, pp. 518-540. *In:* L.E. Cronin (ed.) *Estuarine Research.* Academic Press, New York.

Thayer, G.W., P.L. Parker, M.W. LaCroix and B. Fry. 1978. The stable carbon isotope ratio of some components of an eelgrass, *Zostera marina,* bed. *Oecologia 34:*1-12.

Twilley, R.R., W.M. Kemp, K.W. Staver, W.R. Boynton and J.C. Stevenson. 1983. Effects of nutrient enrichment in experimental estuarine ponds containing submerged vascular plant communities, Chap. 8. *In:* W.M. Kemp et al. (eds.) Submerged Aquatic Vegetation in Upper Chesapeake Bay. U.S. Environmental Protection Agency. NTIS, Springfield, VA.

Valderrama, J.C. 1981. The simultaneous analysis of total nitrogen and total phosphorus in natural waters. *Marine Chem. 10:*109-122.

Verhoeven, J.T.A. 1979. The ecology of *Ruppia*-dominated communities in Western Europe. I. Distribution of *Ruppia* representatives in relation to their autecology. *Aquat. Bot. 6:* 197-268.

Ward, L.G., W.M. Kemp and W.R. Boynton. 1984. The influence of waves and seagrass communities on suspended sediment dynamics in an estuarine embayment. *Marine Geol.* (In press).

Weinstein, M.P. and H.A. Brooks. 1983. Comparative ecology of Nekton residing in a tidal creek and adjacent seagrass meadow: community composition and structure. *Mar. Ecol. Progr. Ser. 12:*15-27.

Westlake, D.F. 1978. Rapid exchange of oxygen between plant and water. *Verh. Intern. Verein. Limnol. 20:*2363-67.

Wood, E.J.F., W.E. Odum and J.C. Zieman. 1969. Influence of Seagrasses on the productivity of coastal lagoons, pp. 495-502. *In:* Laguna Costeras. UN Simposio Mam. Simp. Intern. Lagunas Costeras. (Nov. 1967) Mexico, DF.

Zieman, J.C. and R.G. Wetzel. 1980. Productivity in seagrasses: Methods and rates, pp. 87-118 *In:* D.C. Phillips and C.P. McRoy (eds.), *Handbook of Seagrass Biology: An Ecosystem Perspective*. Garland STPM Press, New York.

THE SEAGRASS FILTER: PURIFICATION OF ESTUARINE AND COASTAL WATERS

Frederick T. Short
Catherine A. Short

Jackson Estuarine Laboratory
University of New Hampshire
Durham, New Hampshire

Abstract: Seagrasses can provide a "filtering" mechanism in estuarine waters by trapping suspended sediments and taking up dissolved water column nutrients. These two processes are discussed from the perspective of water filtration by seagrasses in an effort to establish the plants' benefit to the estuarine system. Previous examinations of such processes have stressed environmental influences on seagrass plants, overlooking the impact that seagrasses may have on the environment. Our approach to the concept of seagrass as a filter has been to examine previous work and combine it with results of measurements of suspended sediment and dissolved nutrient removal in culture tank systems with and without seagrasses. In manipulation experiments, suspended sediment removal was measured by the increase in light penetration, and varied according to added sediment type. Nutrient addition and subsequent depletion in the water column of the culture tanks was measured to determine seagrass community uptake rates. These rates were then extrapolated to a somewhat eutrophic coastal environment for evaluation of potential nutrient removal by seagrasses. A synopsis of these filtering experiments and other studies indicates that seagrass communities remove material of natural or human origin from estuarine waters, but excessive loading of nutrients or suspended material upsets the balance of the seagrass ecosystem, promoting degradation of the seagrass beds and loss of the filtering mechanism.

ISBN 0-12-405070-0

INTRODUCTION

Seagrass meadows form the basis of many estuarine and coastal ecosystems. These seagrass communities are best known for their roles as nurseries for coastal fish and invertebrate populations, as supporters of complex trophic food webs, and as suppliers of large quantities of detrital organic material to estuarine and offshore environments (Wood, Odum and Zieman 1969; Phillips and McRoy 1980). Most studies of seagrass ecology have investigated environmental influences on these plants (McRoy and Helfferich 1977); few investigations have examined the effects of seagrasses on the environment (Den Hartog 1983). The possible effects of seagrasses as a "filter" for both suspended particulate materials and nutrients present in the water column have never been quantitatively investigated. A review of the literature combined with field data and experimental culture work leads us to postulate that seagrasses have a decided influence on the environments which they inhabit. It is possible and indeed constructive to compare these influences to a filtering process.

The filtering mechanism a seagrass meadow supplies to the water flowing through it is in many ways parallel to secondary and tertiary water treatment processes. Secondary treatment entails removal of suspended particulate material, while tertiary treatment involves removal of dissolved nutrients from water. As early as 1944, Jackson observed that "a large amount of flocculent material ... was strained out of the inflowing water by the eel grass" in Great Bay, New Hampshire, prior to the eelgrass wasting disease of the 1930's. And nutrient removal was acknowledged by Cambridge (1979) who suggested that "the seagrass ecosystem acts as ... a filter ...."

Seagrass beds have long been depicted as effective mechanisms for baffling current flow and increasing sedimentation (Ginsburg and Lowenstam 1958; Swinchatt 1965; Wood et al. 1969; Davies 1970; Scoffin 1970; Brasier 1975; Wolfe, Thayer and Adams 1976). Although authors continue to make this point and to reference earlier work, we were unable to find anything more than observations documenting the influence of seagrass. A possible exception is the early paper of Molinier and Picard (1952), which provides a detailed description of Mediterranean seagrass beds and the sediment characteristics associated with bed formation and destruction. That seagrasses play a major role in sedimentation is strongly supported by numerous observations and by circumstantial evidence, but it is a difficult role to quantify.

The ability of seagrasses to remove dissolved nutrients from estuarine waters has not been extensively investigated. The early work of Raymont (1947) suggested the efficiency of nutrient removal by seagrass leaves; he was forced to stop broadcast fertilization of Scottish lochs because it "encouraged ... a heavy growth of seaweed and *Zostera*." Several authors have examined nutrient uptake by seagrass leaves in chamber experiments and determined that the leaves take up both nitrogen and phosphorus from the surrounding water as a function of nutrient concentration (McRoy and Barsdate 1970; Penhale and Thayer 1980; Iizumi and Hattori 1982; Thursby and Harlin 1982). Recent experimental nutrient enrichment of the water column in seagrass beds resulted in an increase of *Zostera marina* biomass (Harlin and Thorne-Miller 1981). However, the extent to which seagrass leaves affect the water column nutrient environment remains an open question.

A review of the literature shows that few studies have directly investigated seagrass as a filter or presented quantitative data concerning the filtering capacities of seagrasses. The aim of this paper is to present the results of experimental manipulations designed to test the potential capacity of the seagrass ecosystem to function as a filter in the estuarine environment. Our approach is to examine the impact of seagrass on the environment rather than the more usual investigation of environmental influences on seagrass.

METHODS

Seagrasses were grown in tank culture in Ft. Pierce, Florida, to provide controlled simulation of field conditions. Several 0.5 x 0.7 x 2.2 m cement outdoor tanks were filled with 15 cm of sediment from the Indian River. The seagrasses *Halodule wrightii* Aschers. and *Syringodium filiforme* Kütz. were then transplanted in monoculture into the tanks in water 20 cm deep. Constant low-nutrient water inflow and circular current were maintained in the tanks. *Halodule wrightii* and *S. filiforme* were successfully established and maintained in culture for 15 months prior to use in these manipulation experiments. Although the tanks were small, they provided a controlled environment for testing the impact of seagrasses on suspended particulate loads and nutrient enrichments.

The seagrasses and tanks were kept relatively free of macroscopic epiphytes by large numbers of grazing snails and amphipods. The *H. wrightii* tank had an average midwater current velocity of 3.3 cm sec^{-1}, a density of 7350 shoots m^{-2}, and dry wt biomass of 169 g m^{-2} (leaves) and 230 g m^{-2}

(roots plus rhizomes). Shoots averaged 2.5 cm in height, 1.3 mm in width, with ca 4 leaves per shoot. The *S. filiforme* tank had a comparable current of 6.8 cm sec^{-1}, a shoot density of 2200 m^{-2}, and dry wt biomass of 90 g m^{-2} (leaves) and 85 g m^{-2} (roots plus rhizomes). Shoots averaged 24 cm in height, 0.8 mm in width, with ca 2 leaves per shoot. An unvegetated control tank had an average midwater current velocity of 10.5 cm sec^{-1} and a mud bottom comparable to the seagrass tanks.

Sediment Removal

Experimental manipulations were used to examine the removal of suspended sedimentary material from the water column by seagrasses. Three types of sediment were added in measured quantities during separate experiments to seagrass and control tanks: sand-silt and silt-clay sediment from the Indian River; pure clay-size material in the form of dry red potters' clay; and terrestrial organic silt material. Water inflow, but not circular current, was turned off during the experiments.

Sediments were added to the experimental tanks by creating a slurry of the appropriate grain-size material and then quickly pouring this mixture into each tank. Removal of suspended sediments from the water column of both culture and control tanks was determined using two methods for measuring light attenuation in the water column. First, light levels just below and 10 cm below the surface were measured with a spherical Biosphere sensor in units of quanta sec^{-1} cm^{-2}. Second, the irradiance on a horizontal surface just below the water surface and just above the sediment surface (18 cm depth) was measured in μE sec^{-1} m^{-2} with a Li-Cor submersible photosensor. In both cases measurements were made in a grass free area of the tanks to avoid the shading influence of the plants. From these measurements the percent surface light was calculated. Extinction coefficients m^{-1} were calculated from the light gradient and water depth.

Nutrient Removal

In a similar experiment, nitrogen and phosphorus were added to the water in the form of ammonium chloride and sodium phosphate. Again, the addition was made to both culture and control tanks. Nutrients were added to the high current area of the tank to obtain maximum dispersal, and water in the tanks was then mixed by hand. Water samples

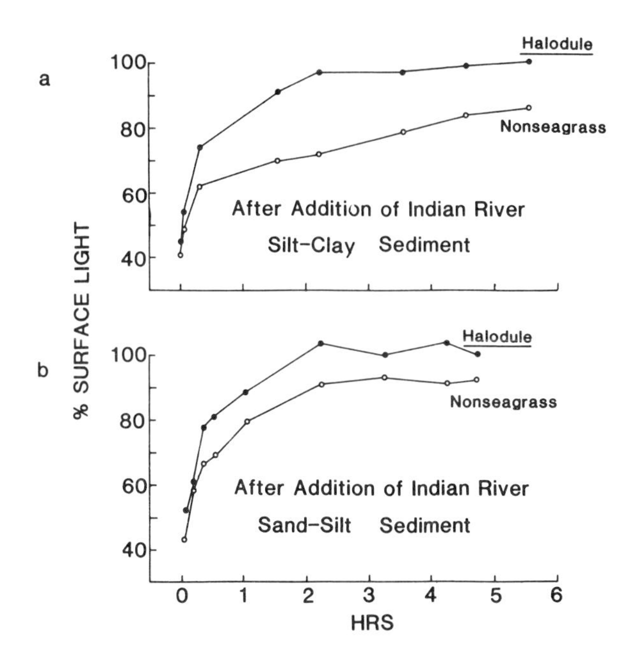

*FIGURE 1. Percent of surface light reaching 10 cm depth after addition of Indian River silt-clay sediment (a) and sand-silt sediment (b) in seagrass (*Halodule wrightii*) and nonseagrass tanks.*

for nutrient analysis were collected after two minutes and then at designated time intervals, filtered and frozen immediately for later analysis of ammonium and phosphate on a Technicon AutoAnalyzer (Zimmermann, Price and Montgomery 1977).

RESULTS

Sediment Removal

The impact of seagrasses in decreasing the suspended sediment load was clearly demonstrated in the culture systems. The seagrass extended through the water column to the surface, thereby providing a large sediment-trapping capacity. Light transmission measured over time after addition of various sediment types in a *Halodule wrightii* tank and an unvegetated control tank increased more rapidly in the former. A seagrass culture tank of *H. wrightii* to

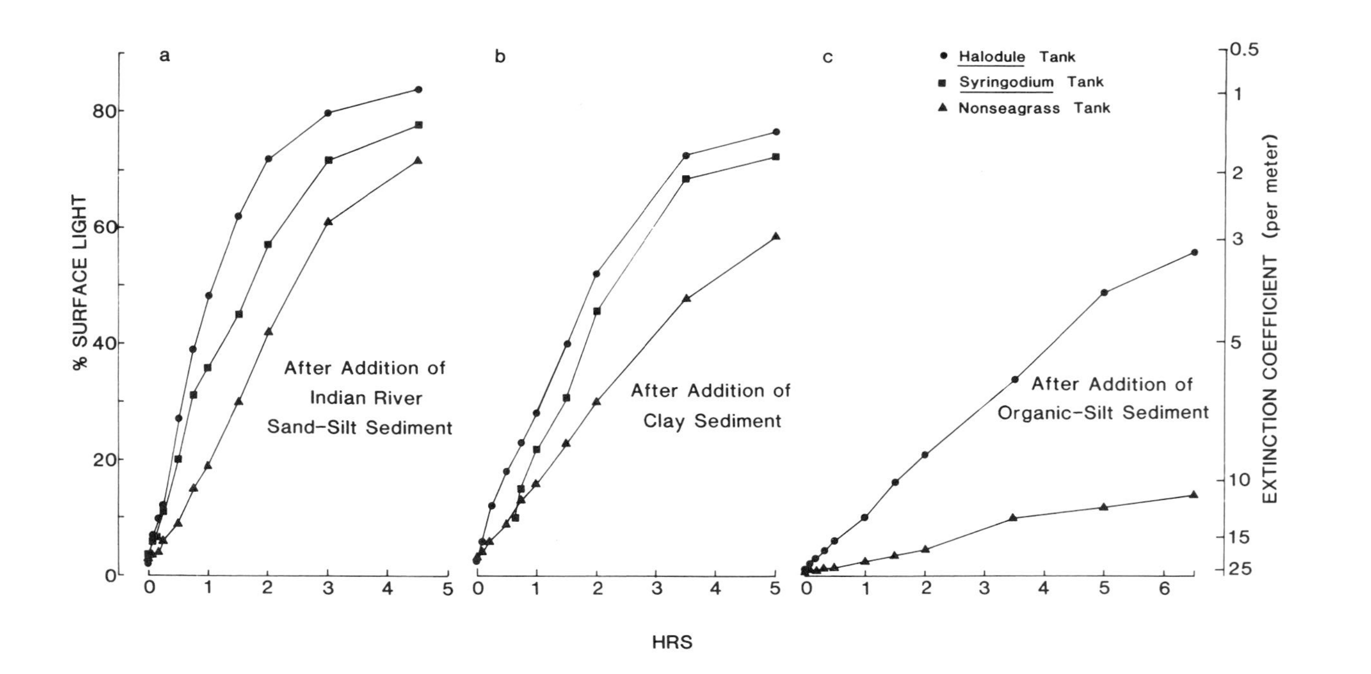

FIGURE 2. Percent of surface light reaching an unvegetated area of bottom, and the extinction coefficient for a Halodule wrightii *tank, a* Syringodium filiforme *tank and an unvegetated tank after addition of Indian River sand-silt sediment (a), clay sediment (b), and organic-silt sediment (c).*

which Indian River silt-clay was added showed a 14% increase in light transmission over the control tank; with the addition of sand-silt material a 10% increase in light transmission was observed (Fig. 1).

In a similar experiment, light extinction at the sediment surface was compared in three tanks containing *Halodule wrightii*, *Syringodium filiforme* and no vegetation, respectively (Fig. 2). A more rapid decrease in turbidity was evident in tanks of both seagrass species, although in all cases *H. wrightii* had a greater impact. Suspended Indian River sand-silt was removed in the presence of seagrasses more rapidly than in the unvegetated control tank, and after six hours, the extinction coefficient in the *H. wrightii* tank was reduced to 1 m^{-1} while the two other tanks

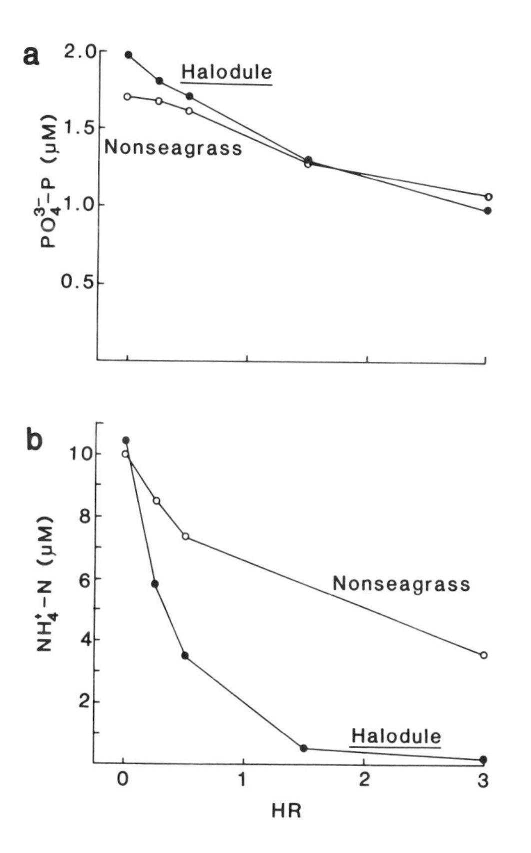

FIGURE 3. Nutrient removal from the water column by the seagrass community. Decrease with time in phosphate concentration (a) and ammonium concentration (b) for the Halodule wrightii *tank and the nonseagrass control tank.*

remained higher (Fig. 2). Addition of uniform-size clay particles to these three tanks resulted in a greater reduction in turbidity for both *H. wrightii* and *S. filiforme* than for the unvegetated control (Fig. 2b). The greatest difference in removal of suspended sediments from the water column occurred when a terrestrial organic fine silt material was added. With this mixture, the turbidity in the unvegetated control tank was more than three times that of the *H. wrightii* tank (Fig. 2c).

Nutrient Removal

Phosphate removal from the water column was similar for both the *H. wrightii* and the unvegetated control tanks (Fig. 3a). Ammonium concentrations in the water column decreased more rapidly in the seagrass tanks than in the control tank (Fig. 3b). Nitrogen removal from the water column of the *H. wrightii* culture tank was measured on several days, with different initial ammonium concentrations (Fig. 4b). The data for resulting measurements of concentration change, together with existing data for seagrass abundance, made it possible to calculate uptake rates at average concentrations (Fig. 4a). Ammonium uptake by *H. wrightii* was linearly related to concentration in the water column up to ca 20 μM NH_4^+ - N as demonstrated by the correlation coefficient of 0.97.

DISCUSSION AND CONCLUSIONS

Suspended Sediment Removal

The suspended material in estuarine waters includes both inorganic sediments and particulate organic material. These are the products of phytoplankton production, terrestrial run-off, and resuspension by wind driven currents and by tidal effects on the shore and bottom. Human activities contributing to suspended loads include damming (Pérès and Picard 1975), dredging (Zieman 1975; Lot 1977), removal of terrestrial vegetation in drainage areas, agricultural runoff and other sources of erosion (Chesapeake Bay Study 1982). Whatever the source of sediments to an estuary, the distribution of material appears related to the presence or absence of seagrasses. Studies show that seagrass sediments contain a higher percentage of fine particulate material than nonseagrass sediments (Lynts 1966; Kenworthy, Zieman and Thayer 1982;

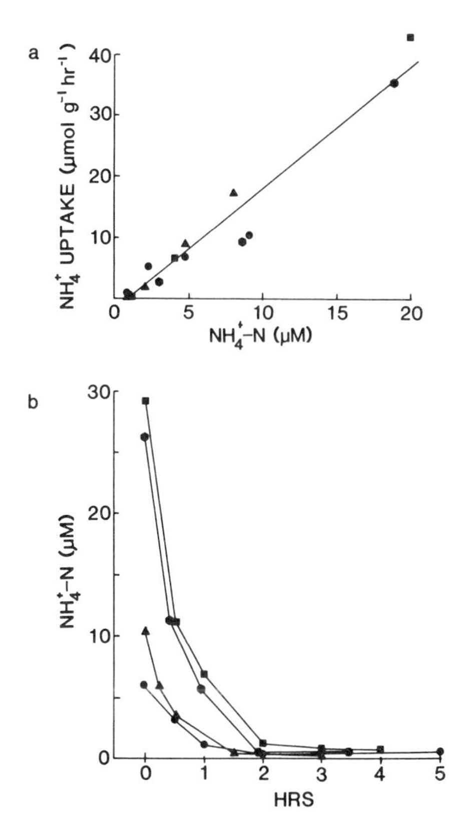

FIGURE 4. Ammonium removal from the water column by the seagrass community. (a) Uptake rates at average ammonium concentrations calculated from (b) decreases in ammonium concentration with time in the Halodule wrightii *tank. Each of the four substrate depletion curves (different symbols in b) represent individual experiments starting at different concentrations. The interval between each pair of points is used to calculate the uptake rate which corresponds to the same symbol in (a). The best fit linear regression for ammonium uptake (y) vs. concentration (x) was $y = 2.00\,x - 2.03$ and the correlation coefficient was 0.97.*

Hoskin 1983) and that seagrass meadows accumulate sediments at a faster rate (Wolfe et al. 1976), although current speed also plays a role in sedimentation rates (Short, Nixon and Oviatt 1974; Fonseca et al. 1982), as does seagrass density (Scoffin 1970). Seagrasses also maintain accumulated sediment, reducing resuspension of particulate material in the water column (Ginsberg and Lowenstam 1958). Seagrasses bend on exposure to currents, reducing current speeds to near zero at the sediment surface and preventing resuspension (Scoffin 1970).

However, seagrasses do not represent a perfect filter for suspended sediments. Storms may rapidly suspend more material than seagrass beds can handle while at the same time damaging the beds themselves (Wolfe et al. 1976). Damming of rivers may result in buildup of sediments that were previously flushed during spring flooding, a buildup that may ultimately destroy seagrass areas (Pérès and Picard 1975). Consistent overloading of the water column with suspended particulate material will decrease available light, thereby limiting plant productivity and diminishing the seagrass bed (Burkholder and Doheny 1968). Seagrasses are affected by the low light levels in turbid waters; as a result of human activities, increased turbidity in the Chesapeake Bay has contributed to the decrease of seagrass density and total area of coverage in the last 15 years (Chesapeake Bay Study 1982; Kemp et al. 1983; Orth and Moore 1983).

In this study, we measured suspended sediment removal using seagrass culture experiments. Our initial examination of these processes showed a distinct difference in removal of suspended material when seagrasses were present, indicated by increased light penetration (Fig. 1). Additionally, seagrasses had a greater influence on some sediment types than others. The removal rates of suspended sediment material indicated by light levels and extinction coefficients measured at the sediment surface showed differences between sediment types and between the two species of seagrass tested and a non-seagrass tank (Fig. 2). The initial rates of sediment removal by both seagrass species were greater than that of the culture tank with no seagrass when either Indian River sand-silt or a suspension of uniform-size potters' clay was added to the tanks (Fig. 2a, b).

The addition of a terrestrially-derived organic silt sediment showed a dramatic difference in sediment removal between the seagrass and nonseagrass situations (Fig. 2c). This last experiment may be similar to conditions created by transport of suspended materials into an estuary via terrestrial runoff. Similarly, the first series of experiments simulated resuspension of bottom sediments due to wind and current mixing. Suspended sediment removal by *Halodule wrightii* was greater than by *Syringodium filiforme* due to differences in shoot density and leaf morphology. *Halodule wrightii* had greater density and a flat blade-like leaf, providing a greater surface area for sediment trapping. Both the lower density and cylindrical leaf blade of *S. filiforme* contributed to its less effective trapping of sediment particles.

The degree of suspended sediment removal from the water column can be influenced by factors other than seagrass shoot density and morphology. The presence of macroscopic epiphytes on seagrass leaves increases the surface area available for the settlement of particles. The sediment trapping effect of seagrasses has an added dimension in species such as *H. wrightii* which are vertical in the water column during daylight when lacunal spaces in the plant are gas-filled, but lie horizontal just above the sediment surface at night. The daily cycle of sediment trapping by upright seagrass leaves and disappearance of sediments from the leaves overnight was strikingly evident in the seagrass cultures.

Dissolved Nutrient Removal

Studies of seagrass nutrient uptake demonstrate the ability of the plants to remove nutrients from the water surrounding the leaves and the roots; in the leaves, concentration-dependent nutrient uptake removes both nitrogen and phosphorus directly from the water column (McRoy and Barsdate 1970; McRoy and Goering 1974; Penhale and Thayer 1980; Iizumi and Hattori 1982; Thursby and Harlin 1982). Additionally, two major factors interact to increase the effectiveness of dissolved nutrient removal by seagrass leaves. First, leaf morphology determines the extent of leaf surface area interacting with the water column. Second, water movement through the seagrass bed increases the volume of water contacting the leaves.

Epiphytic algal growth on seagrass leaves provides an expanded surface area for nutrient absorption. In turn, increased nutrient loads correlate with greater epiphyte growth. However, too great an epiphyte load is detrimental to seagrasses (Bulthius and Woelkerling 1983; Sand-Jensen and Borum 1983). Just as overloading affects the plants' ability to function as a filter of suspended particulate matter, so do excessive nutrients stress the plants (Kemp et al. 1983).

Although it has been suggested that seagrasses take up heavy metals, relatively little work has been done on the mechanisms or rates of uptake (Brinkhuis, Penello and Churchill 1980; Fabris, Harris and Smith 1982). Seagrass beds may represent the largest biological reservoir of metals in the ecosystems which they inhabit (Drifmeyer et al. 1980; Smith, Kozuchi and Hayasaka 1982). Concentrations of metals in seagrass plants are related to concentrations in both the sediment and water column (Parker 1966; Lyngby and Brix 1982).

The use of experimental culture tanks in this study allowed measurement of nitrogen removal from the water column by seagrass leaves and their associated community. The decrease in ammonium in the water column of the *H. wrightii* culture tank was consistently a function of ammonium concentration; ammonium uptake rates by this seagrass assemblage showed a strong linear correlation with average ammonium concentrations. This measure of nitrogen uptake for a tropical seagrass is used later to estimate the potential nutrient removal by seagrass beds in an estuarine system. Measurement of dissolved phosphorus removal from the culture systems showed no discernable difference between the seagrass tank and the control tank (Fig. 3). The absence of phosphorus removal from the water column may be attributable to higher levels of phosphate (ca 10.0 μM PO_4^{3-} - P) observed in the interstitial water in these tanks (F. Short pers. obs.). We restricted this study to removal of water column nitrogen in the form of ammonium.

Manipulation experiments of this kind create an opportunity to extrapolate findings about seagrass filtration to estuarine environments. We realize the inherent problems in extrapolations of small experimental studies to large estuarine areas, but lacking large-scale investigations, such projections do provide an estimate of the capacity of seagrass to act as a nutrient filter. To examine the influence of seagrass communities in the somewhat eutrophied, sub-tropical Indian River Lagoon, Florida, we combined existing data collected during the 1976 Indian River Coastal Zone Study (Harbor Branch Foundation, Ft. Pierce, Florida) with our seagrass culture results. Using nitrogen uptake rates for the seagrass community obtained from the culture experiments (Fig. 4) and seasonal nutrient data (Mahoney and Gibson 1983), we calculated seasonal nitrogen uptake rates for the community (Table 1). Combining these with estimates of seagrass abundance (F. Short pers. obs.; N. Eiseman, Ft. Pierce, FL, unpubl.) and distribution in the Indian River (Thompson 1978) gives an estimate of potential nitrogen removal by the Indian River seagrass community. Removal is greatest in early summer during the maximum seagrass growth period and lowest throughout the winter and early spring (Table 1). The seagrass community present in

Table 1. Potential nitrogen removal by the seagrass community in Indian River Lagoon, Florida

Time	*Lagoon biomass*[a] *(metric tons dry wt.)*	*Lagoon*[b] *NH_4^+ - N* *(μM)*	*Potential*[c] *NH_4^+ removal* *(metric tons N)*
Nov-Jan	1400	8.0	590
Feb-Apr	3300	4.0	600
May-Jul	4400	8.0	1860
Aug-Oct	2800	6.0	840
			3890

a. F. Short, pers. obs. and Thompson 1978; b. Mahoney and Gibson 1983; c. Calculated from Fig. 4.

the Indian River in 1976 had the potential to remove 3890 metric tons of nitrogen. This is a large amount of nitrogen to remove from the water column, but represents only 11% of the nitrogen added to the Indian River water column that year by a sewage plant centrally located at Vero Beach.

The input of nitrogen to the Indian River, amounting to almost 35,000 metric tons in 1976, has increased steadily to nearly double that value in 1983 as the result of increased sewage discharge (Vero Beach Water and Sewer Department, pers. comm.). Increased nutrient load has been accompanied by a local decrease in seagrass abundance in the central Indian River and a virtual disappearance of seagrass beds in the vicinity of Vero Beach. In our opinion the increased nutrient load in this poorly flushed area of the Indian River has resulted in conditions unfavorable to seagrass growth. The exclusion or decline of the seagrass nutrient filtering mechanism from the estuary brought on by excess nutrient loads magnifies the problem of eutrophication.

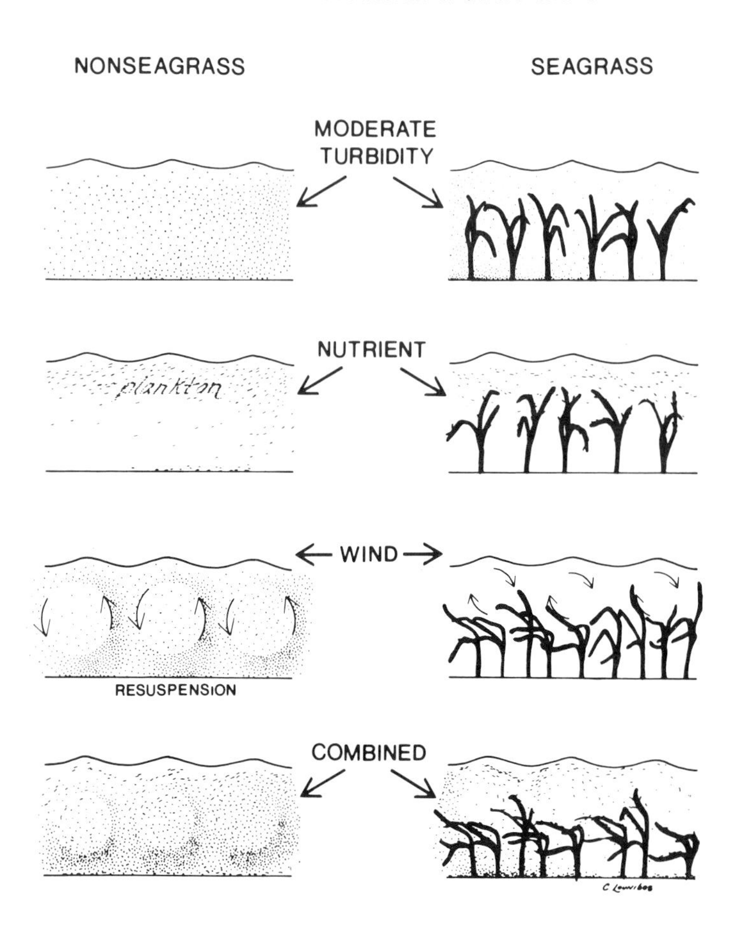

FIGURE 5. Summary diagram illustrating influences of seagrass on the estuarine environment. Conditions are diagrammed for nonseagrass and seagrass systems. Moderate turbidity throughout the water column vs. seagrass removal of suspended sediment; nutrient addition producing a plankton bloom vs. seagrass/epiphytes/few plankton; resuspension of sediments vs. circulation subdued by seagrass; and the combination of factors producing a turbid system vs. a purified seagrass environment.

The individual mechanisms of suspended sediment removal and dissolved nutrient uptake have been shown to be effective and important factors in purification of estuarine waters (Fig. 5). In combination, these two filtering mechanisms can function to help maintain a healthy estuarine environment as well as to enhance the seagrass beds themselves. The fine-grained sedimentary material removed from the water by seagrasses is deposited on the sea bottom between seagrass shoots. Such material is protected from resuspension by the seagrasses and becomes a source of organic carbon and fine-grained mud for the seagrass benthos, ultimately stimulating the rate of decomposition and increasing the nutrients available in the sediments (Iizumi, Hattori and McRoy 1982; Short 1983b). This change in seagrass sediment conditions has been shown to correlate with a change in plant morphology (Short 1983a), resulting in conditions more favorable to both sedimentation and nutrient uptake from the water column.

ACKNOWLEDGMENTS

We thank Harbor Branch Foundation, Inc., for the opportunity to work in the Indian River area of Florida. Thank you to John Ryther for his support and encouragement, to Carl Zimmermann for operation of the autoanalyzer, and to Art Mathieson, Dennis Hanisak and Mark Fonseca for comments on the manuscript. We are grateful to Jim Mullins, Marie Polk, and Janis Marshall for preparation of the manuscript and to Charisa Lounibos for the seagrass drawing. Harbor Branch Foundation contribution no. 361 and Jackson Estuarine Laboratory contribution no. 164.

REFERENCES CITED

Brasier, M.D. 1975. An outline history of seagrass communities. *Palaeontology 18:* 681-702.

Brinkhuis, B.H. W.F. Pennello and A.C. Churchill. 1980. Cadmium and manganese flux in eelgrass *Zostera marina*. Metal uptake by leaf and root-rhizome tissues. *Mar. Biol. 58:* 187-196.

Bulthius, D.A. and W.J. Woelkerling. 1983. Biomass accumulation and shading effects of epiphytes on leaves of the seagrass, *Heterozostera tasmanica*, in Victoria, Australia. *Aquat. Bot. 16:* 137-148.

Burkholder, P.R. and T.E. Doheny. 1968. The biology of eelgrass (with special reference to Hempstead and South Lyster Bays, Nassau County, Long Island, New York). Contribution #3 from the Department of Conservation and Waterways, Town of Hempstead, Long Island. Contribution #1227 from the Lamont Geological Observatory, Palisades, NY.

Cambridge, M.L. 1979. Cockburn Sound Environmental Study. Technical report on seagrass. Department of Conservation and Environment, Western Australia. Report no. 7, June, 1979. 100 p.

Chesapeake Bay Study. 1982. Chesapeake Bay program technical studies: a synthesis. U.S. Environmental Protection Agency, Washington, D.C. U.S. General Printing Office: 1983-606-490. 634 p.

Davies, G.R. 1970. Carbonate bank sedimentation, Eastern Shark Bay, Western Australia, pp. 85-168. *In:* B.W. Logan, G.R. Davies, J.F. Read and D.E. Cebulski (eds.), *Carbonate Sedimentation and Environments, Shark Bay, Western Australia.* The American Association of Petroleum Geologists, Tulsa, OK.

Den Hartog, C. 1983. Structural uniformity and diversity in *Zostera*-dominated communities in Western Europe. *Mar. Tech. Soc. J. 17:* 6-14.

Drifmeyer, J.E., G.W. Thayer, F.A. Cross and J.C. Zieman. 1980. Cycling of Mn, Fe, Cu, and Zn by eelgrass, *Zostera marina* L. *Amer. J. Bot. 67:* 1089-1096.

Fabris, G.J., J.E. Harris and J.D. Smith. 1982. Uptake of cadmium by the seagrass *Heterozostera tasmanica* from Corio Bay and Western Port, Australia. *Aust. J. Mar. Freshw. Res. 33:* 829-836.

Fonseca, M.S., J.S. Fisher, J.C. Zieman and G.W. Thayer. 1982. Influence of the seagrass, *Zostera marina* L., on current flow. *Est. Coast. Shelf Sci. 15:* 351-364.

Ginsburg, R.N. and H.A. Lowenstam. 1958. The influence of marine bottom communities on the depositional environment of sediments. *J. Geol. 66:* 310-318.

Harlin, M.M. and B. Thorne-Miller. 1981. Nutrient enrichment of seagrass beds in a Rhode Island coastal lagoon. *Mar. Biol. 65:* 221-229.

Hoskin, C.M. 1983. Sediment in seagrasses near Link Port, Indian River, Florida. *Fla. Sci. 46:* 153-161.

Iizumi, H. and A. Hattori. 1982. Growth and organic production of eelgrass (*Zostera marina* L.) in temperate waters of the Pacific Coast of Japan. III. The kinetics of nitrogen uptake. *Aquat. Bot. 12:* 245-256.

Iizumi, H., A. Hattori and C.P. McRoy. 1982. Ammonium regeneration and assimilation in eelgrass (*Zostera marina*) beds. *Mar. Biol. 66:* 59-65.

Jackson, C.F. 1944. Physical and biological features of Great Bay and the present status of its marine resources. No. 1. A biological survey of Great Bay, New Hampshire, by the Marine Fisheries Commission. 61 p.

Kemp, W.M., W.R. Boynton, R.R. Twilley, J.C. Stevenson and J.C. Means. 1983. The decline of submerged vascular plants in Upper Chesapeake Bay: Summary of results concerning possible causes. *Mar. Soc. Tech. J. 17:* 78-89.

Kenworthy, W.J., J.C. Zieman and G.W. Thayer. 1982. Evidence for the influence of seagrasses on the benthic nitrogen cycle in a coastal plain estuary near Beaufort, North Carolina (USA). *Oecologia 54:* 152-158.

Lot, A. 1977. General status of research on seagrass ecosystems in Mexico, pp. 233-245. *In:* C.P. McRoy and C. Helfferich (eds.), *Seagrass Ecosystems: A Scientific Perspective.* Marcel Dekker, New York.

Lyngby, J.E. and H. Brix. 1982. Seasonal and environmental variation in cadmium, copper, lead, and zinc concentrations in eelgrass (*Zostera marina* L.) in the Limfjord, Denmark. *Aquat. Bot. 14:* 59-74.

Lynts, G.W. 1966. Relationship of sediment-sized distribution to ecologic factors in Buttonwood Sound, Florida Bay. *J. Sed. Pet. 36:* 66-74.

Mahoney, R.K. and R.A. Gibson. 1983. Phytoplankton ecology of the Indian River near Vero Beach, Florida. *Fla. Sci. 46:* 212-232.

McRoy, C.P. and R.J. Barsdate. 1970. Phosphate absorption in eelgrass. *Limnol. Oceanogr. 15:* 6-13.

McRoy, C.P. and J.J. Goering. 1974. Nutrient transfer between the seagrass *Zostera marina* and its epiphytes. *Nature 228:* 173-174.

McRoy, C.P. and C. Helfferich. 1977. *Seagrass Ecosystems: A Scientific Perspective.* Marcel Dekker, New York. 314 p.

Molinier, R. and J. Picard. 1952. Recherches sur les herbiers de phanérogames marines du littoral méditerranéen français. *Ann. Inst. Océanogr. 27:* 157-234.

Orth, R.J. and K.A. Moore. 1983. Chesapeake Bay: an unprecedented decline in submerged aquatic vegetation. *Science 222:* 51-53.

Parker, P.L. 1966. Movement of radioisotopes in a marine bay: cobalt-60, iron-59, manganese-54, zinc-65, sodium-22. *Publ. Inst. Mar. Sci. Tex. 11:* 102-107.

Penhale, P.A. and G.W. Thayer. 1980. Uptake and transfer of carbon and phosphorus by eelgrass (*Zostera marina* L.) and its epiphytes. *J. Exp. Mar. Biol. Ecol. 42:* 113-123.

Pérès, J.M. and J. Picard. 1975. Causes of decrease and disappearance of the seagrass *Posidonia oceanica* on the French Mediterranean Coast. *Aquat. Bot. 1:* 133-139.

Phillips, R.C. and C.P. McRoy. 1980. *Handbook of Seagrass Biology: An Ecosystem Perspective.* Garland STPM Press, New York. 353 p.

Raymont, J.E.G. 1947. A fish farming experiment in Scottish sea lochs. *J. Mar. Res. 6:* 219-227.

Sand-Jensen, K. and J. Borum. 1983. Regulation of growth in eelgrass (*Zostera marina* L.) in Danish coastal waters. *Mar. Tech. Soc. J. 17:* 15-21.

Scoffin, T.P. 1970. The trapping and binding of subtidal carbonate sediments by marine vegetation in Bimini Lagoon, Bahamas. *J. Sed. Pet. 40:* 249-273.

Short, F.T. 1983a. The seagrass, *Zostera marina* L.: Plant morphology and bed structure in relation to sediment ammonium in Izembek Lagoon, Alaska. *Aquat. Bot. 16:* 149-161.

Short, F.T. 1983b. The response of interstitial ammonium in eelgrass (*Zostera marina* L.) beds to environmental perturbations. *J. Exp. Mar. Biol. Ecol. 68:* 195-208.

Short, F.T., S.W. Nixon and C.A. Oviatt. 1974. Field studies and simulation with a fine grid hydrodynamic model, pp. 1-27. *In: An Environmental Study of a Nuclear Power Plant at Charlestown, Rhode Island,* Univ. of R.I. Mar. Tech. Rpt. 33, VI-B.

Smith, G.W., A.M. Kozuchi and S.S. Hayasaka. 1982. Heavy metal sensitivity of seagrass rhizoplane and sediment bacteria. *Bot. Mar. 25:* 19-24.

Swinchatt, J.P. 1965. Significance of constituent composition, texture and skeletal breakdown in some recent carbonate sediments. *J. Sed. Pet. 35:* 79-90.

Thompson, M.J. 1978. Species composition and distribution of seagrass beds in the Indian River Lagoon, Florida. *Fla. Sci. 41:* 91-96.

Thursby, G.B. and M.M. Harlin. 1982. Leaf-root interaction in the uptake of ammonium by *Zostera marina*. *Mar. Biol. 72:* 109-112.

Wolfe, D.A., G.W. Thayer and S.M. Adams. 1976. Manganese, iron, copper, and zinc in an eelgrass (*Zostera marina*) community, pp. 256-270. *In:* C.E. Cushing, Jr., (ed.), *Radio-ecology and Energy Resources*. Ecol. Soc. of Am. Spec. Pub. no. 1. Dowden, Hutchinson, and Ross, Inc., Stroudsburg, PA.

Wood, E.J.F., W.E. Odum and J.C. Zieman. 1969. Influence of sea grasses on the productivity of coastal lagoons. Mem. Simp. Intern. Lagunas Costeras. UNAM-UNESCO, pp. 495-502.

Zieman, J.C. 1975. Tropical seagrass ecosystems and pollution, pp. 63-74. *In:* E.J.F. and R.E. Johannes (eds.), *Tropical Marine Pollution*. Elsevier Oceanography Series, no. 12. Elsevier, Amsterdam.

Zimmermann, C.F., M. Price and J.R. Montgomery. 1977. Operation, methods and quality control of Technicon AutoAnalyzer system for nutrient determinations in seawater. Harbor Branch Foundation, Inc. Tech. Rpt. No. 11, Ft. Pierce, FL.

BIOLOGICAL AND PHYSICAL FILTERING IN ARID-REGION ESTUARIES: SEASONALITY, EXTREME EVENTS, AND EFFECTS OF WATERSHED MODIFICATION

Joy B. Zedler

Biology Department
San Diego State University
San Diego, California

Christopher P. Onuf

Marine Science Institute
University of California
Santa Barbara, California

Abstract: Both biological and physical processes lead to the removal of dissolved and suspended materials from water that enters estuaries. Our understanding of these "filtering functions" in southern California estuaries is summarized in three concepts: I. In this semi-arid region, ecosystems are highly seasonal in their structure and functioning, so that both the biological and physical "filters" differ for wet and dry seasons. II. Extreme events such as floods substantially modify estuarine structure and functioning. They cause a decrease in biological functioning and dominance by physical phenomena. The effects of storm events differ for channel and marsh habitats. III. When extreme events coincide with major human disturbances in the watershed, estuarine functioning is altered, and the changes can have long-term consequences for the ecosystem. This summary developed from comparisons of three tidally-flushed systems, Tijuana Estuary, Mugu Lagoon, and Colorado Lagoon, for which data sets span years with and without major floods. The three concepts are working hypotheses that need to be tested; they should not yet be regarded as region-wide conclusions.

ISBN 0-12-405070-0

INTRODUCTION

Southern California is a region of highly seasonal rainfall and streamflow (Fig. 1). Coastal ecosystems that are dominated by marine water throughout the dry season become "intermittent estuaries" during the period of substantial freshwater input (approx. January-April). Our review of ecological studies found strong seasonal patterns in both physical and biological events that follow--and appear to be driven by--winter rainfall. The high degree of seasonality in turn made it difficult to characterize estuarine "filtering" without differentiating wet- and dry-season functions. We describe the annual progression of events and then select wet and dry months to explore estuarine "filters." In our view, filtering includes both biological (consumption) and physical (sedimentation) removal of particulate matter from fresh and marine waters, as well as the concentration of dissolved materials by organisms (e.g. nutrient absorption) or substrates (salt accumulation).

MEAN MONTHLY RAINFALL AT SAN DIEGO
Mean for 1931-1973 (25 cm /yr)

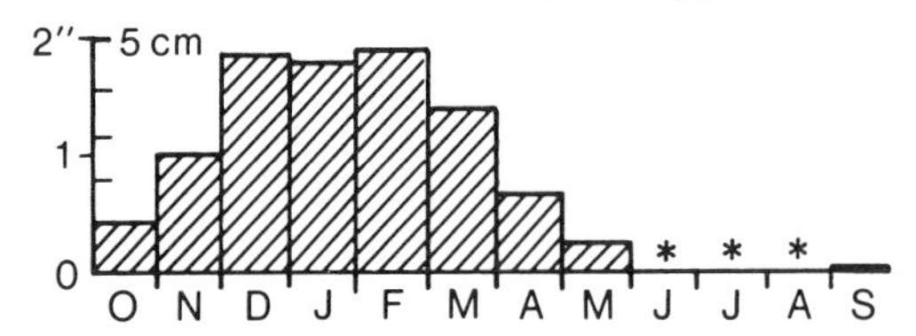

MEAN MONTHLY STREAMFLOW AT TIJUANA ESTUARY
Mean for 1937-1978

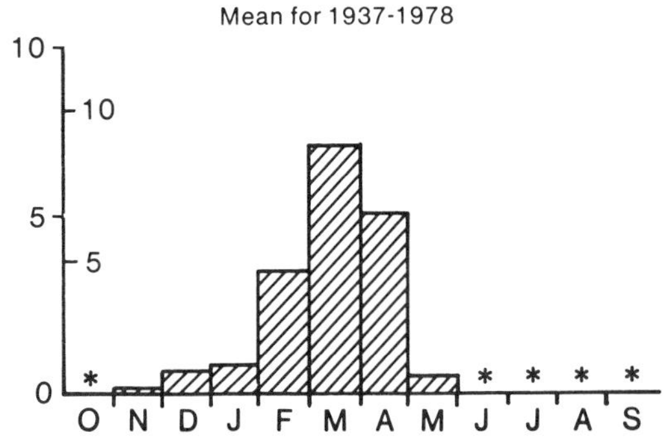

FIGURE 1. Rainfall data for San Diego (on San Diego Bay) and streamflow into Tijuana Estuary (USGS data for Tijuana River, Nestor Gage). Vertical scale for streamflow is in thousand acre-feet (left) and million cubic meters (right). Asterisks = months with averages below 0.1 units.

Ecological records for southern California's estuaries are incomplete, and we extrapolate from three systems to obtain our regional model. Generalization is justified by the similarity in attributes that have been examined for two or more systems. Tijuana Estuary (32°33'N;117°05'W) and Mugu Lagoon (34°06'N;119°05'W) are small in size (440ha and 940ha, respectively) and similar in physiography (approx. 65% intertidal marsh, 20% subtidal habitats, and 10% sand and mudflats). Colorado Lagoon (33°45'N;118°10'W) is a small (6ha) arm of Long Beach Harbor with an urban watershed and tide-gate control of water levels for recreational swimming. This lagoon has no marsh or intertidal flats. The patterns of inorganic nitrogen at Tijuana Estuary (Winfield 1980) agree with those at Colorado Lagoon (Kremer and Kremer 1980). The species composition of large benthic fauna is similar at Tijuana Estuary and Mugu Lagoon (Peterson 1975). Five of the six most common species of fish are the same for both Mugu Lagoon (Onuf and Quammen 1983) and Tijuana Estuary (W.S.White, US Fish and Wildlife Service, pers. comm.). Three of the six most common species at Colorado Lagoon are also among the six most common at Mugu Lagoon and Tijuana Estuary. However, the overwhelming numerical dominant at Colorado Lagoon (northern anchovy, *Engraulis mordax*) is rare at the other locations (Allen and Horn 1975). Shorebird composition and seasonal occurrences are similar at Tijuana Estuary (Boland 1981) and Mugu Lagoon (Quammen 1980). Algal productivity is seasonal and substantial at Tijuana Estuary (Zedler 1980), Colorado Lagoon (Kremer and Kremer 1980), and Mugu Lagoon (Shaffer and Onuf 1983). Vascular plant composition compares well at Tijuana Estuary (Zedler 1977) and Mugu Lagoon (Onuf et al. 1979), except that *Spartina foliosa* is not widespread at Mugu Lagoon. We draw most heavily from our own studies at Tijuana Estuary and Mugu Lagoon, described more fully in Zedler (1982a) and Onuf (in press).

SEASONAL DIFFERENCES IN ESTUARINE FILTER FUNCTIONS

Watershed hydrology drives seasonal patterns and appears to control both structure and function in downstream estuaries. Winter rainfall triggers a number of events, as coastal streams transport fresh water and nutrients to the region's estuaries (summarized in Fig. 2). Maximum concentrations of nitrate, nitrite and phosphate coincided with winter rains at Colorado Lagoon (Kremer and Kremer 1980). Inorganic nitrogen showed ten-fold increases of nitrate and nitrite concentrations at Tijuana Estuary in 1978 (Winfield

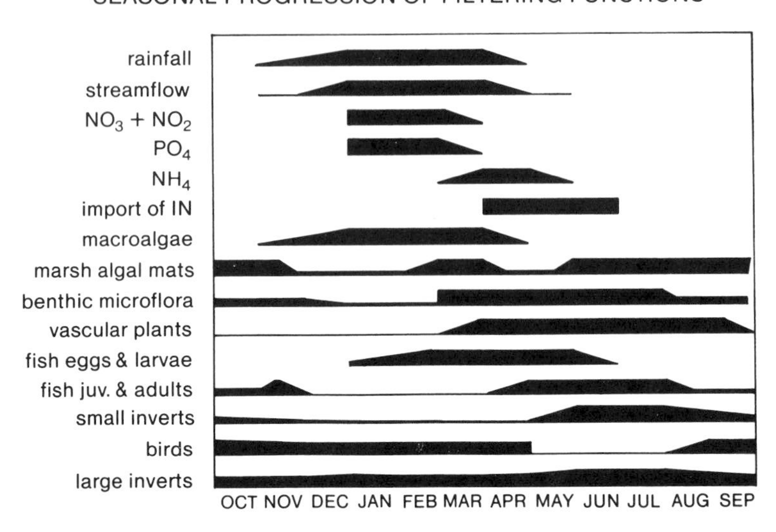

FIGURE 2. Data on estuarine attributes (see text) are presented semi-quantitatively. The USGS water year (Oct. 1 to Sept. 30) depicts the wet season as one unit. January and July characterize wet- and dry-season filtering functions (see Fig. 3). IN=inorganic nitrogen.

1980) and Colorado Lagoon in 1979 (Kremer and Kremer 1980), when winter peaks were compared to dry season averages. Ammonium peaks occurred one month later than the oxidized forms of nitrogen at Tijuana Estuary (Winfield 1980), suggesting that the estuary transforms inorganic nitrogen to its reduced form.

The influx of inorganic nutrients is then followed by peaks in algal productivity, most thoroughly documented at Colorado Lagoon in 1979 (Kremer and Kremer 1980), where macroalgae (predominantly *Enteromorpha* spp.) had their highest biomass immediately after the maximum concentration of nitrates, nitrites, and phosphates. Green macroalgae produce floating mats in winter at Tijuana Estuary also (pers. observ. and Nordby 1982), and green algae had a winter peak abundance in the marsh soil algal mats there as well (Zedler 1982b). Productivity data for these soil algal mats (in 1977; Zedler 1980) and for benthic microalgae in Mugu Lagoon intertidal and subtidal channels (in 1978; Shaffer and Onuf 1983) suggest that their maximum productivity occurs later than for macroalgae, although this may be an artifact of studying different processes in different years.

Vascular plant growth is most rapid during the spring and summer months at both Tijuana Estuary (Winfield 1980) and Mugu Lagoon (Onuf in press). Growth rates of vascular plants, along with those of the algal mats under the marsh canopy and benthic microalgae in the channels suggest high rates of nutrient uptake during this period. Winfield's (1980) data on nutrient import from tidal creeks to the salt marsh support the existence of such a filter function during spring and summer. The highest imports of total dissolved inorganic nitrogen were measured in June 1977 (200gN ha^{-1} per high tide) and April 1978 (250gN ha^{-1} per high tide; Fig. 5 in Winfield 1980).

The winter rainfall and nutrient pulses are closely linked with biological filtering functions occurring in the spring. However, later events are more difficult to assign to single causes. Kremer and Kremer (1980) noted a winter peak of macroalgae followed by a summer maximum for phytoplankton at Colorado Lagoon. The first seasonal pattern is probably driven by the winter nutrient pulse, but the phytoplankton bloom may result from decomposition of macroalgae and nutrient recycling, from competitive interactions among algal groups as light and temperature increase, from grazing on macroalgae and recycling of nutrients through animals, or from reduced tidal circulation during summer months when the lagoon is used for recreational swimming. Hydrology drives the estuarine ecosystem, but the control mechanisms are complicated.

Strong seasonal patterns are also present among several of the estuarine animal communities. Fishes lay eggs in Tijuana Estuary primarily in March, April, and May (Nordby 1982). For topsmelt (*Atherinops affinis*), a species that attaches its eggs to *Ulva* spp. and to *Enteromorpha* spp., spawning and macroalgal abundance coincide. Fish larvae also reach maximum abundance in spring, and since most are herbivorous at this life stage, feeding activities imply another filtering function. At Mugu Lagoon, juvenile and adult fish are present in very low numbers from December to March and increase rapidly to a sharp peak in June (Onuf and Quammen in press). Invertebrates fall into two distinct categories (Fig. 2). Small, short-lived forms all show strong seasonal patterns; together their abundance increases three-fold during summer. Larger, generally long-lived invertebrates are weakly seasonal, and species peaks are not in synchrony (M.L. Quammen, Univ. California, Santa Barbara, pers. comm.). Peterson's (1982) observations on two species of clams, however, suggest that growth and reproductive effort may be higher in summer. Williams (1981) found rapid growth of mussels placed in Tijuana Estuary; his experimental work indicated that phytoplankton and dissolved organic matter, rather than vascular plant detritus, were utilized by mussels.

These temporal patterns suggest that the life cycles of several animal groups are adjusted to the period of highest primary productivity, whether directly to optimize consumption or indirectly to increase predation on herbivores. In contrast, water-related birds are abundant between October and April with greatest density in December. They are rare from May to September and essentially absent in June (Fig. 2; Boland 1981; Onuf and Quammen in press). This exception is discussed later as an example of how predator-prey interactions complicate the analysis of filtering functions.

At the same time that biological functions undergo seasonal patterns, physical filters also operate and vary through the year. Winter rains have an important leaching effect on intertidal substrates, which accumulate salts at other times of the year. Seasonal changes in soil salinity are most pronounced in the upper salt marsh (Figs. 11 & 12 in Zedler 1982a), where tidal inundation is less frequent and where soils become brackish during the wet season. Leaching allows salt marsh plants to establish seedlings by stimulating germination and growth, thereby fostering biological filtering of nutrients. In contrast, salt concentration, a physical filtering process, predominates in the dry season and reduces halophyte growth. Plants continue to grow in summer, but the nutrients they absorb may be recycled within the estuarine-tidal creek system, rather than imported from the ocean. Winfield (1980) concluded that only 6% of the nitrogen required by marsh plants (vascular and algal combined) was imported from the ocean or from the deeper channels to the tidal creek/marsh system at Tijuana Estuary.

The interplay between physical and biological filtering processes also exists within estuarine channels. Winter streamflow transports sediments to the channels and settling occurs in areas of quiet water. If sedimentation (a physical filtering function) is sufficiently heavy, biological filtering functions are impaired as benthic invertebrates become smothered with silt and as habitat for water-column species is reduced in volume. At other times, organisms may increase sedimentation in channels. Algae help to stabilize sediments by secreting mucus, and suspension feeders remove and concentrate materials in fecal pellets.

To summarize the variety of seasonal estuarine functions as a simple model, we contrast wet and dry season events (Fig. 3) and identify the major physical and biological filters that dominate the two time periods. During the wet season, the estuary removes nutrients and sediments from the watershed. Estuarine animals may obtain particulate organic carbon (POC) from both stream and tidal waters, but data are insufficient to determine if there is a net gain in POC (Mauriello and Winfield 1978; Onuf et al. 1979).

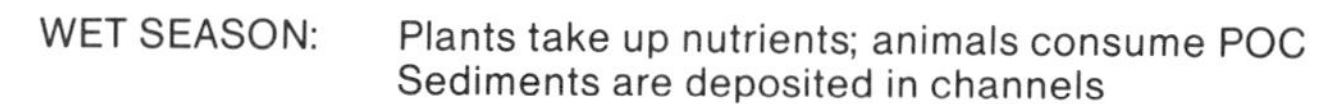

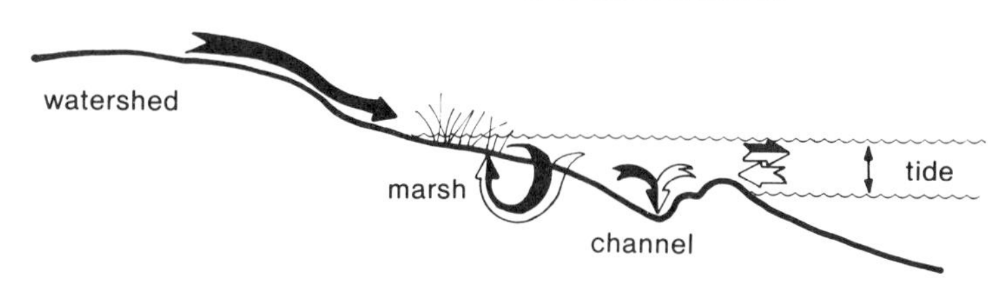

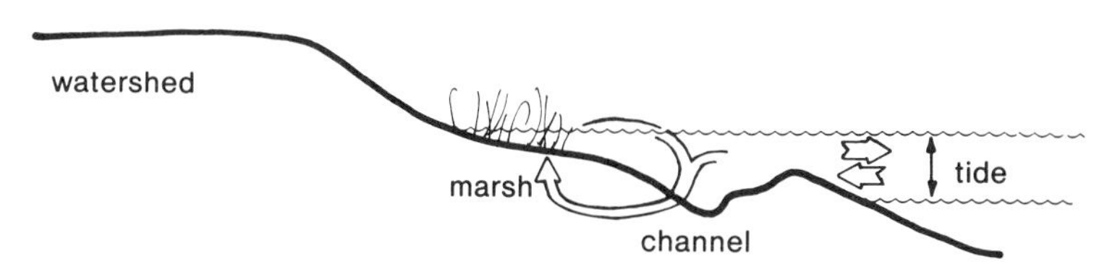

FIGURE 3. Wet and dry seasons: Sources from which materials can be filtered, the types of biological filtering, and the nature of physical filters. POC=particulate organic carbon. Solid arrows=fresh water; clear arrows=marine water.

During the dry season, removal of materials from the watershed ceases as stream flow drops to zero. The ocean provides a source of both inorganic and organic materials, but the quantities filtered from tidal waters are not known. Consumption of particulate organic matter probably reaches a peak with high densities of fishes and invertebrates, but the majority of foods probably are derived from within the estuary. The filtering function may be predominantly one of nutrient uptake and recycling rather than substantial trapping of imported particulate organic matter.

Through the year, the influence of the nutrient pulse can be traced qualitatively through progressively higher trophic levels. However, late in the year when several trophic levels are at work simultaneously, it becomes difficult to determine which processes drive the system. We know of one situation that suggests a governing role for predators, rather than seasonal nutrient pulses. As noted above, the activites of water-related birds appear to be out of phase with the rest of the system. Quammen (1984) demonstrated that in intertidal areas with fine substrates, the feeding activities of birds (from fall to spring) can actually reverse the seasonal patterns in small invertebrates. We presume that this reversal would have an impact down to the base of the food web. Birds are "out-of-step" with the hydrologic pulse in the estuary because they are responding to conditions on their northern breeding grounds as well as in their winter feeding areas. Complexities and uncertainties notwithstanding, we believe that the strongest driving force is the highly seasonal rainfall that dictates the timing of critical inputs to the estuary.

EFFECTS OF EXTREME EVENTS

During the past decade, southern California's annual rainfall and streamflow shifted from below average to well above average during the flood years of 1978 and 1980 (Fig. 4). At Mugu Lagoon, the storms were similar in magnitude and substantial in their effect in both years (Onuf and Quammen 1983; Onuf in press), whereas at Tijuana Estuary, the major flood of 1978 was dwarfed by the "hundred-year event" of 1980.

Our understanding of the effects of flooding on southern California estuarine function comes primarily from detailed observations on salt marsh vegetation at Tijuana Estuary (Zedler 1982a; in press) and at Mugu Lagoon (Onuf in press), and five years' data on fish populations at Mugu Lagoon (Onuf and Quammen 1983). Additional information summarized in Zedler (1982a) and Onuf (in press) allows comparison of flood- and non-flood years. In attempting to generalize our observations of flood effects, we rapidly concluded that the nature and degree of impact differed substantially for channel and intertidal marsh habitats (Table 1).

Channels

The studies of Onuf and associates on channels suggested that flooding is capable of eliminating both plant and animal communities. Two processes (sedimentation and salinity reduction) are indicated, with the former being more important. An average of 13 cm of silt was deposited in subtidal areas of the lagoon in 1978 and another 7 cm in 1980. An entire eelgrass bed was eliminated and the bottom changed from predominantly sandy to predominantly muddy substrate. The combined sedimentation events reduced low tide volume of Mugu Lagoon by 40%.

Responses of the channel fauna were variable, but there were patterns of change that related to life history type (among invertebrates; Onuf in press) and habitat utilization (among fishes; Onuf and Quammen 1983). Except for gastropods, the large, long-lived invertebrates were reduced in density

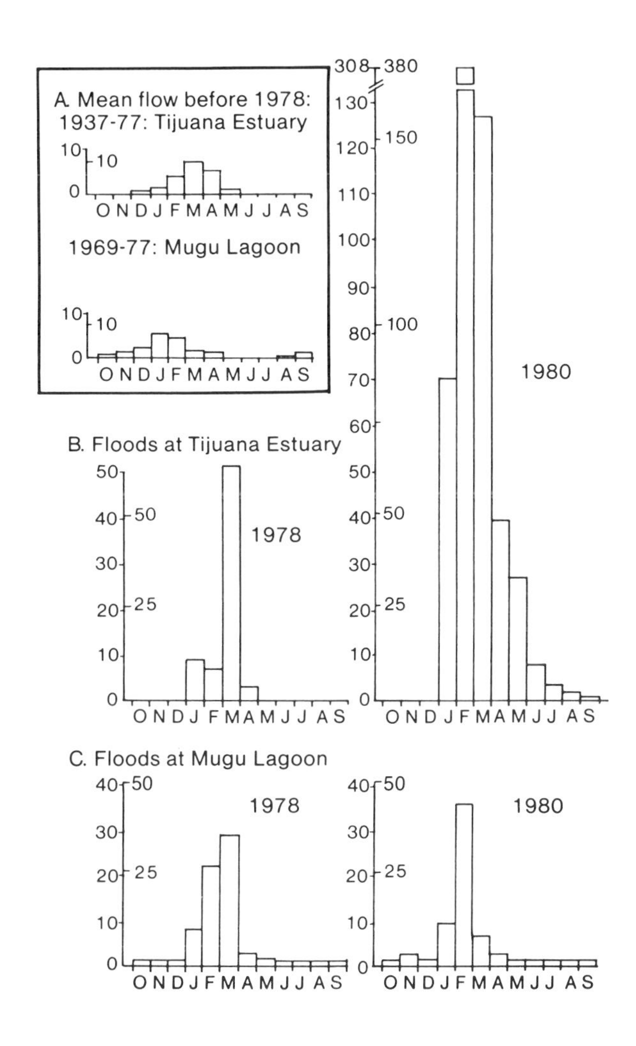

FIGURE 4. Monthly hydrographs of streamflows into Tijuana Estuary (Nestor Gage) and Mugu Lagoon (Calleguas Creek, Camarillo Hospital Gage). All data are from USGS records and are plotted on the same scale. Vertical scales are thousand acre-feet (left) and million cubic meters (right).

TABLE 1. Effect of extreme flood events on channels and on intertidal marshes.

Physical disturbance	*Biological change*
	A. Channels
Sedimentation	Invertebrates smothered
Cumulative sedimentation	Reduced habitat for water-column fishes
Reduced water salinity	Sensitive species eliminated
	B. Salt marsh
Reduced soil salinity (and increased nutrients?)	Increased vascular plant growth; increased establishment of cordgrass

but maintained similar relative abundances. The small, short-lived invertebrates had an opposite response. Overall abundance increased rather than decreased after the major storms, although abundances of the major groups fluctuated erratically among years. Furthermore, the species composition changed drastically in all but the gastropod group. Species that were common before 1978 became rare or absent; species that became common after the 1978 flood had been rare or absent before it (Onuf and Quammen in press). The responses of fishes seemed to vary with the habitat utilized. Species characteristically found in the water column (topsmelt and shiner surfperch) sustained major declines that have persisted, presumably in response to reduced lagoon volume. Curiously, the bottom dwelling fishes appeared to be little affected by the radical shift from sandy to muddy sediments (Onuf and Quammen 1983).

Sediment deposition was probably responsible for the abrupt compositional changes among small invertebrates, and for most of the persistent changes at the Lagoon. However, the coinciding salinity reduction was an important cause of mortality of sensitive species such as sand dollars (*Dendraster excentricus* and bubble snails (*Bulla gouldiana*). Similar salinity effects were seen after the 1978 flood at Tijuana Estuary; populations of both sand dollars and purple-hinged clams *(Nuttallia [=Sanguinolaria] nuttallii)* were eliminated and had not recovered by 1983 (D. Dexter, San Diego State Univ., pers. comm.). Heavy mortality of bubble snails occurred at the Sweetwater River after reservoir release in 1983 (pers. observ.). In both cases, fresh water dominated the main channels during peak flow.

In estuarine channels, major floods reduce biological filtering processes by reducing or eliminating populations of both plants and animals. At the same time, the physical process of sedimentation increases dramatically. As discussed elsewhere (Onuf in press), sedimentation is a cumulative process. The floods of 1962 and 1969 had no lasting impact on the eastern arm of Mugu Lagoon, even though they were similar or greater in magnitude than the floods of 1978 and 1980. More sediments were available in 1978 because of erosion in the watershed. Also, the 1978 flood filled major portions of the central basin, so that sediment-laden waters then spread out over the eastern arm. Sedimentation was significantly greater in amount and area, apparently because of watershed modification (discussed later), but we believe the types of effects on the estuary were characteristic of flooding in general.

Salt Marsh

By contrast, intertidal marsh vegetation was stimulated by the hundred-year flooding at Tijuana Estuary, as low-marsh soils were leached of salts (Zedler 1982a; 1983). Normally these soils have nearly constant hypersaline conditions (40-45ppt) because of frequent tidal influence and concentration of salts by evaporation and cordgrass transpiration. By April 1980, the soil salinity was brackish (15ppt), and cordgrass biomass was 40% higher in August 1980 than in non-flood years. Onuf (in press) noted a similar stimulation of pickleweed (*Salicornia virginica*) following the 1978 flood at Mugu Lagoon. In both estuaries, the increase in halophyte biomass was restricted to the year of flooding, indicating a positive but short-lived change in biological functioning.

A longer-term effect on the Tijuana Estuary salt marsh was the broadened distribution of cordgrass that occurred through accelerated vegetative expansion and unusual seedling establishment. It appears that flooding is important to the regional, as well as local, distribution of cordgrass (pers. observ.). To the extent that less productive communities are invaded by this C-4 grass, overall nutrient filtering ability of an estuary may be enhanced.

Sedimentation was also measurable and significant in the lower marsh (mean accretion approx. 5cm), but few effects could be attributed to the deposition of particulate material. The most obvious effects were small patches of dead cordgrass underneath piles of debris filtered from the watershed. These patches caused short-term reductions in productivity.

The effects of flooding can be summarized as a conceptual model (Fig. 5) for comparison with the wet-season model of estuarine filtering. Channels lose their biological ability to filter both inorganic and organic materials if either plants or animals are eliminated. At the same time, physical processes dominate as large volumes of sediment accumulate. From our vegetation studies, it appears that biota of intertidal marshes are less negatively affected by flooding than are channel populations. The period of time that sediment-laden fresh water dominates the intertidal zone is shorter than in channels, reducing the potential for both sedimentation and salinity reduction--although floating debris is more likely to accumulate at higher elevations. The period of time that biological filtering functions are curtailed is also brief, and reduced nutrient uptake during flooding may be compensated for by later growth spurts. However, it is not clear if biological filtering increases overall, because of the interaction between vascular plants and understory algal mats (Zedler 1980, 1982a). Increased growth of overstory vegetation should reduce light available to soil algae, thereby decreasing epibenthic productivity (not measured in 1980).

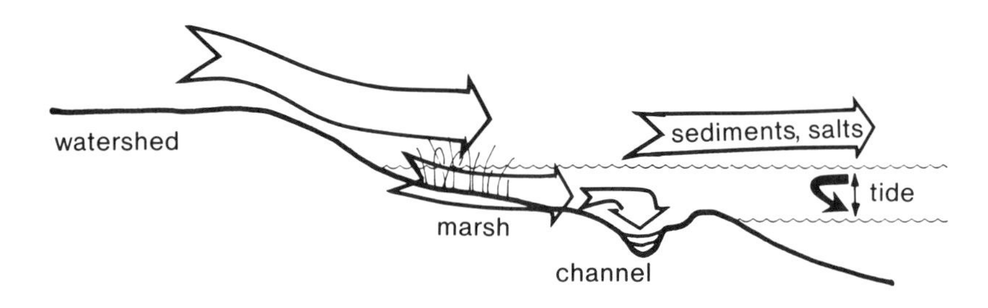

FIGURE 5. Effects of flooding on biological and physical filtering processes. Clear arrows = fresh water; solid arrows = marine water.

LONG-LASTING IMPACTS OF EXTREME FLOODING AND WATERSHED MODIFICATION

In a highly disturbed landscape, such as southern California, it is difficult to separate normal flood impacts from those augmented by human disturbances. Ideally we should examine flooding before and after major modifications to a lagoon's watershed. Selected features of Mugu Lagoon were studied in the 1960's, so that some historical comparisons are possible. For Tijuana Estuary, the scarcity of historical data have made it necessary to draw upon comparisons with more disturbed systems, such as the nearby San Diego River. These comparisons suggest a third concept, that the combination of extreme flooding and major watershed disturbance can have catastrophic (that is, major and long-lasting) consequences for the estuary. In some cases, filtering ability may be irreversibly impaired.

In developing this concept, our different research focuses have again indicated complexities, and we conclude that the nature of the effect differs for the habitat under consideration (Table 2). Intertidal marshes seem more sensitive to changes in the salinity regime: prolonging the period of freshwater influence can eliminate natural salt marsh vegetation and allow invasion of other plant communities. Channel and lagoon habitats are especially subject to damage from sedimentation: disturbed watersheds can release substantially more material for deposition in the estuary. Two examples led to these generalizations:

Prolonged release of fresh water from reservoirs in the San Diego River watershed extended the 1980 flood flows well beyond the normal wet season. Flows were constricted within a flood control channel, which concentrated the effects of freshwater. In this situation, the 1980 floodwaters had an effect much greater than at Tijuana Estuary. Salt marsh vegetation (primarily *Salicornia virginica*) died and *Typha domingensis* invaded. The mortality of halophytes probably resulted from the extended period of inundation. However, the invasion event was triggered by low soil salinities, here reduced to 0ppt. Laboratory experiments (P. Beare, San Diego State Univ., pers. comm.) show that *T. domingensis* requires lower salinities for germination than are tolerated by older individuals. In the field, the *T. domingensis* has survived for four years, during which time the salinities have averaged around 20-25ppt in spring to 30-35ppt in summer. Perennial rhizomes continue to resprout in salinity conditions that are too stressful for establishment of *T. domingensis*.

TABLE 2. Effect of extreme flood events on watersheds that have substantial human modifications.

Habitat	*Disturbance*	*Physical change*	*Biological effect*
Channels	Hillside development and accelerated erosion rates	Sedimentation reduces lagoon volume	Loss of fishes (e.g. topsmelt, shiner surfperch)
		Subtidal habitat becomes intertidal	Loss of eelgrass beds
Salt marsh	Reservoir drawdown prolongs streamflow beyond wet season	Long period of inundation	Loss of halophyte cover and slow recovery
		Long period of low salinity	Invasion and persistence of non-halophytes

Thus, a very unusual and brief event initiated a long-term shift in intertidal marsh vegetation. Its effect on plant productivity has not been measured. However, because the variations in soil salinity cause periodic mortality of *Typha dominguensis*, plant growth is held below its potential. Upstream populations of *T. dominguensis* had up to 5 times the biomass of intertidal populations (pers. observ.). Both the development of soil algal mats and the recolonization by halophytes are inhibited by the *T. dominguensis* canopy (whether live or dying), so that overall nutrient filtering must be impaired by the invading vegetation.

Extensive disturbance in the watershed of Mugu Lagoon released enormous quantities of sediment, which filled its central basin and extended the area of sedimentation well into the eastern arm of the lagoon. Onuf (in press) compared the sedimentation events of 1962, 1969, 1978 and 1980 floods and found much greater impacts of flooding in the latter two years. Using records of hillside development within the watershed of Calleguas Creek, he found that 60% of all development had occurred between 1969 and 1979, and that later developments occurred more often on steep slopes. Areas with "high rates of accelerated erosion" were identified by Steffen (1982) and suggested as responsible for 40% of the sediment deposited in Calleguas Creek.

Flooding after watershed disturbance increased both the amount of sedimentation and the area of deposition. The central basin of Mugu Lagoon changed from subtidal to intertidal habitat as material was filtered from the floodwaters of 1978. The main channel changed from sandy to muddy bottom habitat as the area of flood influence expanded beyond the central "catchment basin" of Mugu Lagoon. Overall, the lagoon's low-tide volume decreased by 40%.

It is clear that the filtering ability of an estuary changes as that system shrinks in volume. Physical processes depend on suitable physiography. As deep channels are filled, more sediments will be transported to the ocean. Biological processes (food consumption, nutrient uptake) also change. As densities of fish and of the large, mostly filter-feeding invertebrates decline, there will be less filtering of organic particles. Reduced areas of eelgrass suggest lower nutrient uptake rates. Whether alternative plant and animal communities will develop to resume these processes at comparable rates is not known. Even if they do, the functions have been interrupted, thereby decreasing the system's long-term average filtering rates.

Overall, the estuary's filtering ability has a threshhold that is exceeded when floods coincide with major changes in watershed land use and hydrology. Normal flooding has a beneficial and short-lived effect on marsh vegetation--soil salinity is decreased and vascular plant growth is briefly stimulated. But when inundation is prolonged well beyond the wet season, salt marsh vegetation dies and plant productivity drops as new communities develop and decline. Likewise, channels and subtidal basins can accommodate normal sediment input but are rapidly filled when erosion is accelerated within the watershed. The filtering processes are interrupted in both marshes and channels, and the changes are persistent. Filling of basins and channels may be irreversible.

CONCLUSION

This review is a first attempt to identify filtering functions in southern California's highly variable estuaries. Due to the paucity of appropriate studies, our generalizations have been qualitative, rather than quantitative. We have hypothesized ecosystem responses to seasonal pulses of freshwater and nutrient influxes; we have discovered how extreme events can modify species composition and productivity; and we have documented the far-reaching influences of disturbance within the watershed. The changes brought about by modifying hydrology and land use can be catastrophic for both marsh and channel biota. It now remains to test these hypotheses experimentally, to determine recovery rates for the species lost to extreme flooding, and to develop management practices that will reverse the negative impacts of disturbance in the watershed and lessen the chances for human-caused catastrophes.

ACKNOWLEDGMENTS

We thank John W. Day, Jr., who catalyzed the writing of this paper by suggesting the ERF Symposium presentation. We are grateful to the many researchers whose data contributed to our understanding of seasonal patterns and filtering functions. The California Sea Grant College Program supported research on Tijuana Estuary (R/CZ-33c and R/CZ-51) and Mugu Lagoon (R/CZ-33a and R/CZ-52; laboratory facilities provided by R. Holmes, UCSB). In addition, the US Fish and Wildlife Service supported our Profiles on southern California salt marshes and on Mugu Lagoon. Special thanks to Millicent Quammen, James N. Kremer, V. S. Kennedy, and anonymous reviewers for improving the manuscript.

REFERENCES CITED

Allen, L. G. and M. H. Horn. 1975. Abundance, diversity and seasonality of fishes in Colorado Lagoon, Alamitos Bay, California. *Estuar. Coastal Mar. Sci. 3*:371-380.

Boland, J. M. 1981. Seasonal abundances, habitat utilization, feeding strategies and interspecific competition within a wintering shorebird community and their possible relationships with the latitudinal distribution of shorebird species. M.S. Thesis, San Diego State U., San Diego, CA. 78 pp.

Kremer, J. N. and P. Kremer. 1980. Ecology of a small tidal lagoon under the influence of urban recreational use. Univ. of Southern California Sea Grant Annual Report 1979-80, Los Angeles, CA. 16 pp.

Mauriello, D. and T. P. Winfield. 1978. Nutrient exchange in the Tijuana Estuary. Coastal Zone '78, Volume III:2221-2239. Am. Soc. Civil Engr., New York.

Nordby, C. 1982. Comparative ecology of ichthyoplankton within and outside Tijuana Estuary. M.S. Thesis, San Diego State Univ., San Diego, CA. 101 pp.

Onuf, C. P. In press. The Ecology of Mugu Lagoon: an Estuarine Profile. U.S. Fish and Wildlife Service, Office of Biological Services. Washington, DC.

Onuf, C. P. and M. L. Quammen. 1983. Fishes in a California coastal lagoon: effects of major storms on distribution and abundance. *Mar. Ecol. Prog. Ser. 12*:1-14.

Onuf, C. P. and M. L. Quammen. In press. The biota: distribution and abundance, Chapter 4. *In*: C. P. Onuf, The Ecology of Mugu Lagoon: an Estuarine Profile. U.S. Fish and Wildlife Service, Office of Biological Services, Washington, DC.

Onuf, C. P., M. L. Quammen, G. P. Shaffer, C. H. Peterson, J. W. Chapman, J. Cermak and R. W. Holmes. 1979. An analysis of the values of central and southern California coastal wetlands, pp. 186-199. *In*: P. W. Greeson, J. R. Clark, and J. E. Clark (eds.), *Wetland Functions and Values: the State of our Understanding*. American Water Resources Association, Minneapolis, MN.

Peterson, C. H. 1975. Stability of species and of community for the benthos of two lagoons. *Ecology 56*:958-965.

Peterson, C. H. 1982. The importance of predation and intra- and inter-specific competition in the population biology of two infaunal suspension-feeding bivalves, *Protothaca staminea* and *Chione undatella*. *Ecol. Mongr.* *52*:437-475.

Quammen, M. L. 1980. The impact of predation by shorebirds, benthic feeding fish and a crab on the shallow living invertebrates in intertidal mudflats of two southern California lagoons. Ph.D. Dissertation, U. California, Irvine. 131 pp.

Quammen, M. L. 1984. Predation by shorebirds, fish and crabs on invertebrates in intertidal mudflats: an experimental test. *Ecology* (In press).

Shaffer, G. P. and C. P. Onuf. 1983. An analysis of factors influencing the primary production of benthic microflora in a southern California lagoon. *Neth. J. Sea Research 17*:126-144.

Steffen, L. J. 1982. Mugu Lagoon and its tributaries: geology and sedimentation. USDA Soil Conservation Service. Davis, California. 73 pp.

Williams, P. 1981. Detritus utilization by *Mytilus edulis*. *Estuar. Coastal Shelf Sci. 12*:739-746.

Winfield, T. P. 1980. Dynamics of carbon and nitrogen in a southern California salt marsh. Ph.D. Dissertation, Univ. California, Riverside, CA., and San Diego State Univ., San Diego, CA. 76 pp.

Zedler, J. B. 1977. Salt marsh community structure in the Tijuana Estuary, California. *Estuar. Coastal Mar. Sci.* *5*:39-53.

Zedler, J. B. 1980. Algal mat productivity: comparisons in a salt marsh. *Estuaries 3*:122-131.

Zedler, J. B. 1982a. The ecology of southern California coastal salt marshes: a community profile. U.S. Fish and Wildlife Service, Biological Services Program. Washington, DC. FWS/OBS-81/54. 110 pp.

Zedler, J. B. 1982b. Salt marsh algal mat composition: spatial and temporal comparisons. *Bull. Southern California Acad. Sci.* *8*:41-50.

Zedler, J. B. 1983. Freshwater impacts in normally hypersaline marshes. *Estuaries 6*:346-355.

MANAGEMENT IMPLICATIONS

ASSESSING AND MANAGING EFFECTS OF REDUCED FRESHWATER INFLOW TO TWO TEXAS ESTUARIES

Nicholas A. Funicelli[1]

National Coastal Ecosystems Team
U.S. Fish and Wildlife Service
Slidell, Louisana

Abstract: While a major objective of coastal zone management plans is to maintain the value and use of estuaries, decisions affecting inflow, a critical component of estuaries, are often based on considerations of water use inland. The functions of estuaries as "filters" for inland based signals is often ignored. For effective management of both inland-water resources and estuarine freshwater inflow, methods for assessing the actual effects of inland perturbations are needed. Because no single method exists, the U.S. Fish and Wildlife Service used a set of evaluative methods to assess the effects of reduced freshwater inflow on estuaries in Nueces-Corpus Christi and Matagorda Bays and to develop management plans for ensuring fish and wildlife productivity. The set included (1) studying the effects of environmental changes on key species, (2) regression analysis between inflow and fishery harvest data, (3) compilation and analysis of a nutrient budget, and (4) comparison of historical and projected deltaic marsh inundations. Proposed management plans to address the findings of these assessment methods involved the release of freshwater during spring and low inflow, the rerouting of selected wastewater to the river delta, and the diversion of in-channel water to inundate deltaic marshes. These plans all depend upon compromises between inland and coastal needs.

[1]Present address: Florida Fishery Research Station, P.O. Box 1669, Homestead, Florida.

ISBN 0-12-405070-0

INTRODUCTION

In defining estuaries, Pritchard (1967) called freshwater inflow an essential part of estuarine functioning. Freshwater can be thought of as a conduit for critical "deliverables" such as nutrients and sediments to the estuarine ecosystem. Inflow also effects salinity control and gradient, delta formation, and hydrologic regimes, depending upon the "filtering" activity of the estuary.

Although the value and use of estuaries for municipal and industrial development are widely recognized, the importance of freshwater inflow to estuaries is neither widely recognized nor broadly understood by managers. Consequently, while a major objective of coastal zone management is maintaining the value and use of estuaries, few programs effectively address freshwater inflow. Instead, decisions affecting freshwater distribution are often made by inland planners and are based on inland considerations. Overcoming this situation is not simple. To relate inland water-resource management to coastal management adequately, methods for assessing the actual effects of altered freshwater inflow to estuaries are needed.

Such assessment methods were part of recently concluded studies in Texas by the U.S. Fish and Wildlife Service in Nueces-Corpus Christi Bay (Henley and Rauschber 1980) and Matagorda Bay (Wiersema, Armstrong and Ward 1982). The methods were used to determine the effects on estuaries of reduced amounts of freshwater inflow from present and proposed upland reservoirs and to develop management plans for ensuring fish and wildlife productivity in bay estuaries. A description and evaluation of these methods and related studies as well as management plans based on their results follow.

METHODS FOR FORECASTING ENVIRONMENTAL EFFECTS OF MODIFYING FRESHWATER INFLOW

A single method for predicting the exact amounts of freshwater inflow needed to sustain or enhance a given biological productivity level has not been developed. To develop a management plan therefore, the Nueces-Corpus Christi and Matagorda inflow studies used and evaluated several approaches to assessing the effects of freshwater inflow: responses of key species; impact on the commercial fishery harvest; effects on nutrient inflow; and inundation of deltaic marsh.

Each method has its shortcomings, but each reflects some aspect of the role of estuaries as filters, in this case, in modifying the incoming signals associated with freshwater drainage.

Key Species

The most common method of analyzing the effect of altering freshwater inflow to estuaries is to select "key species". All species selected in both Matagorda and Nueces-Corpus Christi Bays were considered to be major components of the bay ecosystem, representative of various habitats within the bay, and/or commercially important.

Key species included three seagrasses that dominate the submerged rooted vegetation along the lower half of the Texas coast: *Thalassia testudinum*, *Halodule beaudettei*, and *Ruppia maritima*. The emergent grasses selected are dominant species of the brackish and salt marshes of the Texas coast; *Spartina alterniflora*, *S. patens*, *Batis maritima*, *Salicornia virginica*, *S. bigelovii*, *S. europaea*, *S. utahensis*, *Monanthochloe littoralis*, *Distichlis spicata*, and *Juncus roemerianus*. The commercial invertebrate catch of Texas is dominated by the white shrimp (*Penaeus setiferus*), brown shrimp (*P. aztecus*), blue crab (*Callinectes sapidus*) and oysters (*Crassostrea virginica*), so all were selected. The finfish chosen are important commercially and/or recreationally (*Cynoscion nebulosus*, *Sciaenops ocellatus*, *Pogonias cromis*) or represent an important trophic component such as filter feeder (*Brevoortia patronus*, *Anchoa mitchilli*), demersal inhabitant (*Paralichthys lethostigma*), or mid-level consumer (*Fundulus similis*).

Species tolerances and preference as outlined in the scientific literature were compared with predicted changes in temperature, salinity, dissolved oxygen, depth, and substrate caused by reduced freshwater inflow. Although the Matagorda Bay study revealed that most finfish and shellfish species were most abundant at salinities of 14-19 parts per thousand (ppt), the salinity tolerance levels for key species were so broad that reduced freshwater inflows would not raise salinities above the tolerance limits of most species. However, it was found that upon completion of an upland reservoir during extended low inflow periods, upper Nueces Bay could become hypersaline (above 40 ppt) and thus negatively impact key shellfish and finfish species.

The key species approach did not consider any biological or community processes and therefore has some obvious limitations. Managers will need to obtain an integrated picture of estuarine function in order to make better predictions concerning effects of altered freshwater inflow.

Fishery Harvest

The primary reason that finfish and shellfish have been used to illustrate the effects of freshwater inflow on estuaries involves the availability of commercial fishery statistics. However, ecologists and planners legitimately argue that effects of freshwater inflow should be viewed from a broader perspective than harvest and should include other physical, chemical, and biological parameters. Nevertheless, given the relative lack of ecosystem data, fishery statistics represent a considerable data base that can be exploited. Broad relations between freshwater inflow and quality and quantity of estuarine habitats and their biota have been identified through regressing commercial fishery harvest with one or more variables. Regressions are easy to comprehend and are used as basic components to build comprehensive simulation models (Armstrong and Gordon 1979).

Chapman (1966) used commercial fishery catch data from six Texas estuaries to show the direct relations between freshwater inflow and commercial catch (primarily menhaden, white shrimp, and brown shrimp). He used data to judge the potential effects of the Texas Water Plan, a water development program for the inland rivers of Texas. As a result of this analysis, the Texas Department of Water Resources (TDWR 1980) has conducted extensive studies to determine the freshwater needs of Texas estuaries.

These studies analyzed the influence of freshwater inflows on the Lavaca-Tres Palacios Estuary. A part of the study examined the relation of seasonal freshwater inflow to the harvest of various commercial species of the Matagorda Bay system. TDWR (1980) performed a time series analysis of Lavaca-Tres Palacios estuary. The analysis computed a sequence of multiple linear regression equations in a stepwise manner. Each sequential step utilized the next variable which yielded the largest reduction in the sum of squares error term, which was then added to the equation. The best significant equation was then developed that had the highest multiple correlation coefficient (r), and lowest error sum of squares. TDWR's (1980) typical form of the harvest regression equation was:

$$H_t = a_0 + a_1 Q_{1,t-b_1} + a_2 Q_{2,t-b_2} + a_3 Q_{3,t-b_3} + a_4 Q_{4,t-b_4} + a_5 Q_{5,t-b_5} + e$$

where, a_0 is the intercept harvest value and e is the normally distributed error term with a mean of 0 and the regression variables are:

H_t = annual inshore harvest of a fisheries component in thousands of pounds at year t,

$Q_{1,t-b_1}$ = winter season (January-March) mean monthly freshwater inflow in thousands of acre-feet at year $t-b_1$, where b_1 is a positive integer,

$Q_{2,t-b_2}$ = spring season (April-June) mean monthly freshwater inflow in thousands of acre-feet at year $t-b_2$, where b_2 is a positive integer,

$Q_{3,t-b_3}$ = summer season (July-August) mean monthly freshwater inflow in thousands of acre-feet at year $t-b_3$, where b_3 is a positive integer,

$Q_{4,t-b_4}$ = autumn season (September-October) mean monthly freshwater inflow in thousands of acre-feet $t-b_4$, where b_4 is a positive integer,

$Q_{5,t-b_5}$ = late fall season (November-December) mean monthly freshwater inflow in thousands of acre-feet at year $t-b_5$, where b_5 is a positive integer,

TDWR (1980) studies yielded nineteen (19) statistically significant regression equations that related to eight harvest groups: shellfish, all penaeid shrimp, white shrimp, blue crab, oyster, finfish, spotted seatrout, and red drum (Table 1).

These equations provided tools to assess the effects of the predicted future reduced inflows from the Colorado and Lavaca Rivers to all coastal basins associated with Matagorda Bay. In only one instance (spotted seatrout) did the study fail to produce a statistically significant equation for harvest versus combined inflows from the Lavaca and Colorado Rivers and Coastal Basins. Thus, except for the spotted seatrout, the effects of modifying each river system individually or effects of combined inflow can be measured (Wiersema, Armstrong and Ward 1982).

TABLE 1. Predicted commercial harvest resulting from reduced inflows (after Wiersema, Armstrong and Ward 1982)

			Catch (thousands of Kilograms)	
Organism	*Inflow*[a]	R^2	*Mean*	*Predicted*
Shellfish[d]	LD[b]	68	1378	1432
	CD	NS[c]		
	Comb	50	1378	1335
Penaeid shrimp[e]	LD[b]	75	879	893
	CD	62	879	703
	Comb	67	879	866
White shrimp	LD	46	723	709
	CD	NS		
	Comb[b]	48	723	707
Blue crab	LD	72	355	297
	CD	62	355	257
	Comb[b]	72	355	284
Bay oyster	LD	44	114	121
	CD	NS		
	Comb[b]	51	114	117
Finfish[f]	LD	53	125	165
	CD[b]	73	125	136
	Comb	40	125	154
Spotted seatrout	LD	NS		
	CD[b]	73	55	56
	Comb	NS		
Red drum	LD	50	30	40
	CD[b]	65	30	45
	Comb	42	30	37

[a] *Abbreviations: LD - Lavaca Delta inflow; CD - Colorado Delta inflow; Comb - Total inflow, all rivers and coastal basins*

[b] *Best fit*

[c] *No significant equation*

[d] *Includes blue crab, oyster, all penaeid shrimp.*

[e] *Includes white, brown and pink shrimp*

[f] *Includes croaker, black drum, red drum, flounder, sea catfish, spotted seatrout, sheepshead*

The regression of freshwater inflow with commercial harvest, like other regression analysis, identifies statistical relations between two or more variables, but does not identify the cause for variation. For example, a reduction of freshwater inflow from a new reservoir can be correlated with a reduction in commercial shrimp catch, but the causes may be several: eg. altered salinity regime in the bay system combined with reduced nutrients delivered to the estuary. Thus, a more complete analysis of potential effects is desirable.

Nutrient Inflow

Estuaries have long been regarded as traps (filters) for nutrients (carbon, nitrogen, and phosphorus) receiving them from freshwater inflows and precipitation, from marsh drainage and tidal exchange, and as a result of internal recycling. Inorganic nutrients, after incorporation in seagrass, emergent vegetation and phytoplankton, eventually become part of the detrital food chain. Nutrients are indefinitely recycled between the water column and sediments until exchanged with the ocean (Armstrong 1982; Wiersema, Armstrong and Ward 1982).

The importance of freshwater inflows and other external nutrient sources on estuarine functioning may be assessed by compiling a nutrient budget and accounting for spatial and temporal loading of nutrients. In this way, the relative contribution of each source to inputs of particulate and dissolved organic and inorganic forms of carbon, nitrogen, and phosphorus may be determined. The consequences of perturbations such as freshwater inflow alteration (and presumably nutrient-input alterations) may be thereby evaluated in terms of departures from previous input budgets (Wiersema, Armstrong and Ward 1982).

In the Matagorda Bay study, the nutrient budget was compiled by determining the nutrient input from (1) gauged and ungauged freshwater inflows; (2) adjacent saltwater, brackish water, and freshwater; (3) tidal exchange of the system at existing passes; and (4) direct precipitation on the bay surface (Wiersema, Armstrong and Ward 1982). Use of these nutrients in the bay was gauged by measuring the uptake and release of nutrients by phytoplankton through primary productivity (Wiersema, Armstrong and Ward 1982).

In Matagorda Bay, the total input of organic carbon from all external sources was 81.6 million kg/yr; of nitrogen, 4.33 million kg/yr; and of phosphorus, 1.54 million kg/yr. The dominant source of organic carbon (about 93% of the total) was freshwater inflow; about 7% was derived from marsh drainage (6.6% from normal tidal exchange and 0.2% from floods); 0.5% from wastewater; and less than 0.1% from tidal exchange and precipitation. For nitrogen, the dominant source was freshwater inflow (83%); precipitation accounted for almost 12%; marshes, 2% and waste discharges, 3.4%. Almost 85% of the phosphorus came from freshwater inflows; 9% from marshes; 3% from waste discharges; just over 2% from precipitation; and much less than 0.1% from tidal exchange (Wiersema, Armstrong and Ward 1982).

The Matagorda Bay study concluded that freshwater inflows contributed most of the nutrients to Matagorda Bay, primarily in the dissolved inorganic form. Marshes contributed less than 10% of any nutrient, most of which was from normal tidal exchange. The contribution of nutrients via flood inundation appeared small. Based on aerial extent of seagrasses in Matagorda Bay the seagrass contribution to the exchange of nutrients is believed to be quite small (Wiersema, Armstrong and Ward 1982).

Nutrient Utilization

Nutrient concentrations measured in water samples in Matagorda Bay showed that organic carbon levels average less than 10 ppm and never exceeded 15 ppm. Nitrite nitrogen concentrations were usually less than 0.01 ppm. Nitrate concentrations were less than 0.02 ppm (except after a May flood). Phosphorus concentrations never fell below 0.01 ppm (Wiersema, Armstrong and Ward 1982).

Bay-wide primary productivity was estimated by Wiersema, Armstrong and Ward (1982) at 0.48 ± 0.46 g $C/m^2/d$ and 0.64 ± 2.94 g $O_2/m^2/d$. These values were used in the Matagorda Bay study to estimate biomass production and nutrient uptake and release in the nutrient budgets.

Carbon in Matagorda Bay arrives externally from freshwater inflow, as well as being cycled through primary production of macrophytes, seagrasses, algae and phytoplankton. Losses to the system include deposition in bay sediments as well as transport to the open Gulf, and respiratory loss.

Wiersema, Armstrong and Ward (1982) concluded their nutrient budget exercise for Matagorda Bay by stating nitrogen is limiting primary production in the bay. They supported this speculation by citing low observed nitrogen vs phosphorus levels in the bay. Despite this observation and the small magnitude of external nutrients (freshwater inflow) delivered to Matagorda Bay, they further observed that the majority of external nutrient supplies are inorganic and thus readily assimilable while the bay's internal nutrient pool involves various sinks which remove nutrients (tidal and non-tidal export, sedimentation and denitrification). Removal of external nutrients would not immediately affect this internal nutrient budget because of its relatively small amounts, however in time, the nutrient pool would be depleted because of these various sinks. This reliance of the bay on external nutrients emphasizes the importance of freshwater inflows.

Deltaic Marsh Inundation

Another method for assessing the effects of altered fresh-water inflow is a predictive one that compares the numbers of historical freshwater inundations on a deltaic marsh with numbers of projected inundations expected from some alteration of a river's flow. In the Nueces-Corpus Christi study, projected river flows below the proposed reservoir were calculated (based on 1958-1979 conditions) and compared with historical (actual) flows for the same period. The effects of full reservoir use compared to historical freshwater inundation events (Table 2) revealed that operating an upland reservoir would reduce the annual freshwater inundation frequency by about one event a year. During the May-June period, the reservoir reduced the number of inundation events from 13 (historically) to 4 floodings for the 22-year period. For the October period, delta floods were reduced from 11 to 7. Temporally, the loss of freshwater marsh inundations were most severe in the spring when many commercially important estuarine organisms use the habitat as nursery areas.

The seriousness of these projected losses of spring inundations, based on the deltaic marsh inundation evaluation is corroborated by the study's commercial harvest evaluation. For 1962-78, a positive correlation was found between spring overbanking (flooding) and the commercial catch of white shrimp, a species considered to be a fall to early winter inhabitant of the estuary. Using the results of both methods, Henley and Rauschuber (1980) reported on this relationship but they did not determine a specific cause and effect; rather they speculated that this relationship could be caused by several factors such as lower salinity habitat as well as nutrient flushing of the marsh.

TABLE 2. Comparison of historical and projected freshwater inundations.

	Historical freshwater inundation (1958-1979)	*Projected freshwater inundation*
Average flow duration (Days)	9.8	13
Average one-day flood volume (Thousands acre-feet)	17.0	19.4
Number of occurrences	50	26
Occurrences/Year	2.3	1.2

MANAGEMENT PLANS

The above studies indicated that reduced freshwater inflows to estuaries would directly reduce riverborne nutrients and freshwater inundations of deltaic marshes and would indirectly cause impacts on habitat (i.e., absence of freshwater inundation flushing) and result in commercial fisheries losses. Management recommendations to reduce the effects of reduced freshwater inflows to estuaries must be aimed at increasing nutrient flows to estuaries and increasing freshwater inundation of deltaic marshes.

Releasing freshwater from storage to the estuarine system is an emotional and sensitive issue. The benefits are obvious: freshwater delivers nutrients to the estuary and, if delivered in quantity, causes marsh inundation. But releases of freshwater to estuaries defeat the very purpose for which the reservoir was built. Therefore, management compromises between inland and coastal planners are needed. Freshwater releases could be delivered during the critical springtime when postlarval and juvenile forms of commercially important shellfish are known to use the marsh or during below-average antecedent river inflows to the estuary.

About 60% of water from municipal and industrial reservoirs is eventually returned as wastewater to the estuary, but return flows are usually discharged far from secondary and tertiary bays (deltaic marshes) where they would naturally have entered. In Corpus Christi Bay, the two major sewage treatment plants (Allison and Broadway) are located on

Corpus Christi Bay far removed from Nueces River. Rerouting selected treated discharges back to the river delta can partially replace nutrients lost to the marsh from reduced flooding, provide a continuous source of water, and transport nutrients from the deltaic marshes to the estuarine system.

Diversion channels enable normal river flows (nonflooding events) to inundate deltaic marshes by diverting normal in-channel river flows into estuarine marshes. This diversion allows available low to average flows to deliver freshwater and its constituents to marsh areas that would not naturally be inundated by these flows. Diversion of the Colorado River into Matagorda Bay is an authorized Corps of Engineers project. Project benefits were calculated using the fishery harvest method. Diversion channels have been recommended from the Nueces river to Nueces Bay by the National Marine Fisheries Service. Henley and Rauschuber (1980) recommended diversion as partial mitigation for Choke Canyon Reservoir.

CONCLUSIONS

Although a single method to ascertain impacts on estuaries from altered freshwater inflows in coastal areas has not been determined, a series or set of methods allows the resource manager to develop a management plan to compensate for losses. The key species approach is traditional within environmental impact assessment, but may have to be replaced by more integrated approaches. The nutrient budget approach deals with the basic raw materials of all biological productivity, and therefore, upland carbon losses can be tied to productivity of estuarine organisms. Using both the fishery harvest and the historical versus projected marsh inundation approach has the advantage of immediate transition to economic impact. In all cases, managers need to understand the various ways in which the filtering functions of estuaries can be affected by manipulation of freshwater inflow.

ACKNOWLEDGMENTS

Principal contractors for these USFWS funded studies were Henningson, Durham and Richardson for the Nueces-Corpus Christi study and Espey, Huston and Associates and the University of Texas at Austin for the Matagorda Bay study. Drs. Wiley Kitchens and Norman Benson provided technical reviews. My sincere thanks to Gaye Farris for her editorial diligence and to Dawnlyn M. Harris and Jayne Dayton for typing the manuscript. And a special thanks to Dr. Victor Kennedy for his many comments and suggestions.

REFERENCES CITED

Armstrong, N.E. 1982. Responses of Texas estuaries to freshwater inflows, pp. 103-120. *In:* V.S. Kennedy (ed.), *Estuarine Comparison*. Academic Press, New York.

Armstrong, N.E. and V.N. Gordon. 1979. Nutrient exchange studies on the seagrasses of Texas. Report CRWR-161. Center for Research in Water Resources, U. of Texas, Austin, Texas.

Chapman, C.R. 1966. The Texas Basins Project, pp. 83-92. *In:* R. Smith, A.H. Swartz and W.H. Massman (eds.), *A Symposium on Estuarine Fisheries*. Spec. Pub. 3. American Fisheries Society.

Henley, D.E. and D.G. Rauschuber. 1980. Studies of freshwater needs of fish and wildlife resources in Nueces-Corpus Christi Bay area, Texas. Phase 4 Rep. to the U.S. Dept. of Interior, Fish and Wildlife Service. Albuquerque, N.M. pp. 373-398.

Pritchard, D.W. 1967. What is an estuary: physical viewpoint, pp. 3-5. *In:* G.H. Lauff (ed.), *Estuaries*. American Assoc. Adv. Sci. Pub. 83.

Texas Department of Water Resources (TDWR). 1980. Lavaca-Tres Palacios Estuary: a study of the influence of freshwater inflows. Doc. No. LP-106, Texas Department of Water Resources, Austin, TX. chapter VII (VIII-1-45)

Wiersema, J.M., N.E. Armstrong and G.H. Ward. 1982. Studies of the effects of alterations of freshwater inflows into Matagorda Bay area, Texas: Phase 3 final report. Division of Ecological Services, U.S. Fish and Wildlife Service, Albuquerque, N.M.

DETERIORATION OF COASTAL ENVIRONMENTS IN THE MISSISSIPPI DELTAIC PLAIN: OPTIONS FOR MANAGEMENT

Donald F. Boesch

Louisiana Universities Marine Consortium
Chauvin, Louisiana

John W. Day, Jr.
R. Eugene Turner

Center for Wetland Resources
Louisiana State University
Baton Rouge, Louisiana

Abstract: Coastal environments in the Mississippi Deltaic Plain of southeastern Louisiana are undergoing rapid change as a result of natural decay of Mississippi delta lobes, channelization of river flow, and dredging activities. As a result, over 100 km^2/yr of coastal wetlands are converted to open water or uplands--85% of the total U.S. loss rate. Of the options available to reduce the deterioration of coastal wetlands, management of new delta accretion and regulatory control and rehabilitation of channelized wetlands are likely the most effective. Attempts to stabilize coastal barriers should also be made because of the potential for very rapid erosion in their absence. Controlled diversions of the river above the active deltas will be limited in capacity by the presence of populated areas and roads and are unlikely to contribute significantly to new wetland accretion.

ISBN 0-12-405070-0

INTRODUCTION

The Mississippi Deltaic Plain consists of coastal lands formed by delta-building processes during the short period (approximately 7,000 years) since the Holocene transgression of sea level. It occupies much of southeastern Louisiana and contains by far the largest area of coastal wetlands in the United States (9,000 km^2 or about 25% of U.S. total). Highly productive estuarine and nearshore ecosystems are associated with the region. In 1982, 7.7×10^8 kg of commercial fish and shellfish were landed in coastal Louisiana ports, representing 27% of total U.S. marine landings (National Marine Fisheries Service 1983). Most of this catch is comprised of species which spend all or part of their life cycle in the estuarine-wetland complex.

The coastal and estuarine ecosystem of the Mississippi Deltaic Plain has functioned as an effective "filter" between the river and the sea during this period of relatively stable sea level. Much of the river's sediment load was captured and built new coastal landforms which in essence defined the estuaries and established conditions for extensive proliferation of wetlands. Utilization of the nutrients contributed by the river has supported the tremendous production of animal biomass within the coastal and nearshore environments. However, recent changes in the coastal environments suggest that for one reason or another they are becoming less efficient filters and that sediment fluxes are being affected.

Although the Mississippi Deltaic Plain has been long known as an area of rapid change in coastal environments as a result of delta lobe growth and decay (Russell 1936), systematic comparisons of maps and aerial imagery from periods during the last three decades have documented a surprisingly large and accelerating rate of wetland deterioration and coastal habitat change. Wicker (1980) estimated a loss of coastal wetlands of 83 km^2/yr in the Deltaic Plain between 1955 and 1978. Furthermore, map comparisons for various intervals during this century indicate a geometric increase in this rate from a maximum of about 17 km^2/yr in 1913 to an extrapolated 102 km^2/yr in 1980 (Gagliano, Meyer-Arendt and Wicker 1981). This represents approximately 85% of the coastal wetland loss presently experienced in the United States.

In addition to the conversion of wetland habitat to open water, the character of wetland and estuarine habitats in the Deltaic Plain is being modified by saltwater intrusion as the wetlands and barrier islands nearest the Gulf of Mexico deteriorate. Thus the areal extent of tidal freshwater and transitional (fresh to slightly brackish) habitats has particularly diminished (Wicker 1980). Large bays near the coast have grown in size and increased in salinity. Waterbodies in previously fresh marshes and swamps have opened as salinity has killed intolerant vegetation.

Here we will examine the causes of the accelerated modification of the estuarine ecosystems of the Mississippi Deltaic Plain. We will assess the options available to reduce the rate of wetland loss and to regulate salinity in the component estuarine basins of the region. Additional background and discussion of these issues is included in the papers in Boesch (1982).

CAUSES

Natural Deltaic Processes

The Mississippi Deltaic Plain has been created by cycles of delta construction and abandonment as the Mississippi River has switched its course to the sea. Delta switching occurred at a frequency of approximately once every 1,000 years over the last 7,000 years. As a delta lobe was extended to a point of decreased efficiency of channel flow, the flow was gradually captured by a more efficient route to the sea. Seven major delta lobes, each with multiple distributaries, can be recognized extending from Vermilion Bay to Mississippi Sound to the east (Fig. 1). Presently, most flow debauches through the Balize Delta, the familiar "birdfoot" delta extending into the Gulf of Mexico. However, the U.S. Army Corps of Engineers allows about 30% of the flow to travel down the Atchafalaya River, a 307 km shorter, and thus more efficient, route. Without the artificial control structure at the junction of the Atchafalaya and Mississippi rivers, the Atchafalaya would have captured most of the river flow by the mid 1970's (Fisk 1952). A subaqueous delta began to be deposited at the mouth of the

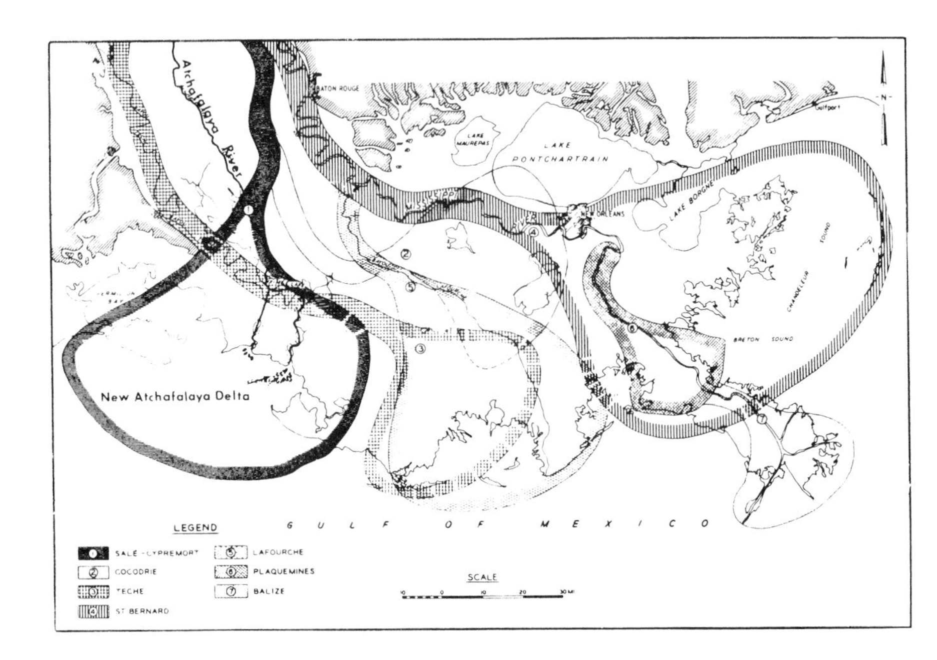

FIGURE 1. Major delta lobes that have constructed the Holocene Mississippi River deltaic plain (modified from Kolb and Van Lopik (1966). Note the location of the most recent lobe in the Mississippi River delta complex, the Atchafalaya delta.

Atchafalaya River in the early 1950's and a subaerially exposed delta has rapidly prograded into Atchafalaya Bay since 1973 (van Heerden and Roberts 1980).

As delta lobes grow due to fluvial processes, wetlands establish between distributaries and contribute to accretion by peat accumulation and trapping of fine suspended sediments (Fisk 1960). Upon abandonment of a delta lobe or distributary channel, there is no longer a source of fluvial sediments for accretion, the effects of sediment dewatering and compaction and regional subsidence become dominant, and the associated wetlands deteriorate (Boesch et al. 1983). This deterioration is caused by lateral erosion along shorelines by waves and by the sinking of the marsh surface as a result of subsidence due to compaction of thick unconsolidated sediments and downwarping of underlying formations. Despite the very rapid subsidence rate, the marsh is not immediately inundated, because it is able to trap and accrete resuspended sediments and organic debris. Thus, for a while the marshes continue to grow upward as their sedimentary foundation subsides and may actually advance over previously subaerial land of the natural levee (Fisk 1960). However, eventually the rate of sediment accretion in the marsh falls behind the rate of subsidence. Protective barrier islands also subside, exposing the erodable marsh deposits to waves. Rapid destruction of wetlands results.

The rapid subsidence in the Mississippi Deltaic Plain effectively results in an apparent local sea-level rise of approximately 12 mm/yr near the coast, which is ten times the rate of eustatic sea-level rise and approximately four times the locally apparent rise along the U.S. east coast (Boesch et al. 1983). In order to prevent submergence, wetlands have to accrete sediments and organic detritus at a rate at least equal to the local sea-level rise. This is apparently accomplished for east coast marshes and even for Louisiana marshes along tidal waterbodies (Table 1); however backmarshes, tens of meters removed from tidal waterbodies, have an apparent deficit in accretion under most salinity regimes in Louisiana. As the deficiency of sediment accretion compared with the subsidence-induced rise in sea-level increases, "die back" of marsh grass results and shallow, open ponds develop (Mendelssohn, McKee and Patrick 1981). The loss of grass further reduces sediment trapping and increases the accretionary deficit.

Subsidence also affects wetland deterioration indirectly by hastening the demise of the sand barriers which mark the outer flank of the delta lobes. After an active delta lobe is abandoned, marine processes transport delta-front sands into arcuate, flanking barrier islands which enclose interdistributary bays (Penland and Boyd 1982). The erosion of the barrier islands results in direct wave attack on the easily erodable, fine wetland sediments and in enhanced erosion by currents as a result of the increased tidal prism in the bays.

Because of these processes, wetland deterioration can be expected in all of the inactive delta lobes of the Mississippi Deltaic Plain. However, the net overall rate of land gain in the Deltaic Plain must have averaged 5 to 6 km^2/yr since the end of the Holocene transgression in order to

TABLE 1. Summary of marsh accretion rates measured in coastal Louisiana and along the U.S. east coast.

Location	*Marsh type*	*Marsh accretion rate (mm/yr)*	*Mean sea-level rise (mm/yr)*	*Reference*
Louisiana Deltaic Plain	Freshwater		11.0	a
	streamside	10.6		
	backmarsh	6.5		
	Intermediate (Spartina patens)			a
	streamside	13.5		
	backmarsh	6.4		
	Brackish (S. patens)			a
	streamside	14.0		
	backmarsh	5.9		
	Saline (S. alterniflora)		13.0	b
	streamside	13.5		
	backmarsh	7.5		
Chenier Plain	Salt-brackish (S. patens)	7.0	12.0	c
Georgia	S. alterniflora	3-5		a
Delaware	S. alterniflora	5.0-6.3	3.8	a
New York	S. alterniflora	2.5-6.3	2.9	a
Conn.	S. alterniflora	8-10	2.5	a
	S. patens	2-5		
Mass.	S. alterniflora	2-18	3.4	d

a. Hatton et al. 1983; b. DeLaune et al. 1978; Baumann 1980; c. Baumann and DeLaune 1982; d. Redfield 1972.

account for its present size. The net land loss rate of 17 km^2/yr which may have existed in the early part of this century was a reversal of the long-term trend, perhaps because the Mississippi River has been building a narrow deep-water delta (the Balize Delta) and depositing much of its alluvium near the edge of the continental shelf rather than dispersing it in a broadly triangular, shallow-water delta as for the previous lobes.

The presently estimated overall net rate of wetland loss of over 100 km^2/yr, although certainly not unusual during transgression, has probably not been experienced since the construction of the Holocene Mississippi Deltaic Plain began. The rapid acceleration of the loss rate since the mid-century seems largely a result of the direct and indirect effects of human activities.

Alteration of Flow Regimes

Soon after European settlement of southeastern Louisiana, construction of artificial levees for flood protection began in 1719 and continued to the point that the Mississippi River is effectively leveed to within 17 km of the Head of the Passes, where the active distributaries join. The Atchafalaya River is also confined by flood control levees, however it fills a much larger flood plain. Even prior to leveeing, the Mississippi River flowed mainly through the Balize Delta where much sediment was lost to deep water. As a consequence, even though there has been active progradation of that delta, there was probably little or no net wetland accretion within the Deltaic Plain. However, floods regularly overtopped the river's natural levees or formed breaching crevasses. This resulted in broad dispersal of sediments within the Deltaic Plain, including portions of naturally deteriorating delta lobes. The great flood of 1882, for example, inundated most of the Deltaic Plain (Elliot 1932), including the abandoned delta lobes.

Presently, other than the Atchafalaya River, the only diversions of Mississippi River flow allowed outside of the confining artificial levees are 1) flood relief flows through the Bonnet Carré spillway into Lake Pontchartrain, and 2) lesser flows down Bayou Lafourche and through navigation locks and several small structures below New Orleans operated to control salinity in adjacent estuaries.

In addition to the lack of new sediments which river flooding previously provided, the control of the Mississippi River's flow has also resulted in increasing salinity in many of the estuarine systems of the Deltaic Plain. This is reflected in alterations of wetland vegetation types (Wicker 1980).

In addition to flood control levees along the Mississippi River and Atchafalaya Basin, other structures erected by man have altered flow and sedimentation regimes on a smaller scale. These include levees to protect inhabited areas from tidal and backwater flooding, levees erected for land drainage and reclamation, spoil banks along dredged channels and canals, and channels themselves.

Channelization

Enormous numbers of ditches, canals, and channels have been dug throughout the wetlands of the Mississippi Deltaic Plain. They include large navigation channels, oil and gas well access canals, pipeline canals, drainage canals, and the narrow ditches which provide trappers access to expansive marshes. A few large navigation canals traverse the estuarine gradient perpendicular to the coast (Mississippi River-Gulf Outlet, Barataria Waterway, Houma Navigation Canal) and enhance saltwater intrusion on a large scale. The Gulf Intracoastal Waterway which runs parallel to the coast affects the flow of fresh water within and between estuarine basins.

Virtually all canals widen with time as a result of bank erosion or flow alteration. Johnson and Gosselink (1982) found that oil and gas access canals widened at an average rate of 1 m/yr and those experiencing boat traffic widen more than twice that fast. Thus, the extensive development of canals presents a legacy of continued wetland erosion and hydrologic modification even without new channelization.

Several attempts have been made over the last 15 years to estimate the impacts of canal development on Louisiana's coastal wetlands (reviewed by Craig, Turner and Day 1980). Detailed and comprehensive estimation of the areal extent of impacts were made possible by the habitat maps of Wicker (1980) based on 1955 and 1978 photoimagery of the entire Mississippi Deltaic Plain. Based on these inventories, Turner, Costanza and Scaife (1982) estimated that the area of canals and associated spoil banks amounts to about 9 % of the total marsh surface. Scaife, Turner and Costanza (1983) demonstrated a significant relationship between coastal land loss and canal density within specific delta lobes and distances from the coast. If the actual canal and spoil bank area is subtracted from the land loss rate, there is a residual loss rate which itself is correlated with canal density. Thus, there may be indirect effects of canals on wetland habitats in addition to the actual direct effects of dredging and filling.

Examination of the patterns of wetland deterioration in areas of extensive canal development provides some insight into the indirect effects of canals. Numerous shallow ponds, evident in 1978 imagery (Fig. 2), have opened in marsh adjacent to canals and spoil banks. Similar pond development is not seen adjacent to natural channels. These ponds seem to be the result of a disruption of the natural marsh hydrology. Spoil banks interfere with overbank flooding of the marsh, disrupting the supply of suspended sediments which subsidize the aggradation necessary to counteract subsidence. Further, spoil banks also may result in impounding standing water over the marsh surface and decrease subsurface flows by gravity compression of the marsh deposits underlying the spoil banks. In either case, marsh grasses succumb to continuous inundation because of the lack of oxygen or presence of high sulfide levels to which roots are exposed (Mendelssohn et al. 1981). These indirect effects of channelization have been estimated to result in the loss of approximately four times as much wetland as the canals themselves (Craig et al. 1980). Consequently, the construction of new canals, the placement of dredged material, canal widening, and the peripheral ponding effect may, according to Scaife et al. (1983), be responsible for 48 to 97 % of the regional land loss.

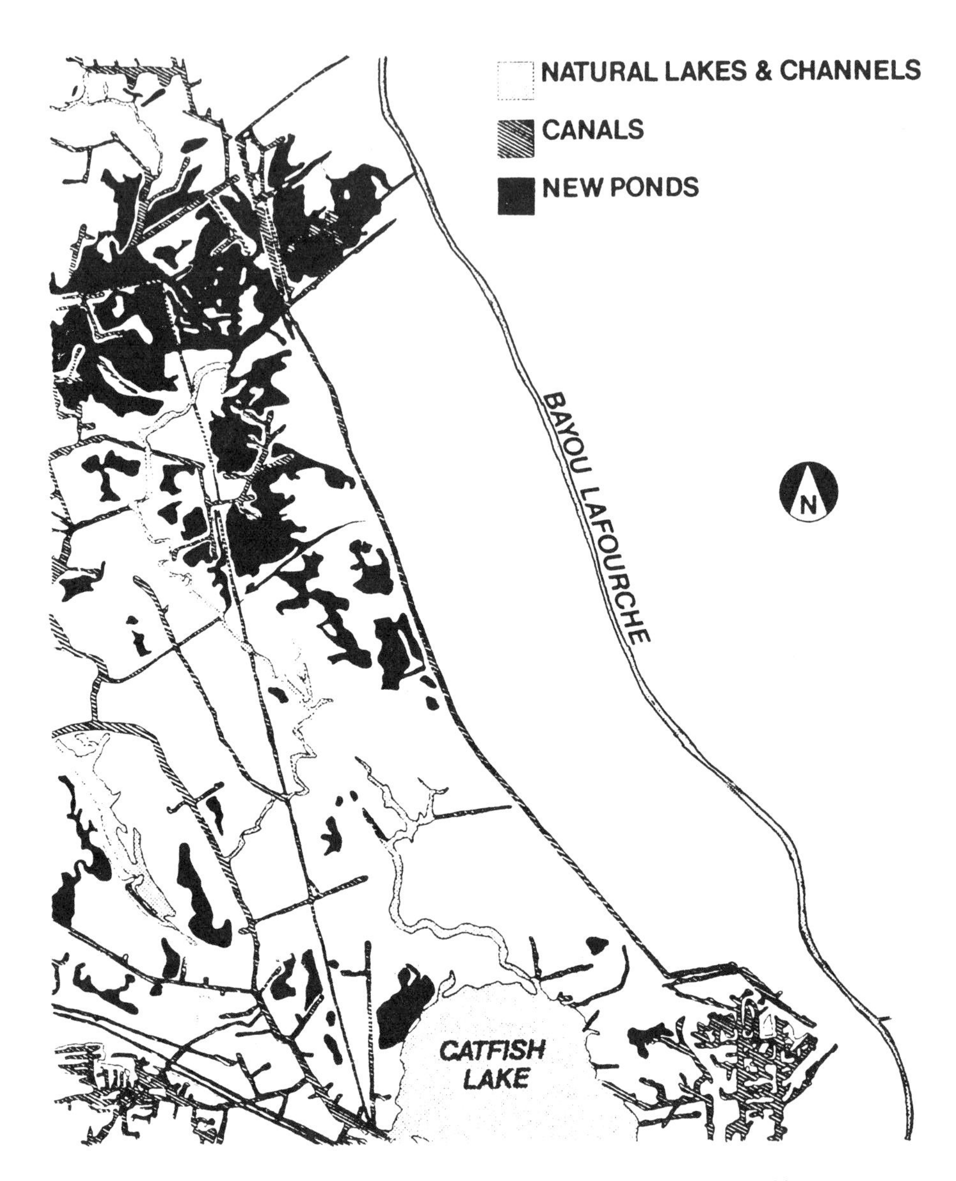

FIGURE 2. Shallow water bodies which formed adjacent to dredged canals between 1969 and 1978 in the vicinity of Golden Meadow, Louisiana (Turner et al. 1982).

OPTIONS FOR MANAGEMENT

An extrapolation of the destructional processes and trends described above projects a bleak future for perpetuation of the coastal ecosystems of the Mississippi Deltaic Plain in their present form. For example, Cleveland, Neill and Day (1981) developed speculative models which predict the marsh area remaining in the Barataria Basin given the effects of subsidence on backmarsh, the lateral erosion of streamside marsh, and the effects of canals. Their models predict the disappearance of 72% of the marsh in the basin after 100 years without the construction of new canals, and the virtual disappearance of marsh after 72 years if canal construction continues at present rates. However, if it is assumed that two new subdeltas are created in the Barataria Basin over a 100 year period as a result of artificial breaches in the Mississippi River levee, total marsh area would actually increase by 15%.

As the filtering activities of the systems are modified by human activities, the coastal ecosystems of southeastern Louisiana will change dramatically in the coming years. Much of what is now marsh will become shallow water habitat and embayments will become larger and more saline. The challenge to environmental management is to preserve the net habitat value as much as is possible. Because of the assumed high value of vegetated wetlands as habitat for fish, shellfish and wildlife; in buffering storms; and in assimilating wastes, one objective of an appropriate management strategy is to minimize the net loss of these wetlands within the Mississippi Deltaic Plain. The options available in order to achieve this objective are 1) management of presently accretional subdeltas; 2) building of flow and water-level control structures; 3) diversion of river water and sediments to maintain salinity levels, build new wetlands and subsidize accretion in decaying marshes; 4) maintenance of coastal barrier islands as a line of defense; and 5) reduction in losses due to channelization by tighter regulatory control and rehabilitation of existing canals.

Various state and federal programs address each of these options in some degree. Notably, the Coastal Environmental Protection Trust Fund was established through state legislative act. Planning of several barrier island-beach stabilization projects, river diversions, and delta accretion management programs is underway (Louisiana Department of Natural Resources 1982). The U.S. Army Corps of Engineers is also undertaking a variety of planning studies concerning river diversions, flood protection, beach protection, and navigation channel dredging which have great implications to estuarine and wetlands management.

Management of Subdelta Accretion

Most active accretion of coastal wetlands is associated with the Atchafalaya River mouth and includes 1) progradation of the Atchafalaya delta *per se* (Roberts, Adams and Cunningham 1980), 2) mud flat progradation along the coast to the west (Wells and Kemp 1981), and 3) infilling of older deteriorating marshes peripheral to Atchafalaya Bay (Baumann and Adams 1982). Day and Craig (1982) estimated that these three processes result in a reduction of land loss rate of 11.9, 1.1, and 4.9 km^2/yr, respectively. Through management of flow and constructive use of material dredged from the navigation channel in the Atchafalaya delta, wetland accretion may be enhanced. A planned extension of a levee along the eastern margin of Atchafalaya Bay for the purpose of controlling backwater flooding will limit the supply of alluvial sediments into the extensive, marshes to the east (Baumann and Adams 1982).

Despite receiving most of the Mississippi River flow, the Balize delta is experiencing very rapid wetland loss (Wicker 1980) in part because the flow is largely confined to the distributaries and minor crevasses are routinely closed by the Corps of Engineers. Because of the rapid subsidence rates (43 mm/yr; Swanson and Thurlow 1973) prodigious accretion is required for the maintenance of interdistributary wetlands in the delta. Through operation of small flow control structures, sediment accretion and marsh re-establishment can be enhanced.

Flow and Water-level Control

Tidal and runoff flow control and water-level regulation by wiers and gates are extensively employed in many Louisiana wetlands by both public wildlife management agencies and private landowners. Water-level regulation is practiced in an attempt to control vegetation type to maximize habitat value for waterfowl or fur- and hide-bearers. Control structures are also employed to reduce salt-water intrusion and for that reason are considered as a tool for protection of wetland habitats.

Such management techniques have evolved from practical experience and their effectiveness and side effects have seldom been carefully evaluated (Wicker, Davis and Roberts 1983). Maintaining water at too high a level can result in destruction of marsh vegetation. Curtailment of tidal flow through limited entrance into impoundments may reduce the supply of suspended sediments needed for continued accretion of wetlands. The effect of control structures on the value of the habitat as a nursery for fishery species is also not well known (Adkins and Bowman 1976).

River Diversions

With the exception of the Bonnet Carré spillway, opened 7 times since 1937, present river diversions can accommodate <100 m^3/sec flow rates. Siphons or gated control structures have been employed. While flows of such magnitude are useful in preventing saltwater intrusion and maintaining desired salinity gradients in relatively small areas, only very large flows (>1000 m^3/sec) are likely to result in dramatic accretion of wetlands. A practical limitation of such large volume diversions is that, except in the lower delta, they would have to pass under roadways, requiring expensive bridges and levees. Controlled diversions along the lower river may be able to reduce land loss by 1 to 3 km^2/yr (Day and Craig 1982).

Barrier Stabilization

The retreat of the coastal strand in the Mississippi Deltaic Plain is the most rapid in the

United States (May, Dolan and Hayden 1983), averaging 4.2 m/yr and reaching values as high as 22 m/yr (Penland and Boyd 1982). Between 1880 and 1980, total coastal barrier area decreased by 41%. These rapidly eroding and subsiding barriers are sand starved, thus their stabilization either by structural means or sand nourishment and revegetation (Louisiana Department of Natural Resources 1982; Mendelssohn 1982) will be difficult and expensive. Unfortunately, there is a lack of quantification of the effectiveness of coastal barriers in reducing wetland deterioration. The uncertainty concerning the future consequences of barrier disappearance to coastal and estuarine environments makes comparison of the effectiveness of barrier stabilization to other management options difficult. However, because rapid acceleration of wetland erosion is a serious possibility, the pilot stabilization efforts underway seem prudent.

Control of Canals

Turner et al. (1982) estimated that if no canals were constructed the wetland loss rate would be 30 to 40 km^2/yr less than it would be over the next 20 years than if the present level of development continued. Canal development is also theoretically the most manageable cause of coastal wetland loss (compared to subsidence, rising sea level, and flow alteration by river levees). By far most new canals are dredged for access for oil and gas development and, because of economic pressures, the rate of canal development has not decreased despite growing concern about their effects (Turner et al. 1982). State and federal regulatory agencies have recently been moderately successful in reducing the length of canals for which permits are requested and in some cases in requiring directional drilling (drilling an oblique rather than vertical well) in lieu of dredging. Backfilling or plugging canals has also been required, but there are unresolved issues regarding the effectiveness of these approaches, primarily because the highly organic dredged material loses much of its volume if placed in subaerial spoil banks and on refilling will not return the sediment surface to the previous grade.

Clearly, regulatory approaches to reducing the impacts of canal construction should be pursued vigorously. The length of canals should be minimized and canals should be plugged or filled after use, particularly where they intersect natural stream channels. Spoil banks which interfere with marsh hydrology should not be left or at least should have frequent gaps. Mitigative attempts should also be made to rehabilitate existing canals which continue to affect wetland deterioration.

Integrated Management

As the causes of the deterioration of coastal environments in the Mississippi Deltaic Plain are multiple, so must be the efforts to reduce the loss of valuable habitats. The various options described here are all recognized and discussed among environmental managers, elected officials, university scientists, environmentalists and the general public. What is lacking is an integration wherein the effectiveness of the options can be evaluated in terms of the same currency. In our view, management of new deltaic accretion and regulatory control and rehabilitation of proposed and existing channelized wetlands are the most effective and affordable options. Coastal barrier stabilization should be pursued because of the potentially catastrophic consequences if these barriers are lost. However, the costs of such stabilization are very high and success is speculative. Controlled river diversions upstream of the active natural distributary systems (the Balize and Atchafalaya Deltas) are constrained in capacity by the presence of populated areas, roads and leveed areas. Thus, their primary role will be in maintaining salinity distribution and subsidizing the sediment budget of deteriorating wetlands rather than in building new wetlands.

ACKNOWLEDGMENTS

Partial support for our efforts has been provided by the National Coastal Ecosystem Team of the U.S. Fish and Wildlife Service through Cooperative Agreement 14-16-0009-81-1016 and by the Ocean Assessments Division of the National Oceanographic and Atmospheric Administration through Grant NA82RAD00008.

REFERENCES CITED

Adkins, G. and P. Bowman. 1976. A study of the fauna in dredged canals of coastal Louisiana. *Louisiana Wildlife and Fisheries Commission Tech. Bull. 18*: 1-72.

Baumann, R. H. 1980. Mechanisms of maintaining marsh elevation in a subsiding environment. M.S. Thesis. Louisiana State Univ., Baton Rouge, LA., 92 pp.

Baumann, R. and R. Adams. 1982. The creation and restoration of wetlands by natural processes in the lower Atchafalaya River system: possible conflicts with navigation and flood control objectives, pp. 8-24. In: Randall H. Stovall (ed.), *Proceedings Annual Conference on Wetlands Restoration and Creation, Vol. 8.* Hillsborough Community College, Environmental Studies Center in cooperation with Tampa Port Authority, Tampa, Florida.

Baumann, R. H. and R. D. DeLaune. 1982. Sedimentation and apparent sea-level rise as factors affecting land loss in coastal Louisiana, pp. 2-13. In: D. F. Boesch (ed.), Proceedings of the Conference on Coastal Erosion and Wetland Modification in Louisiana: Causes, Consequences and Options. U.S. Fish and Wildlife Service, Biological Services Program, FWS/OBS-82/59, Washington, D.C.

Boesch, D. F. (ed.). 1982. Proceedings of the Conference on Coastal Erosion and Wetland Modification in Louisiana: Causes, Consequences and Options. U.S. Fish and Wildlife Service, Biological Services Program, FWS/OBS-82/59, Washington, D.C. 256 pp.

Boesch, D. F., D. Levin, D. Nummedal and K. Bowles. 1983. Subsidence in Coastal Louisiana: Causes, Rates and Effects on Wetlands. U.S. Fish and Wildlife Service, Biological Services Program, FWS/OBS-83/26, Washington, D.C. 30 pp.

Cleveland, C. J., Jr., C. Neill and J. W. Day, Jr. 1981. The impact of artificial canals on land loss in the Barataria Bay Basin, Louisiana, pp. 425-

435. In: W. Mitsch, R. Bosserman and J. Klopatele (eds.), *Energy and Ecological Modelling.* Elsevier Scientific Publ., New York.

Craig, N. J., R. E. Turner and J. W. Day, Jr. 1980. Land losses and their consequences in coastal Louisiana. *Zeitschrift fuer Geomorphologie Suppl. 34*: 225-241.

Day, J. W., Jr. and N. J. Craig. 1982. Comparison of effectiveness of management options for wetland loss in the coastal zone of Louisiana, pp. 232-239. In: D. F. Boesch (ed.), Proceedings of the Conference on Coastal Erosion and Wetland Modification in Louisiana: Causes, Consequences and Options. U.S. Fish and Wildlife Service, Biological Services Program, FWS/OBS-82/59, Washington, D.C.

DeLaune, R. D., W. H. Patrick, Jr. and R. J. Buresh. 1978. Sedimentation rates as determined by ^{137}Cs dating in a rapidly accreting salt marsh. *Nature 275*: 532-533.

Elliot, D. O. 1932. The improvement of the lower Mississippi River for flood control and navigation. U.S. Army Corps of Engineers, Waterways Experiment Station, Vicksburg, MS, Vol. 1, 2, & 3.

Fisk, H. N. 1952. Geological investigation of the Atchafalaya Basin and the problem of Mississippi River diversion. U.S. Army Corps of Engineers, Mississippi River Commission, Vicksburg, MS, Vol. 1, 145 pp.

Fisk, H. N. 1960. Recent Mississippi River sedimentation and peat accumulation, pp. 187-199. In: E. Van Aelst (ed.), Congrès pour l'Advancement des Études de Stratigraphie et de Geologie du Carbonifere, *4th Heerlen*, 1958, Vol.1.

Gagliano, S. M., K. J. Meyer-Arendt and K. M. Wicker. 1981. Land loss in the Mississippi River deltaic plain. *Trans. Gulf Coast Assoc. Geol. Soc. 31*: 295-300.

Hatton, R. S., R. D. DeLaune and W. H. Patrick, Jr. 1983. Sedimentation, accretion, and subsidence in marshes of Barataria Basin, Louisiana. *Limnol. Oceanogr. 28*: 494-502.

Johnson, W. B. and J. G. Gosselink. 1982. Wetland loss directly associated with canal dredging in the Louisiana coastal zone, pp 60-72. In: D. F. Boesch (ed.), Proceedings of the Conference on Coastal Erosion and Wetland Modification in Louisiana: Causes, Consequences and Options. U.S. Fish and Wildlife Service, Biological Services Program, FWS/OBS-82/59, Washington, D.C.

Kolb, C. R. and J. R. Van Lopik. 1966. Depositional environments of the Mississippi River deltaic plain--southeastern Louisiana, pp. 17-61. In: M. L. Shirley (ed.), Deltas in Their Geologic Framework. Houston Geological Society, Houston, Texas.

Louisiana Department of Natural Resources. 1982. Coastal Protection Task Force Report. Louisiana Department of Natural Resources, Baton Rouge, 25 pp.

May, S. K., R. Dolan and B. P. Hayden. 1983. Erosion of U.S. Shorelines. *Eos 64*: 521-523.

Mendelssohn, I. A. 1982. Sand dune vegetation and stabilization in Louisiana, pp. 187-207. In: D. F. Boesch (ed.), Proceedings of the Conference on Coastal Erosion and Wetland Modification in Louisiana: Causes, Consequences and Options. U.S. Fish and Wildlife Service, Biological Services Program. FWS/OBS-82/59, Washington, D.C.

Mendelssohn, I. A., K. L. McKee and W. H. Patrick, Jr. 1981. Oxygen deficiency in Spartina alterniflora roots: metabolic adaptation to anoxia. *Science 214*: 439-441.

National Marine Fisheries Service. 1983. Fisheries of the United States, 1982. National Marine Fisheries Service, National Oceanic and Atmospheric Administration, Washington, D.C., 117 pp.

Penland, S. and R. Boyd. 1982. Assessment of geological and human factors responsible for Louisiana coastal barrier erosion, pp. 14-51. In: D. F. Boesch (ed.), Proceedings of the Conference on Coastal Erosion and Wetland Modification in Louisiana: Causes, Consequences and Options. U.S. Fish and Wildlife Service, Biological Services Program, FWS/OBS-82/59, Washington, D.C.

Redfield, A. C. 1972. Development of a New England salt marsh. *Ecol. Monogr. 42*: 201-237.

Roberts, H., R. Adams and R. Cunningham. 1980. Evolution of sand dominant subaerial phase, Atchafalaya Delta, Louisiana. *Am. Assoc. Petrol. Geol. Bull. 64*: 264-279.

Russell, R. J. 1936. Physiography of the lower Mississippi River delta. Louisiana Geological Survey, Baton Rouge. *Lower Mississippi Delta Geol. Bull. 8*: 3-199.

Scaife, W., R. E. Turner and R. Costanza. 1983. Indirect impact of canals on recent coastal land loss rates in Louisiana. *Environ. Mgmt. 7*: 433-442.

Swanson, R. L. and C. I. Thurlow. 1973. Recent subsidence rates along the Texas and Louisiana coasts as determined from tide measurements. *J. Geophys. Res. 78*: 2665-2671.

Turner, R. E. 1984. Relationships between coastal land losses and canals and canal levees in Louisiana and management options. U.S. Fish and Wildlife Service, Biological Services Program. FWS/OBS-84/ (in press).

Turner, R. E., R. Costanza and W. Scaife. 1982. Canals and wetland erosion rates in coastal Louisiana, pp. 73-84. In: D. F. Boesch (ed.), Proceedings of the Conference on Coastal Erosion and Wetland Modification in Louisiana: Causes, Consequences and Options. U.S. Fish and Wildlife Service, Biological Services Program, FWS/OBS-82/59, Washington, D.C.

van Heerden, I. L. and H. H. Roberts. 1980. The Atchafalaya Delta: rapid progradation along a traditionally retreating coast (South-central Louisiana). *Zeitschrift fuer Geomorphologie Suppl. 34*: 188-201.

Wells, J. T. and G. P. Kemp. 1981. Atchafalaya mud stream and recent mudflat progradation: Louisiana Chenier Plain. *Trans. Gulf Coast Assoc. Geol. Soc. 31*: 409-416.

Wicker, K. M. 1980. Mississippi Deltaic Plain Region Ecological Characterization: A Habitat Mapping Study. A User's Guide to the Habitat Maps. U.S. Fish and Wildlife Service, Office of Biological Services, FWS/OBS-79/07, Washington, D.C., 84 pp.

Wicker, K. M., D. Davis and O. Roberts. 1983. Rockefeller State Wildlife Refuge and Game Preserve; Evaluation of Wetland Management Techniques. Report to Louisiana Department of Natural Resources, Coastal Environments, Inc., Baton Rouge, LA., 92 pp., 8 plates.

AN ENVIRONMENTAL CHARACTERIZATION OF CHESAPEAKE BAY AND A FRAMEWORK FOR ACTION

Virginia K. Tippie

U.S. Environmental Protection Agency
Chesapeake Bay Program
Annapolis, Maryland

Abstract: The scientific findings and management implications of the Environmental Protection Agency's Chesapeake Bay Program are reviewed. They suggest that the Bay has dramatically changed in the last century and this change has accelerated in the last thirty years. Many of the valued living resources of the Bay, such as the submerged aquatic vegetation, shad, striped bass, and oysters are declining. Paralleling this decline, there has been an increase in nutrients and toxic compounds in the Bay. It appears that the Bay acts as a "filter", essentially trapping and recycling pollutants. Thus, to restore the Bay's ecological integrity we must reduce pollutant discharge to the Bay.

INTRODUCTION

Chesapeake Bay is the largest estuary in the United States and biologically, one of the most productive systems in the world. The main-stem of the Bay is situated within the states of Maryland and Virginia. It is 290 km (180 mi) long and 4 to 48 km (5 to 30 mi) wide. The mainstem of the Bay and its tidal tributaries have a surface area of 11,500 km^2 (4,400 mi^2) and a tidal shoreline of 30,000 km (8,100 mi) (Cronin 1982). The Bay's drainage basin is 166,000 km^2 (64,000 mi^2) in extent and includes over 150 rivers, creeks, and branches flowing through portions of six states and the District of Columbia (Fig. 1). Today, over 12.7 million people live in the region and virtually every type of economic activity and land use is found in its watershed.

ISBN 0-12-405070-0

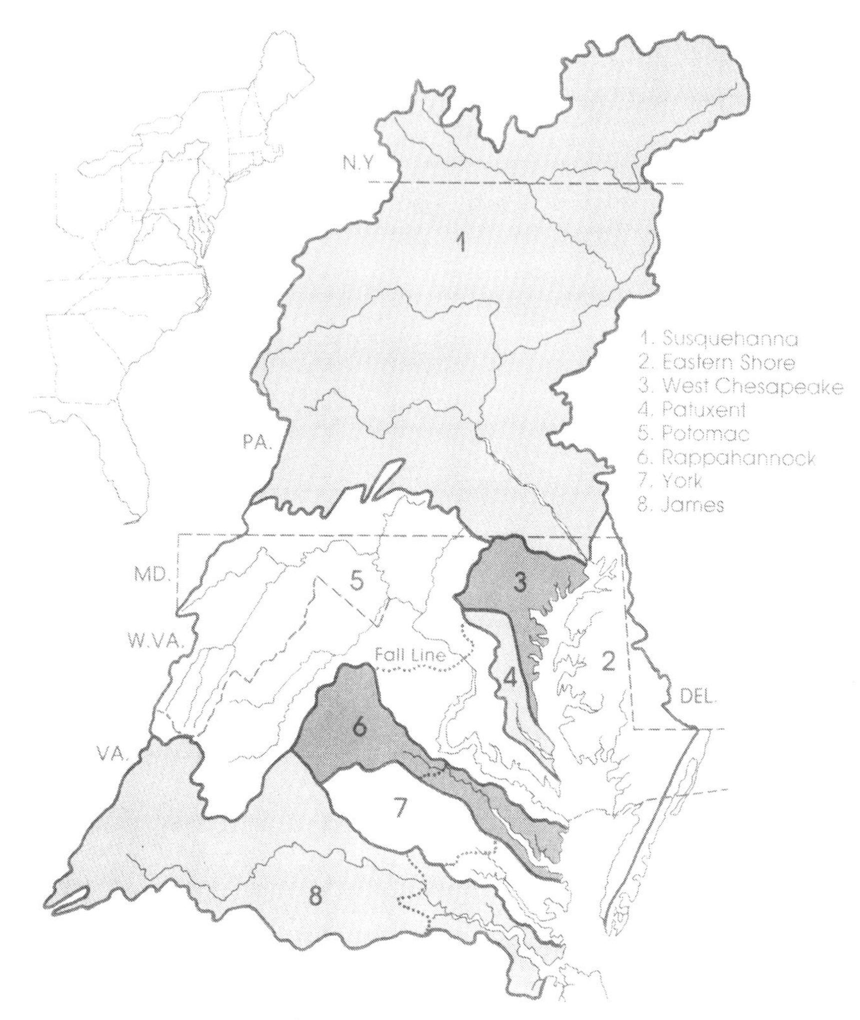

FIGURE 1. Chesapeake Bay drainage basin.

For years, it was believed that the Bay's bounty was endless and that it had an unlimited capacity to assimilate human wastes. However, the very feature that made the Bay so productive, its "filtering" capacity, resulted in the trapping of pollutants and a gradual change in its ecology. Environmental concern over observed changes in the Bay, such as the disappearance of submerged aquatic vegetation (SAV) and declines in important fisheries, prompted the U.S. Congress to authorize an in-depth study of the Bay. This study is known as the Environmental Protection Agency's Chesapeake Bay Program.

This paper discusses the program's purpose and structure, major scientific findings, and management implications. It is hoped that the lessons learned from this study will help us better understand and manage the Chesapeake Bay and other estuarine systems.

PURPOSE AND STRUCTURE

The directives of the U.S. Congress were fairly specific:

- o Assess the principal factors having an adverse impact on environmental quality as perceived by scientists and users;
- o Establish mechanisms for collecting, storing, analyzing and disseminating environmental data;
- o Analyze available environmental data and implement methods for improved data collection;
- o Propose alternative control strategies for long-term protection of the Bay;
- o Evaluate Bay management coordination mechanisms to best assure timely implementation of control strategies. (U.S. Senate Report No. 94-326, 1975).

To address these directives, the program was structured in three phases, namely research, characterization, and management.

To initiate the research phase, state and federal personnel, the scientific community, and citizens from around the Bay were asked to identify and prioritize critical problems. Technical staff then wrote plans of action for the identified high priority critical problems: nutrient enrichment, accumulation of toxic substances, and the decline of submerged aquatic vegetation. Over 40 research projects were funded to address these problems. The individual scientific reports are available through the National Technical Information Service (NTIS) and the research findings are described in a summary report (Environmental Protection Agency 1982).

In the characterization phase of the program, Bay program research and previous efforts were integrated and synthesized. Virtually all current and historical water quality and living resource data on the Bay were compiled in a comprehensive computerized information system. The data were analyzed to determine trends and potential relationships between water and sediment quality and living resources. Positive correlations between water quality parameters and the abundance of certain resources suggested specific management needs. The summary report of this characterization phase was produced by Flemer et al. (1983).

In the management phase of the program, data on the sources and loadings of pollutants were utilized to develop predictive models for evaluating pollution controls. Management strategies identified by state personnel, the scientific community and citizens were evaluated for cost effectiveness and recommendations were developed. The summary report of this management phase was developed by Tippie et al. (1983).

The most significant product of the overall program effort was the regional management ethic it promoted. This ethic was encouraged by the Chesapeake Bay Program Management Committee. The Committee representing the EPA, the state governments, and the citizens of the area, guided the Program's efforts over the years. To assist the management committee in developing the management plan, working groups comprised of citizen leaders in the Bay community and state managers were established. Throughout the characterization and management phases, these teams regularly reviewed findings and conclusions and were involved in developing the management strategies. This process assured that the recommendations of the final management report reflected a general public consensus.

SCIENTIFIC FINDINGS

The Program's scientific investigations and other studies have essentially documented that the Bay has dramatically changed in the last century (Environmental Protection Agency 1982; Flemer et al. 1983). Increasing population growth over time has resulted in major land-use changes and large increases in municipal wastewater which, in turn, have caused substantial increases in pollution loads to the Bay. For many years these activities had a relatively minor impact and many people believed that the Bay had an unlimited capacity to assimilate human wastes. However, we now know that contaminants entering the Bay are not readily flushed out into the ocean. Instead, the Bay is serving as a filter effectively trapping pollutants within the estuary. Over time this process has gradually changed the nature of the Bay in the following ways:

Living Resources

o In the upper Bay, an increasing number of blue-green algal or dinoflagellate blooms has been observed in recent years. In fact, cell counts have increased approximately 250-fold since the 1950's. In contrast, the algal populations in the upper Potomac River have recently become more diverse, with the massive blue-green algal blooms generally disappearing since nutrient controls were imposed in the 1960's and early 1970's in this segment of the Bay watershed. However in the summer of 1983, a large blue-green algal bloom occurred in the Potomac. A task force is presently trying to determine the causative factors;

o Since the late 1960's, submerged aquatic vegetation has declined in abundance and diversity throughout the Bay (Fig.2). The decline is most dramatic in the upper Bay and western shore tributaries. An analysis over time indicates that the loss has moved progressively down-stream, and that present populations are mostly limited to the lower estuary;

o Landings of freshwater-spawning fish such as shad (Fig.3) and alewife have decreased. Striped bass landings, after increasing through the 1930's and 1940's, have also decreased, especially since 1973. Harvests of marine-spawning fish such as menhaden (Fig. 3) and bluefish have generally remained stable or increased. The increased yield of marine spawners and decreased yield of freshwater spawners represent a major shift in the Bay's fishery. Over the 100 year period from 1880 to 1980, marine spawners accounted for 75 percent of the fishery; during the interval from 1971 to 1980, they accounted for 96 percent of the fishery;

o Oyster harvests have also decreased Bay-wide probably due to over-fishing and diseases as well as water quality. Oyster spat set has declined significantly during the 10 year period from 1970 to 1980, as compared to previous years, particularly in the upper Bay and western shore tributaries and some eastern shore tributaries such as the Chester River. In 1980-81, higher spatset occurred in many areas due to increased salinity; however, spatset in the upper Bay and many western shore tributaries remained depressed. The economic impact of the decline in oyster harvest has been somewhat offset by recent increases in the harvest of blue crabs which may be due to increased fishing effort. As a result, the total Bay-wide landings of shellfish have not changed greatly over the last twenty years. However, overall shellfish landings for the western shore have decreased significantly during this period.

Water and Sediment Quality

o Increasing levels of nutrients are entering many parts of the Bay: the upper reaches of almost all the tributaries are highly enriched with nutrients; lower portions of the tribu-

FIGURE 2. General area of submerged aquatic vegetation distribution in 1965 (left) and 1980 (right).

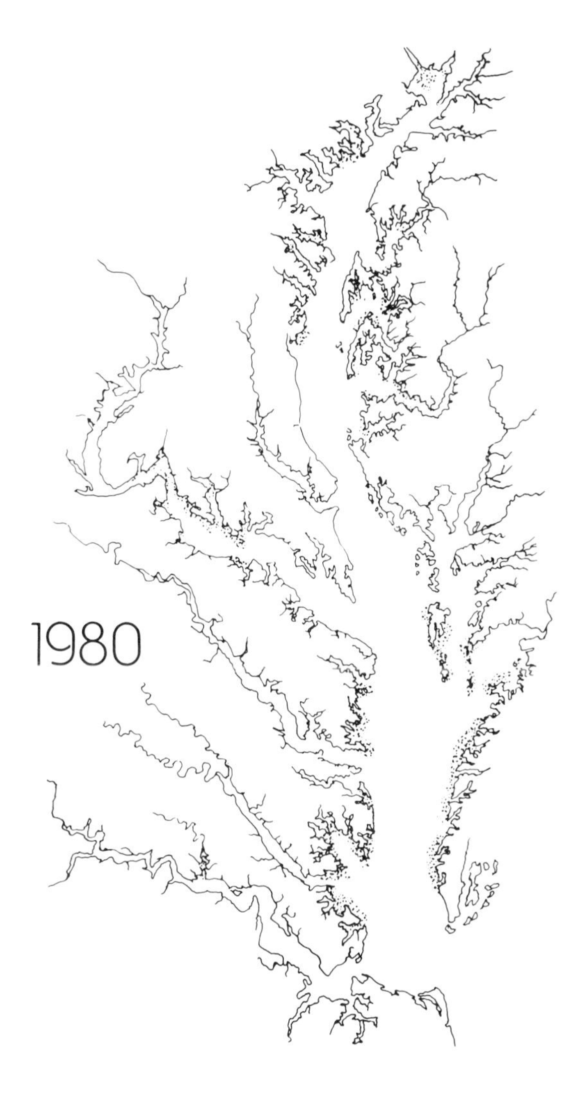

FIGURE 2.

taries and eastern shore embayments have moderate concentrations of nutrients; and the lower Bay does not appear to be enriched (Fig. 4). Data covering 1950 to 1980 indicate that, in most areas, water quality is diminishing, partially because increasing levels of nutrients are entering the waters.

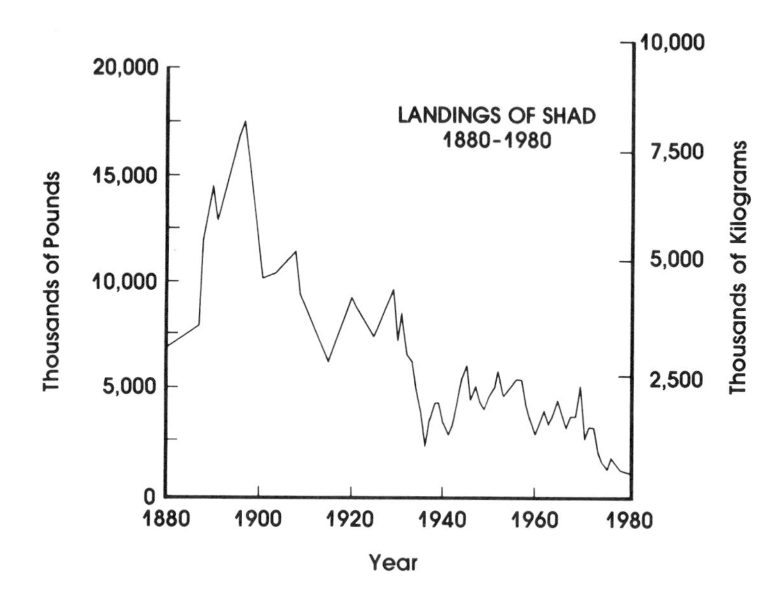

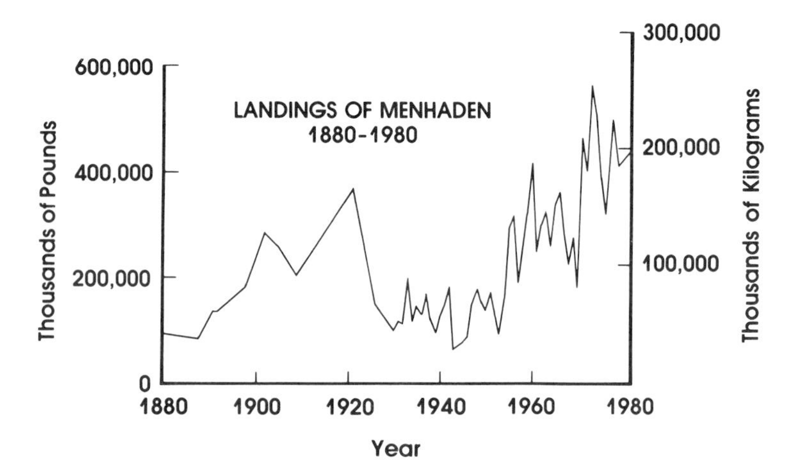

FIGURE 3. Landings of shad (upper) and menhaden (lower) from 1880 to 1980.

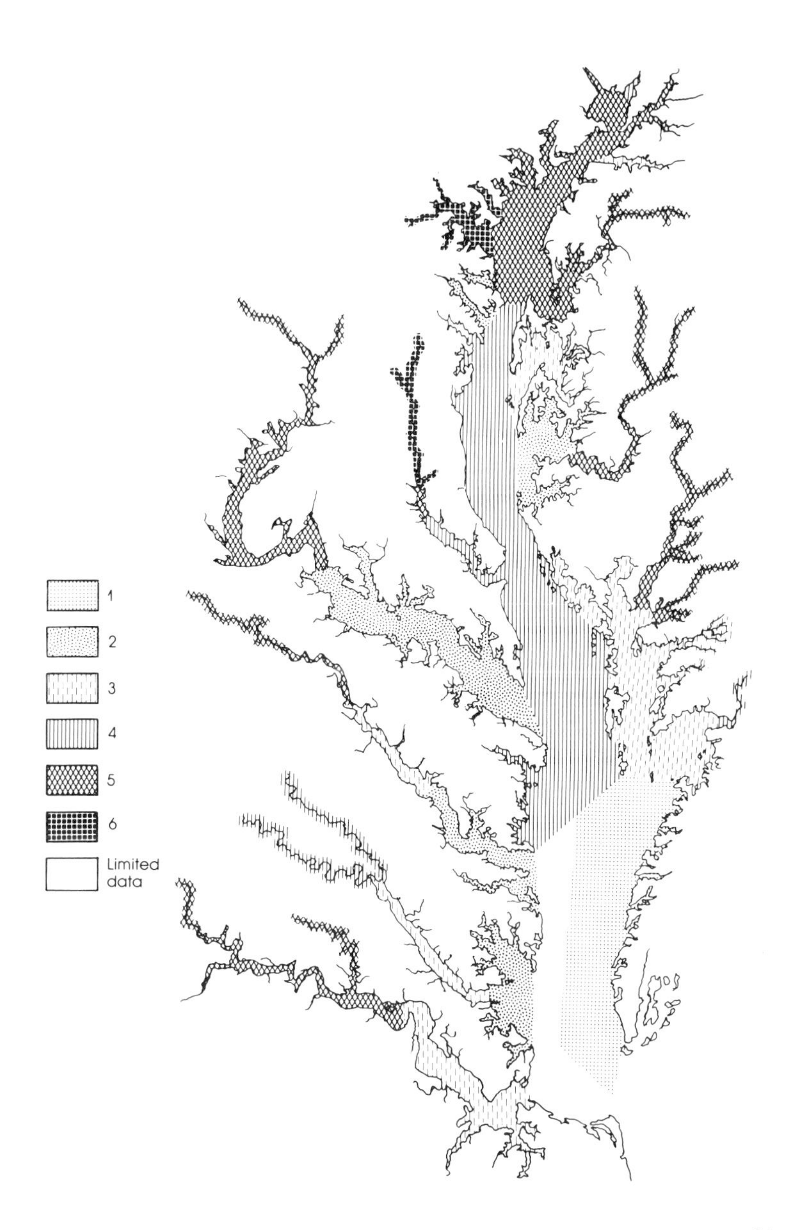

FIGURE 4. Ranking of Chesapeake Bay segments according to the levels of Total Nitrogen (TN) and Total Phosphorus (TP). Rank 1 has the lowest nutrient concentrations; TN (0 - 0.40 mg L^{-1}), and TP (0 - 0.056 mg L^{-1}). Rank 6 has the highest nutrient concentrations TN (1.76+ mg L^{-1}) and TP (0.246+ mg L^{-1}).

However, in the Patapsco, Potomac, and James Rivers (and some smaller areas) there is a relative improvement in water quality; this appears primarily to be due to control efforts in those heavily polluted areas.

o The amount of water in the main part of the Bay which has low or no dissolved oxygen has increased about fifteen-fold

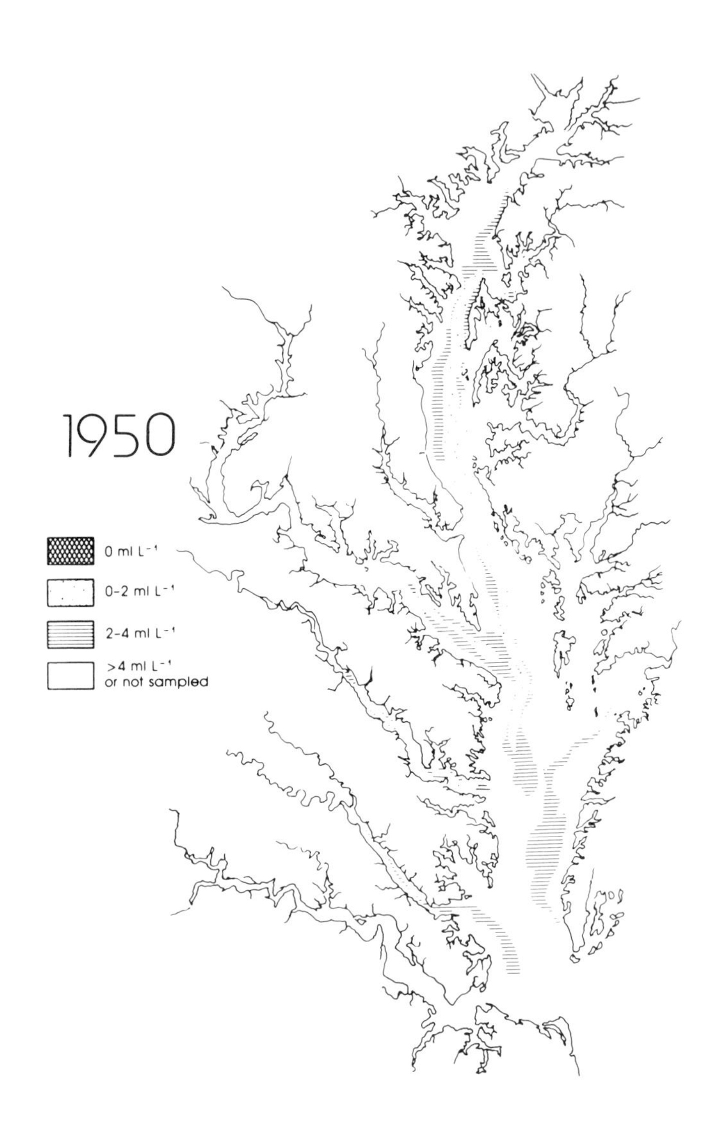

FIGURE 5. Extent of anoxic bottom water in the main stem of Chesapeake Bay in 1950 (left) and 1980 (right).

between 1950 and 1980 (Fig. 5). Currently, from May through September in an area reaching from the Annapolis Bay Bridge to the Rappahanock River, much of the water deeper than 40 feet has no oxygen (Flemer et al. 1983). Dissolved oxygen levels in the Bay appear to be related to increased nutrient levels. High levels of nutrients in the Bay have stimulated growth of

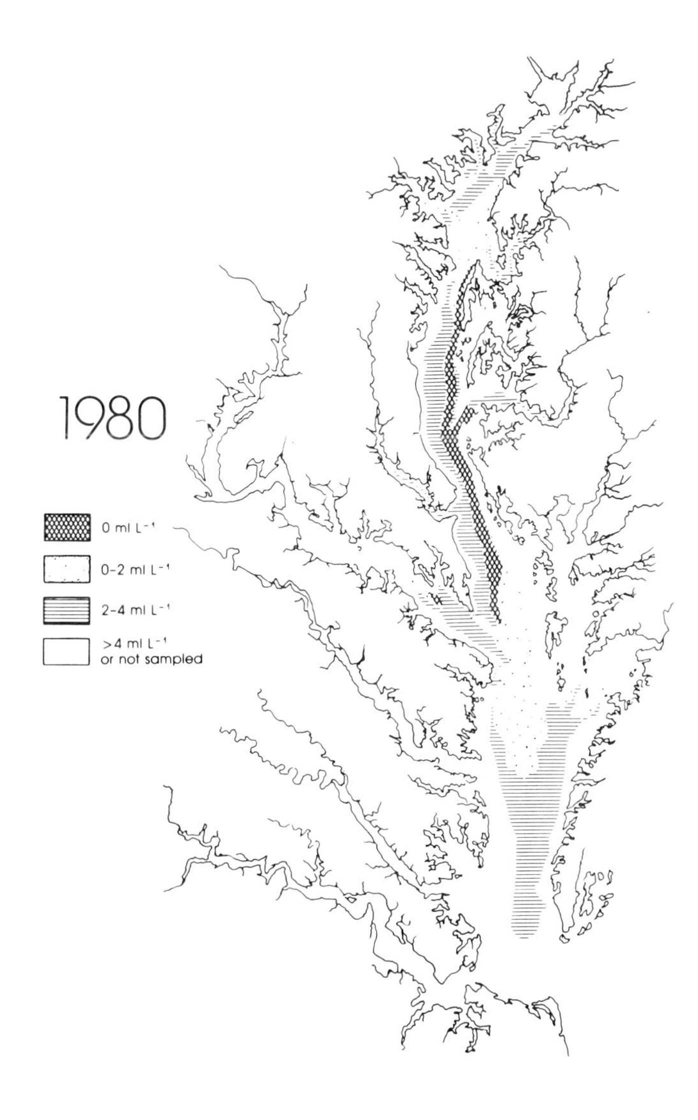

FIGURE 5.

phytoplankton. As the algae die and settle to the bottom, they decay and consume oxygen. Although these processes occur naturally in an estuarine system, they appear to have become far more severe in the Bay in recent years as nutrient inputs have increased.

o High concentrations of toxic organic compounds are in the bottom sediments of the main Bay near known sources such as industrial facilities, near river mouths, and in areas of maximum turbidity. Highest concentrations were found in the Patapsco and Elizabeth Rivers where several sediment samples contained concentrations of organic chemicals exceeding 100 ppm (Environmental Protection Agency 1982). These general patterns suggest that many of these toxic substances adsorb to suspended particles and then accumulate in areas dominated by fine-grained sediments. Benthic organisms located in such areas tend to accumulate the organic compounds in their tissues.

o Many areas of the Bay have metal concentrations in the water column and sediments that are significantly higher than natural (background) levels. In fact, EPA water quality criteria were exceeded in many instances. Also, Bay sediments in the upper Potomac and upper James River, small sections of the Rappahannock and York Rivers, and the upper mid-Bay had high levels of metal contamination (Fig. 6). The most heavily contaminated sediments (with concentrations greater than 100 times natural background levels) occur in the industrialized Patapsco and Elizabeth Rivers (Flemer et al. 1983).

The observed relationships between water quality and resource trends, coupled with laboratory research, has enabled us to begin to identify potential causes and effects. For example, Bay-wide, the areas experiencing significant losses of SAV have high nutrient concentrations in the water column. The high levels of nutrients enhance phytoplankton growth and epiphytic fouling of plants, thus reducing the light reaching SAV to below critical levels. However, it is also probable that high levels of turbidity and herbicides contribute to the SAV problem in localized areas. Reduced diversity and abundance of benthic organisms in some urbanized areas was related to toxic contamination of sediments (Flemer et al. 1983). Low dissolved oxygen (DO) in the summer-time is also a major factor limiting the benthic population, particularly in the upper and mid-Bay (Flemer et al. 1983). The low DO is attributed to increased algal production and decay triggered by nutrient enrichment. Lastly, nutrient enrichment and increased levels of toxicants occur in the major spawning and nursery areas for anadromous fish, as well as in areas experiencing reduced oyster spat. This type of information has been utilized to develop a preliminary Environmental Quality Class-

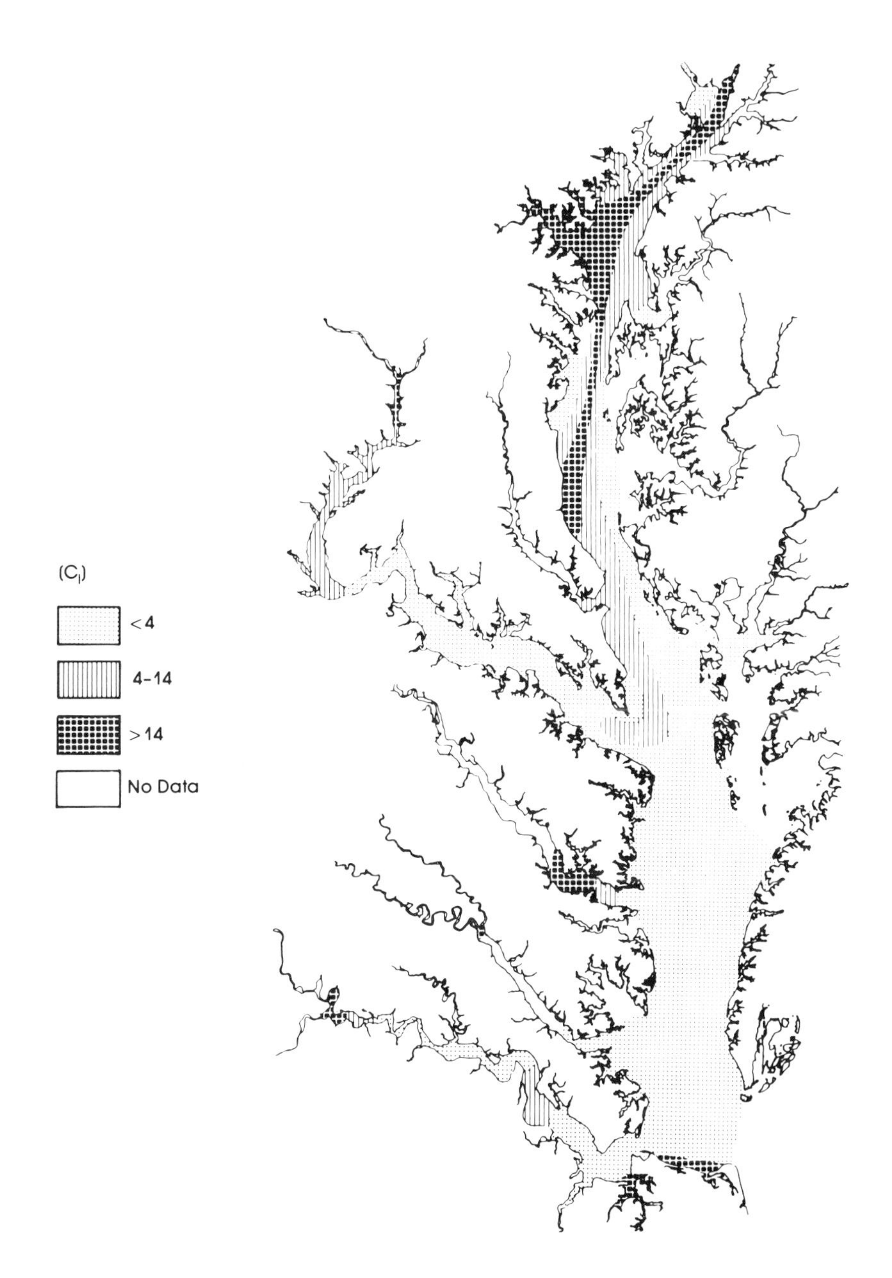

FIGURE 6. Degree of metal contamination in Chesapeake Bay based on the Contamination Index (CI). This index compares surface sediment concentrations of the metals Cd, Cr, Cu, Pb, Ni, and Zn to natural Bay levels.

ification Scheme (EQCS), which relates water quality criteria to the ability to support desired resources or water uses (Table 1).

MANAGEMENT IMPLICATIONS

The Chesapeake Bay Program findings clearly indicate that the Bay is an ecosystem in decline. It is also evident that actions throughout the Bay's watershed can affect the water quality of the tributaries to the Bay. Degradation of the Bay's water and sediment quality can, in turn, affect the living resources. Thus, effective management of Chesapeake Bay must be based on an understanding of and an ability to control both point and nonpoint sources of pollution throughout the Chesapeake Bay basin. To achieve this objective, it is essential that the states and the Federal government work closely together to develop specific management plans that address the basin-wide nutrient and toxic problems identified by the Program. The sources of nutrients and toxic loadings to the Bay and control alternatives are as follows:

Nutrients

Nutrients enter the Bay from point sources, such as sewage treatment plants, and from nonpoint sources, such as agricultural and urban runoff. In general, nitrogen entering Bay waters is contributed primarily by nonpoint sources, which are dominated by cropland runoff loadings. Point sources on the other hand, and especially sewage treatment plants, are the major contributors of phosphorus to Chesapeake Bay. It should be noted that in dry years, point source nutrient discharges tend to be more important than in wet years. In contrast, nonpoint sources, which enter waterways primarily in stormwater runoff, contribute a greater share of total nutrient loading during wet years. Also, different river basins tend to be dominated by different sources, and therefore require different control strategies (Fig. 7). For example, nutrient loading in the Susquehanna River is primarily associated with nonpoint sources, while nutrient loading to the James River is primarily attributable to point sources. The major findings regarding nutrient sources, loadings, and control programs are summarized below:

o The Susquehanna, Potomac, and James Rivers are the major sources of nutrients to the Bay (Fig. 7). This would be expected since they collectively provide 70 percent of the

TABLE 1.
A FRAMEWORK FOR THE CHESAPEAKE BAY ENVIRONMENTAL QUALITY CLASSIFICATION SCHEME

Class	Quality	Objectives	Quality	T_I	T_N	T_P
A	Healthy	supports maximum diversity of benthic resources, SAV, and fisheries	Very low enrichment	1	<0.6	<0.08
B	Fair	moderate resource diversity reduction of SAV, chlorophyll occasionally high	moderate enrichment	1–10	0.6–1.0	0.08–0.14
C*	Fair to Poor	a significant reduction in resource diversity, loss of SAV, chlorophyll often high, occasional red tide or blue-green algal blooms	high enrichment	11–20	1.1–1.8	0.15–0.20
D	Poor	limited pollution-tolerant resources, massive red tides or blue-green algal blooms	significant enrichment	>20	>1.8	>0.20

Note: T_I indicates Toxicity Index;
T_N indicates Total Nitrogen in mg L^{-1};
T_P indicates Total Phosphorus in mg L^{-1}.

*Class C represents a transitional state on a continuum between classes B and D.

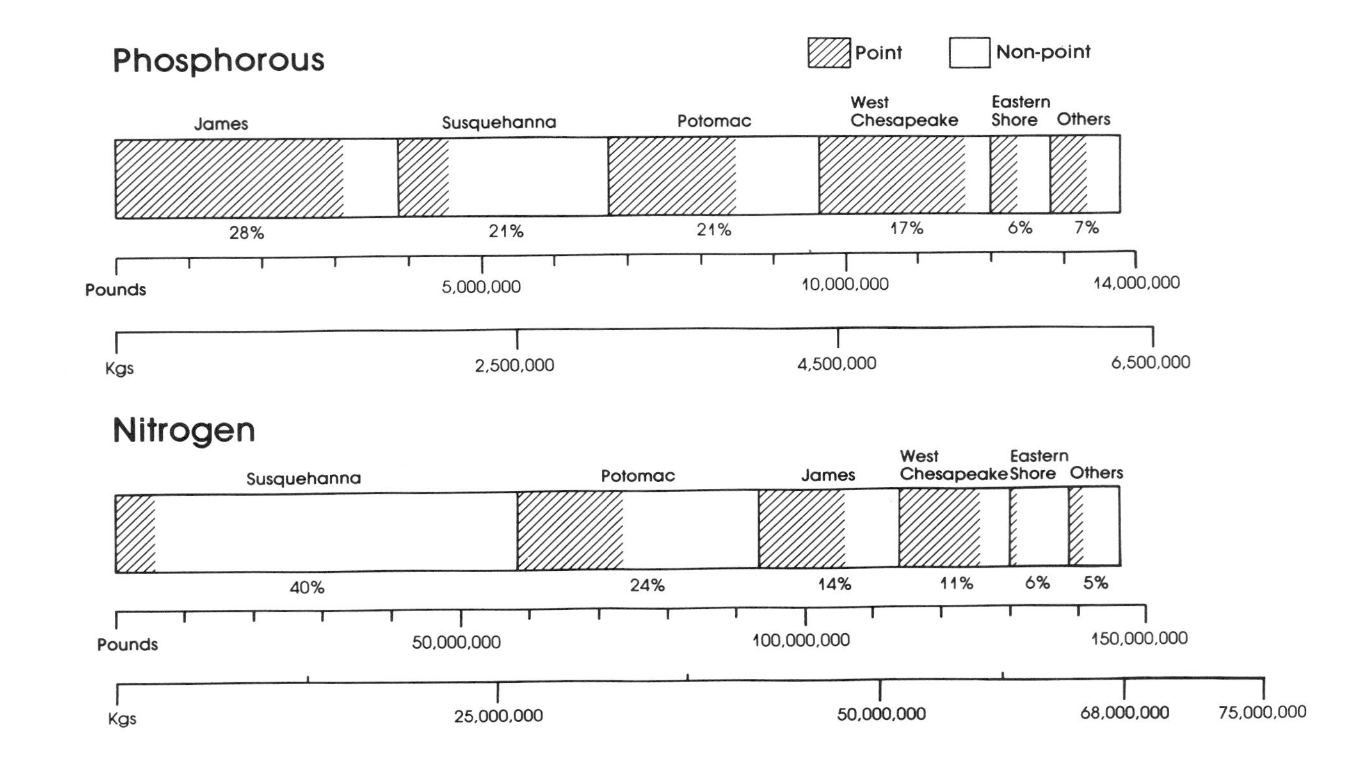

FIGURE 7. Nutrient loading (March to October) by major basin under average rainfall conditions.

freshwater inflow to the Bay. They contribute, respectively, 40, 24, and 14 percent of the nitrogen and 21, 21, and 28 percent of the phosphorus in an average year.

o Runoff from cropland and other nonpoint sources are the major sources of nitrogen to the nutrient-enriched areas in the Bay (Fig. 7). Of the total nitrogen load to the Bay in an average year, nonpoint sources contribute 67 percent whereas point sources contribute only 33 percent.

o Point sources, such as sewage treatment plants, are the dominant source of phosphorus to the nutrient-enriched areas of the Bay (Fig. 7). Of the total phosphorus load to the Bay in an average year, point sources contribute 61 percent whereas nonpoint sources contribute only 39 percent.

o Agricultural runoff control strategies, such as conservation tillage, best management practices, and animal manure waste management, can effectively reduce nutrient loadings from areas dominated by agricultural nonpoint sources (e.g. the Susquehanna River Basin).

o Urban runoff control efforts have been shown to be effective in reducing nutrient loadings to small tributaries located in the Baltimore, MD, Washington, DC, and Hampton Roads, VA, areas.

o Point source control strategies, such as restrictions on nutrient discharges from municipal sewage treatment plants, or limitations on phosphate in detergents, can significantly reduce nutrient loadings to those areas dominated by point sources (e.g., the James and Patuxent River basins).

o Point and nonpoint source controls in combination achieve consistent reductions in pollutant loadings during varying rainfall conditions in all basins.

Toxic Compounds

Toxic compounds enter the Bay from point sources, such as industrial facilities and sewage treatment plants, and from nonpoint sources such as urban runoff, dredged material disposal, and atmospheric deposition. The three major tributaries to the Chesapeake Bay: the Susquehanna, Potomac, and James Rivers, are the major sources of metals (Table 2) and organic compounds to the Bay. Industrial facilities and sewage treatment plants discharging directly to the Bay are significant sources of cadmium, copper, and organic compounds. Urban runoff is an important source of lead, and atmospheric deposition is an important source of zinc to the Bay. The toxics problem is most severe in industrialized areas such as Baltimore, MD, and Norfolk, VA, where the water and sediments have high metal concentrations and many organic compounds.

TABLE 2.
LOADINGS OF METALS FROM MAJOR SOURCES TO THE CHESAPEAKE BAY IN KILOGRAMS/DAY
(PERCENTAGE OF TOTAL LOAD)

	Cd	Cr	Cu	Fe	Pb	Zn
Industry	38 (13)	172 (8)	206 (10)	10,399 (1)	137 (8)	190 (3)
Municipal Wastewater	25 (8)	131 (6)	228 (11)	4,270 (—)	172 (11)	417 (5)
Atmospheric	8 (3)		77 (4)	239 (—)	93 (6)	2,261 (31)
Urban Runoff	19 (6)	30 (1)	25 (1)	2,678 (—)	305 (19)	173 (2)
Rivers	205 (69)	1,510 (72)	1,417 (69)	821,401 (80)	841 (51)	3,958 (54)
Shore Erosion	3 (1)	267 (13)	93 (5)	183,820 (18)	90 (5)	309 (4)
TOTAL	298	2,110	2,046	1,022,807	1,638	7,307

The major findings regarding sources and controls for toxic compounds are summarized below:

o The James, Potomac, and Susquehanna Rivers are the major sources of metals to the Bay. Collectively, they account for 69 percent of cadmium, 72 percent of chromium, 69 percent of copper, 80 percent of iron, 51 percent of lead, and 54 percent of zinc discharged to the Bay system.

o Over 300 organic compounds have been detected in the water and sediments of the Bay; up to 480 organic compounds have been detected in Baltimore Harbor. Most of the compounds detected are toxic and many are priority pollutants which are required to be controlled.

o An analysis of effluent from 20 industries and eight publicly owned treatment works revealed that over 75 percent of the facilities had toxic substances in the effluent. The possible causes of toxicity were metals, chlorine, and chlorinated organic compounds.

o Point source control programs resulted in significant reductions in metal loadings between 1970 and 1980 to areas such as Baltimore Harbor. However, these programs only control a portion of the toxic materials in point source discharges.

o Nonpoint source control efforts, such as urban runoff controls, integrated pest management, and the regulation of dredge spoil disposal, have probably resulted in reduced loadings of toxic compounds to the Bay.

o Toxic pollution control tools, refined or developed by Chesapeake Bay Program research efforts, will help managers address the toxic substance problem. These include the metal contamination index, the toxicity index, an effluent toxicity testing protocol, and an effluent and sediment fingerprinting procedure.

SUMMARY

The Federal government and the states have a variety of point and nonpoint source control programs to reduce loadings to the Bay. However, Chesapeake Bay Program research has shown that much of the Bay is over-enriched with nutrients and that high levels of toxic compounds occur near urbanized areas. This condition appears to be the result of decades of pollution for research has shown that the Bay acts as a filter, essentially trapping and recycling pollutants (Environmental Protection Agency 1982). We can only conclude that additional actions are necessary to reduce the pollutant loadings to the Bay and reverse the deteriorating trend.

To restore and maintain the Bay's ecological integrity will require a tremendous effort. The states must refine and strengthen their permit and enforcement programs. Also, state managers working with citizen groups need to develop basin management plans. Each plan should be targeted to the unique characteristics of the specific basin. For example, in the Susquehanna River basin, efforts should focus on agricultural runoff controls, while in the West Chesapeake basin and James River basin nutrient point source controls and toxic controls should receive priority attention. All of these efforts should be coordinated by a regional policy group to assure that actions in one area of the Bay do not detrimentally affect other sections of the Bay.

Recognizing the need for a cooperative regional approach, the state of Maryland, the Commonwealths of Pennsylvania and Virginia, the District of Columbia, and the Environmental Protection Agency signed an agreement on December 9, 1983. The Chesapeake Bay Agreement of 1983 recognized that EPA and the states share the responsibility for management decisions and resources regarding the high priority issues of the Chesapeake Bay. Accordingly, the agreement established three management mechanisms: a Chesapeake Executive Council of cabinet designees to oversee the implementation of coordinated plans; an Implementation Committee of agency representatives to coordinate technical matters; and a liaison office to advise and support the Council and the committee. Both the states and EPA have committed resources to the effort and projects have been initiated to reduce point and nonpoint source loadings to the Bay. To evaluate the effectiveness of the control efforts, a coordinated Bay-wide monitoring/ research program has also been implemented. Although, this unique federal/state partnership is still in its infancy, it is anticipated that it will be an effective management mechanism. In the future, this experiment in regional watershed management may serve as a model for other estuarine and coastal systems.

ACKNOWLEDGMENTS

Over the years numerous individuals have contributed to the Chesapeake Bay Program effort and it would be impossible to recognize them all. However, a special acknowledgement is given to those individuals who served as principal authors for the final summary reports (see each report). In addition, the Chesapeake Bay Foundation, Chesapeake Research Consortium, Citizens Program for the Chesapeake Bay; District of Columbia

Department of Environmental Services; Maryland Department of Health and Mental Hygiene; Maryland Department of Natural Resources; University of Maryland; Old Dominion University; Pennsylvania Department of Environmental Resources; Susquehanna River Basin Commission; U.S.D.A. Soil Conservation Service; Virginia Institute for Marine Sciences; Virginia Council on the Environment and the Virginia State Water Control Board are gratefully acknowledged for their cooperation, active support, and sustained interest in the Chesapeake Bay Program.

REFERENCES CITED

Cronin, L. E. 1982. A case history and assessment of pollution in Chesapeake Bay. In: Duke, T.W. (ed.). Impact of Man on the Coastal Environment. (EPA/8-82-021). U.S. Environmental Protection Agency, Washington, D.C. 114 p.

Environmental Protection Agency. 1982. Chesapeake Bay Program Technical Studies: A Synthesis. U.S. Environmental Protection Agency, Washington, D.C. 635 p.

Flemer, D. A., G. A. Mackiernan, W. Nehlsen and V. K. Tippie. 1983. Chesapeake Bay: A Profile of Environmental Change. U.S. Environmental Protection Agency Region III, Philadelphia, PA. 200 p. and Appendices.

Tippie, V. K., M. E. Gillelan, D. Haberman, G. B. Mackiernan, J. Macknis and H. W. Wells. Chesapeake Bay: A Framework for Action. U.S. Environmental Protection Agency Region III, Philadelphia, PA. 186 p. and Appendices.

ENVIRONMENTAL IMPACT OF POLLUTION CONTROLS ON THE THAMES ESTUARY, UNITED KINGDOM

Peter Casapieri

Directorate of Scientific Services
Thames Water Authority
Reading, Berks, United Kingdom

Abstract: A brief introduction is given to the historical background relating to the early history of pollution of the Thames estuary. Particular emphasis is placed on the reasons for deterioration and its subsequent dramatic improvement. Legislative measures that were necessary to enable improvements to be made are described. The process of development of water quality objectives and quality standards associated with them in the United Kingdom from the late 1950s to the present day is discussed. The Thames is finally recovering from a hundred years of serious pollution. Over one hundred species of fish are present in the estuary, some in very large numbers, and the river has been shown capable of supporting a salmonid fish population.

> "If I would drink water, I must quaff the maukish contents of an open aqueduct, exposed to all manner of defilement; or swallow that which comes from the river Thames, impregnated with all the filth of London and Westminster - human excrement is the least offensive part of the concrete, which is composed of all the drugs, minerals, and poisons, used in mechanics and manufacture, enriched with the putrefying carcasses of beasts and men; and mixed with the scourings of all the wash tubs, kennels, and common sewers, with the bills of mortality". Tobias Smollett (18th Century)

ISBN 0-12-405070-0

EARLY HISTORY OF POLLUTION IN THE THAMES ESTUARY

Smollett may have exaggerated for the sake of literary effect, but he was an outspoken surgeon and therefore had plenty of opportunity to observe people and places. It was nevertheless clear that even in those early days, serious pollution did occur involving sewage and industrial effluents which discharged into rivers such as the Thames. However, conditions could not have been quite so bad since the Thames remained a good salmon river until the early 1800s, with the last salmon until recent times caught in 1833.

Considerable pollution of the Thames occurred in the 19th Century primarily for two reasons: (i) The population of London expanded during the Industrial Revolution from under 1 million in 1800 to 3 million in 1850 and 8 million in 1900; (ii) ironically, with the invention of the water closet, effluent was sewered to the Thames instead of discharging directly to land as before. Thus, the increasing quantities of human-related wastes were concentrated by the estuary's 'filtering' activities. This led to increased oxygen demand and its consequent depletion.

The deterioration of the Thames was allowed to progress to such an extent over the next 100 years that the river was totally devoid of oxygen for many miles in certain summers. However, long before this time, certain actions were taken at the political level which slowly started the process of improvement. The Thames , during the last century, was not the only major river in the United Kingdom to suffer the consequences of the Industrial Revolution. However, it did have one big advantage in that the Capital city, London, was situated on its banks and when the year of the 'big stink' came in 1858, the Government acted remarkably promptly in passing legislation. It was no doubt spurred on by the fact that wet sacks had to be placed over the windows of the Houses of Parliament to keep out the smell of the river. This, coupled with cholera outbreaks of 1849 and 1854 that were believed to be due to inadequate drainage, was responsible for the construction of a gigantic sewerage scheme for London under the able control of Sir Joseph Bazalgette. This conveyed sewage to the East End of London and at that time beyond what was the city boundary to two large sewage treatment works, Beckton and Crossness (Fig. 1). The system still exists and is used today. Figure 2 shows the average summer levels of dissolved oxygen and the pattern of improvement (1890-1920), deterioration (1940-1965), and then the final improvement in the Thames in the 1970s. Table 1 shows the summary of construction projects that have been carried out along the estuary to allow for restoration of high water quality. Our estimates of these costs at 1980 prices is of the order of 450 million pounds sterling (650 million U.S. dollars).

HISTORICAL LEGISLATION

The United Kingdom has a long history of controlling water pollution since the evident effects of the rapid urbanization of the country after the Industrial Revolution of the 18th and 19th Centuries. Early legislation in 1857 gave authority for the management, improvement, and regulation of the River Thames from Staines (only a few km to the west of London) to the Yantlet line which is the seaward limit in the outer Thames estuary. Later Acts extended still further the powers to the whole fresh-water Thames catchment whilst pollution control, management, and navigation in the estuary were ceded to the Port of London Authority in 1908. Subsequently, in the 1960s this Authority was given additional power to control polluting discharges by the application of flow limitations and restrictions on chemicals and toxic waste discharges to the Thames estuary. Thus, the power was available for setting the consent conditions to a discharge, and indirectly, for establishing quality standards for our rivers. Sadly, what was lacking was the firm definition of any environmental quality objectives by which to set the scene. In recent years, legislation for pollution prevention had been directed largely towards the control of discharges into freshwaters, and only more recently towards control of all discharges into tidal and coastal waters generally.

In England and Wales, the management of all aspects of the water cycle, including water supply, sewage treatment, pollution prevention, resource management, land drainage, and navigation, were looked after by about 1600 separate undertakings. These included municipalities, water companies, river authorities, sea fisheries committees, and many other bodies. This, clearly, was a situation that creaked along and did not allow for any serious integrated water management, certainly not on a national scale. Certain estuaries in the U.K., the Thames included, were fortunate to have pollution control powers and in the case of the Thames this was through the Port of London Authority Acts; however, this was not of general application to estuaries in the U.K. prior to 1974.

The Water Act of 1973 welded most of these undertakings into ten regional water authorities (Fig. 3). Since April 1, 1974, the Regional Water Authorities have been responsible to

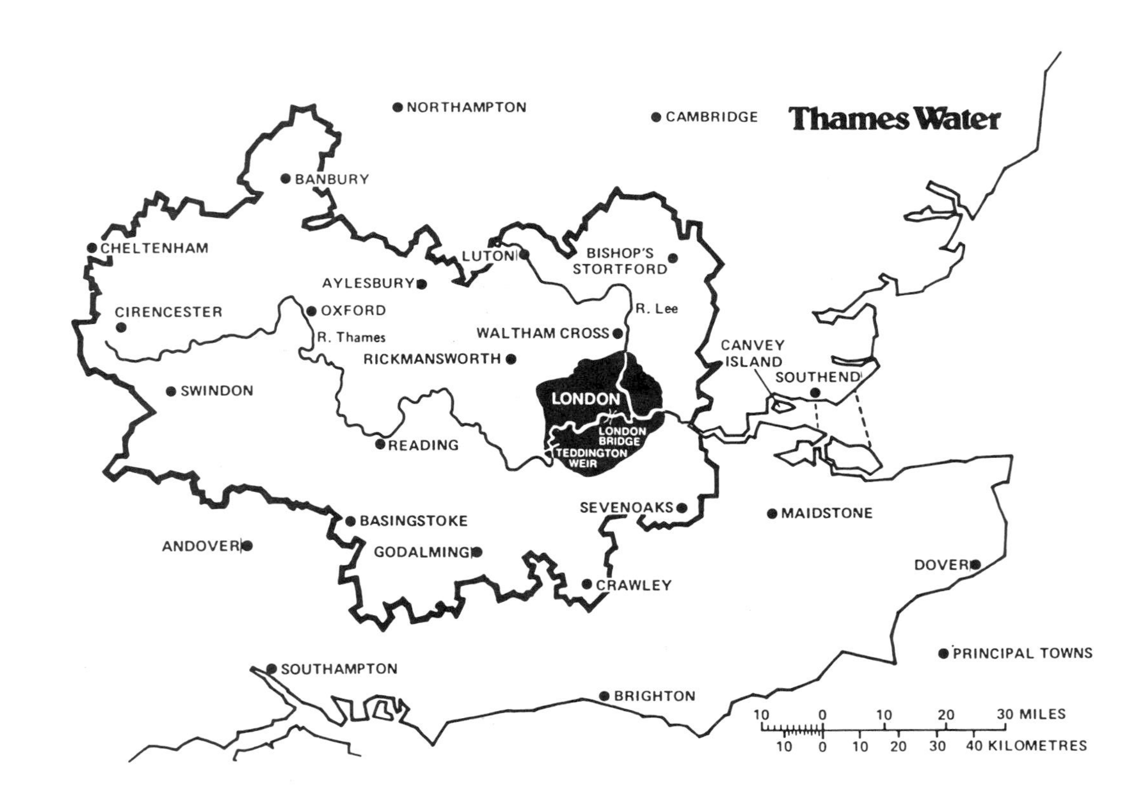

Thames Water
NORTHAMPTON
CAMBRIDGE
BANBURY
CHELTENHAM
LUTON
BISHOP'S STORTFORD
AYLESBURY
CIRENCESTER
OXFORD
R. Thames
R. Lee
WALTHAM CROSS
RICKMANSWORTH
CANVEY ISLAND
SOUTHEND
SWINDON
LONDON
LONDON BRIDGE
TEDDINGTON WEIR
READING
SEVENOAKS
MAIDSTONE
BASINGSTOKE
ANDOVER
GODALMING
DOVER
CRAWLEY
PRINCIPAL TOWNS
SOUTHAMPTON
BRIGHTON
10 0 10 20 30 MILES
10 0 10 20 30 40 KILOMETRES

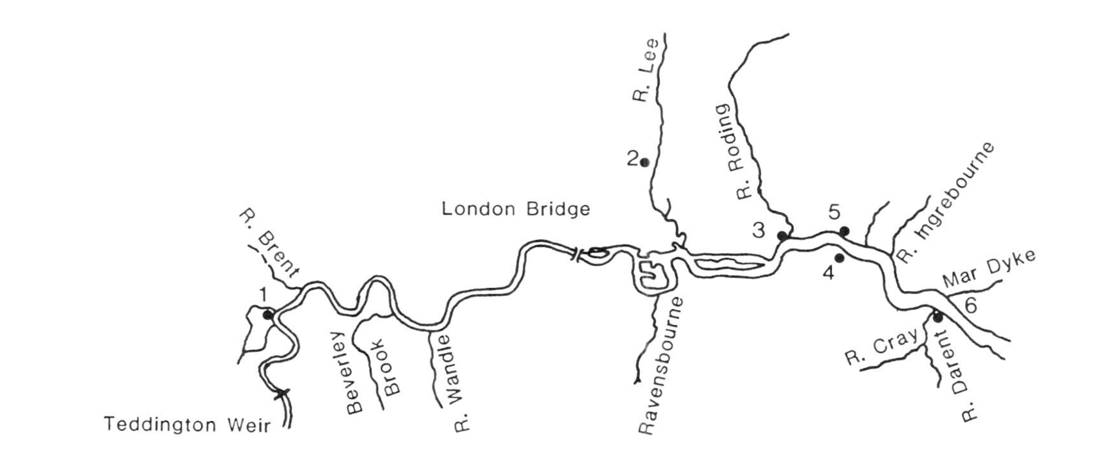

FIGURE 1. *Map of Thames Area (top) with enlargement (bottom) showing major sewage treatment works (STW) in the London area (Nos. 1-6). Average discharge volumes (100 m^3/d) are as follows: 1. Mogden STW (5193); 2. Deephams STW (2065); 3. Beckton STW (10095); 4. Crossness STW (5945); 5. Riverside STW (1093); 6. Long Reach STW (2000).*

the Secretary of State for the Environment for sewerage, sewage treatment, pollution control, water resources, and supply. The Regional Water Authorities also have responsibilities to the Minister of Agriculture, Fisheries and Food for land drainage and fisheries. Thus, the powers and duties previously exercised by a large number of authorities have now passed to a single authority with total control of the water cycle.

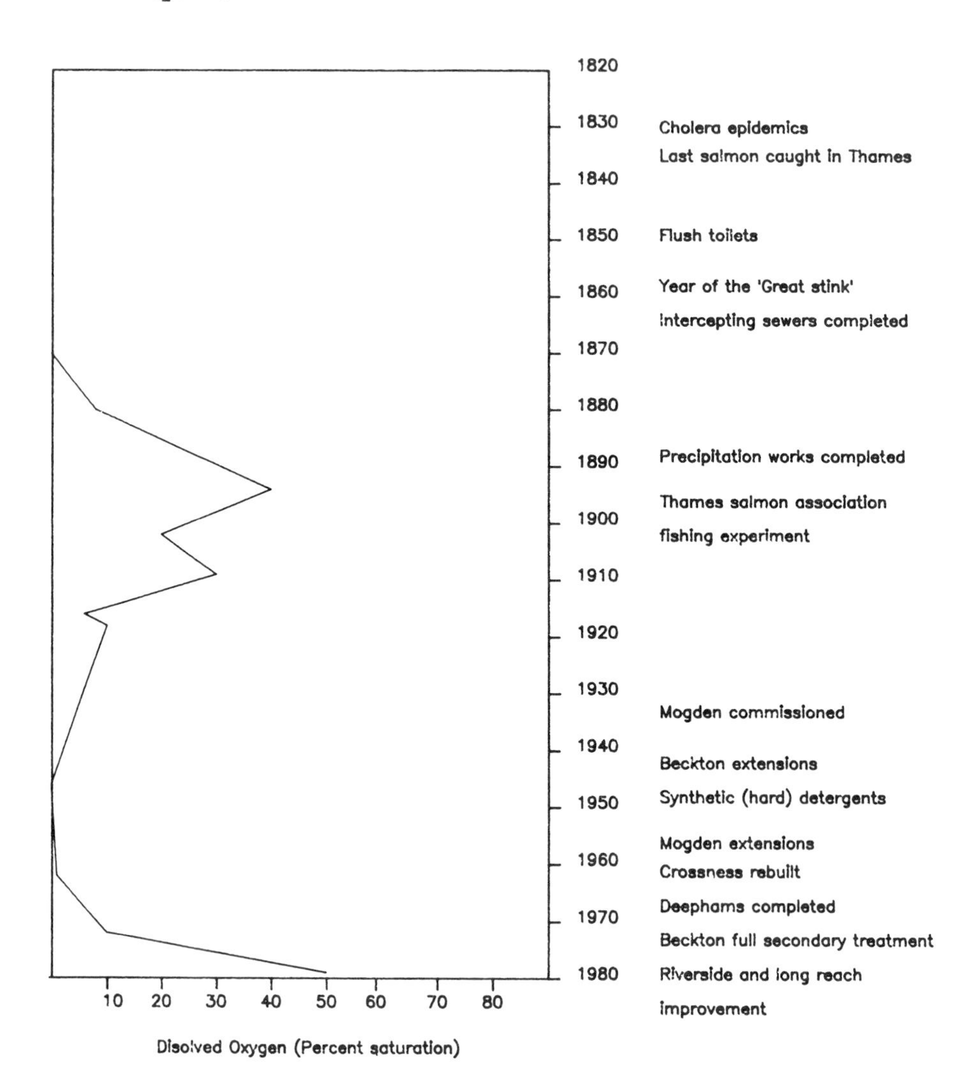

FIGURE 2. Pattern of dissolved oxygen in the tidal Thames at Crossness between 1870 and 1980 (summer quarters).

TABLE 1. Summary of capital works carried out in the restoration of the tidal Thames.

Capital works	*Completion date*	*Capital cost, historic £*
Construction of intercepting sewage overflows, pumping stations to Beckton and Crossness	1865	4.1
Additional sewers, pumping stations	1914	2.5
West Middlesex main drainage scheme; Mogden Sewage Treatment Works (STW)	1935	5.4
Beckton STW, partial secondary treatment, partial sludge digestion	1959	7.0
Hammersmith pumping station	1960	3.0
Extension of West Middlesex drainage and Mogden STW	1961	1.5
East Middlesex drainage scheme and Deephams STW	1965	9.2
Crossness STW replacement providing full secondary treatment	1965-9	11.0
Blackwall Tunnel flood relief	1968	1.0
Beddington STW, rebuilding to give full treatment	1969	4.0
Extension to Deephams	1971	2.8
Lee Valley sludge main	1971	1.1
Beckton STW, full treatment and sludge digestion	1975	26.5
Long Reach STW rebuilt to give secondary treatment	1979	6.4

ENVIRONMENTAL QUALITY OBJECTIVES FOR THE THAMES ESTUARY

The first significant step in the setting of Environmental Quality Objectives for the estuary came as a result of the governmental Committee established under the chairmanship of Professor A.J.S. Pippard. This was as a result of the government being forced to admit that enough pollution was enough. The terms of reference of the

FIGURE 3. *Water authorities in England and Wales.*

Committee were as follows: (i) To assess pollution loads and their sources on the Thames estuary, (ii) to suggest guidelines for controls, and (iii) to look to the future (Pippard 1961). These goals were achieved, and, by working closely with the Thames Survey Committee, one of the by-products was the development of the first really significant and powerful mathematical model of an estuary (Water Pollution Research 1964; Mollowney 1973). This enabled questions to be asked and answered as to what would be the effects on river water quality of changes in pollution loads to the estuary. A more up-to-date version of the model was published in 1978 (Barrett, Casapieri and Mollowney 1978) and a management use was made of it subsequently (Casapieri and Owers 1979).

However, returning to Environmental Quality Objectives, the Pippard report (1961) recommended as a guideline that the minimum objective for the quality of the tidal Thames should be freedom from putrescence. Hardly an earth-shattering event, but it was a beginning. This was interpreted by the Port of London Authority as requiring a minimum figure of 10% saturation dissolved oxygen (ca. 1 mg/l). How this was to be achieved was not made very clear, but it was the beginning of a quality standard.

The Pippard committee went further in the added recommendation that the next stage in the improvement in the estuary would logically be the provision of water quality suitable for the passage of migratory fish. Using the best information at the time, the report considered that it would be necessary during the months of April and May to prevent dissolved oxygen from falling below 30% air saturation (3 mg/l) for nine years out of ten in order to allow for the establishment of a migratory fishery. This represented the beginnings of a management strategy. In the U.K., experience in environmental quality objectives has been derived mainly from the application of the control of discharges into freshwaters, not into estuaries or tidal waters. The first significant break-through came shortly after the reorganization of the water authorities. The National Water Council, through the directors of Scientific Services, set up a working party to report on consent conditions for effluent discharges to freshwater streams (National Water Council 1976). One significant result was the production of a national guideline for environmental quality objectives and associated quality standards for U.K. rivers. A classification of rivers was drawn up and this is shown in Table 2.

The natural sequel to this was to repeat the exercise for tidal and estuarine waters (National Water Council 1978). The latter report provided a very substantial guideline on

TABLE 2. Classification of freshwater river quality and associated quality and standards.

River class	*Quality objectives*	*Quality standards (95 percentile basis)*
1A	Water of high quality suitable for potable abstractions and for all other abstractions; Game and other high-class fisheries; High amenity value; Visible evidence of pollution should be absent	(i) DO saturation >80% (ii) BOD $\leqslant$3 mg/l (iii) Ammonia $\leqslant$0.4 mg/l (iv) For drinking water, it complies with EEC-A2 category (v) Satisfies EIFAC requirements
1B	As for 1A, but cannot be placed in that category because of high proportion of high quality effluent present	(i) DO saturation >60% (ii) BOD $\leqslant$5 mg/l (iii) Ammonia $\leqslant$0.9 mg/l (iv) As for 1A (v) As for 1A
2	Waters suitable for potable supply after advanced treatment; Supporting reasonably good coarse fisheries; Moderate amenity value	(i) DO saturation >40% (ii) BOD $\leqslant$9 mg/l (iii) For drinking water, it complies with EEC-A3 category
3	May be used for low grade industrial abstraction; Waters that are polluted to an extent that fish are absent or only sporadically present; Considerable potential for future use if cleaned up	(i) DO saturation >10% and unlikely to be anaerobic (ii) BOD $\leqslant$17 mg/l
4	Must be cleaned up since these are grossly polluted and are likely to cause nuisance	Inferior to class 3
X	Insignificant watercourses and ditches where the objective is simply to prevent nuisance developing	DO saturation >10%

the environmental quality objectives that are pertinent to estuarine and coastal waters. The working party was conscious of the limited data that existed in general for tidal waters compared to freshwaters and this was reflected in the more limited statements on quality standards. Where a statement on quality standards could be made, then it was made, otherwise only general guidelines were indicated. Nevertheless, it was felt that, taking the nation's tidal waters as a whole, most are substantially unpolluted and, accordingly, the primary objective of pollution control must be to prevent significant deterioration of the present quality. When faced with the prospect that a systematic classification of tidal water quality (i.e. coastal as well as estuarine) would be impracticable at that time, it was decided that the most appropriate course of action would be to outline the basis on which individual water authorities should formulate their own quality objectives for tidal waters under their control (National Water Council 1978). The major factors to be taken into account in formulating quality objectives should be the uses and environmental needs of the tidal waters involved. These objectives are described in Table 3.

TABLE 3. Quality objectives for tidal waters in U.K.

Category	*Quality objectives*
1	Protection of fishery resources including the reproduction, food supply, and habitat of fish involved, for the maintenance of these resources, and to ensure that where these resources are harvested, their quality is suitable for human consumption, although in some circumstances appropriate cleansing treatment may be considered necessary
2	Protection of other particular flora and fauna, and their habitats, for special conservation purposes
3	Protection of abstractions made for industrial, agricultural, or potable supply
4	Protection of bathing and other recreational uses of the waters
5	Avoidance of public nuisance

These broad principles of objectives have also been formulated by the Environmental Quality Objectives sub-group of the Department of the Environment Technical Advisory Committee for Dangerous Substances. A Royal Commission report (Royal Commission 1972) also outlined two principal objectives related to estuarine quality: (i) the quality of water in the estuary should be capable of supporting the passage of migratory fish at all stages of the tide and (ii) the estuary should support on its mud bottom those organisms required to sustain sea fisheries.

As mentioned earlier, the only serious attempt to establish an objective and an associated quality standard for the estuary was described by the Pippard report (1961). This was that the desired dissolved oxygen concentration should be maintained at all times and all places. From the point of view of monitoring and comparative assessment of the estuarine quality, an average value is probably more realistic and in order to assess the mean figure for dissolved oxygen that may be associated with a minimum of 10% saturation, a regression analysis was carried out on data covering the period 1975 and 1976. The quarterly average minimum dissolved oxygen was regressed against the absolute minimum dissolved oxygen (DO), in the respective quarter (Casapieri 1980). The regression analysis showed a highly significant correlation ($r = 0.86$) between the two variables, producing the relationship:

Quarterly average minimum dissolved oxygen (% saturation) = 0.68 x absolute minimum dissolved oxygen + 23.0.

Subsequent examination of quarterly data covering the much longer period of 18 quarters between January 1975 to June 1979 provided a similar relationship:

Quarterly average minimum dissolved oxygen (% saturation) = 0.74 x absolute minimum dissolved oxygen + 23.0.

If a value of 10% is inserted into this equation, then the estimate for the quarterly average minimum is calculated to be 30%. This is in line with the Pippard (1961) recommendations. The objectives and standards for the middle and lower reaches of the estuary were then formulated (Table 4). This shows that a better estuarine water quality is required in the lower reaches, below Canvey Island, to avoid undue constraint on the fish breeding and nursery areas there (Cockburn and Waters 1978).

TABLE 4. Objectives and standards for the middle and lower reaches of the estuary.

Stretch of estuary	*Quality objectives*	*Quality standards*
London Bridge to Canvey Island	Freedom from putrescence and avoidance of public nuisance; Allow passage of migratory fish	Minimum air saturation with dissolved oxygen should be at least 10% (95 percentile) Quarterly minimum average percentage saturation with dissolved oxygen should not fall below 30% in summer quarter
Canvey Island to seaward limit	Quality should be suitable for the whole life cycle of marine organisms including fish	Minimum air saturation with dissolved oxygen should be at least 60% (95 percentile) during the summer quarter

In the case of that part of the tidal Thames upstream of London Bridge and extending to Teddington Weir (30 km), the river may be considered more as a freshwater river and assigned an appropriate classification (see Table 2). Considering the high amenity value of the upper estuary and its potential uses, the objectives proposed were for the stretch from Teddington to Barnes to be given the objective classification of 2, and for Barnes to London Bridge the classification was to be 3. The quality of the first stretch from Teddington to Barnes is dominated by Mogden sewage treatment works, leading to the conclusion that the estuary here could be upgraded to beyond class 2. However, the effluents from treatment works discharging downstream of London Bridge have an additional effect and this causes the Barnes to London Bridge stretch to be assigned an interim objective of class 3. Pending completion of planned capital works at Riverside and Long Reach, it is anticipated that their cumulative effect among others will cause a rising of the standard in the water such that ultimately the objectives pertinent to class 2 will prevail. For creek waters adjacent

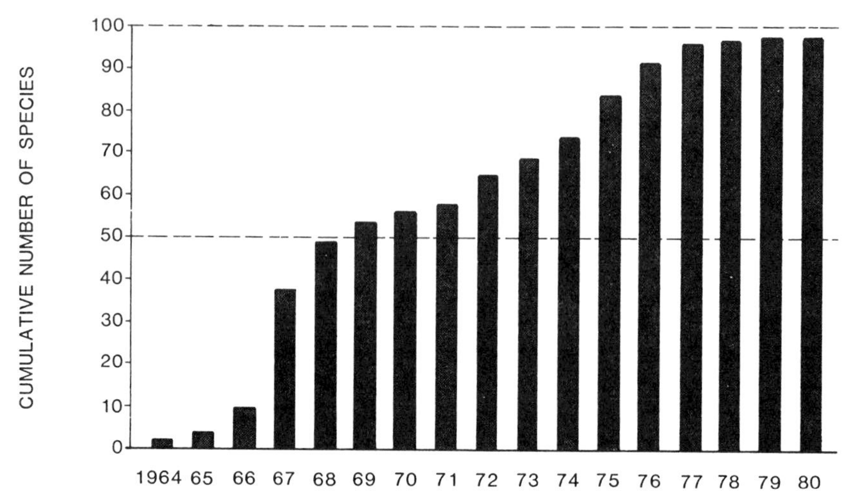

FIGURE 4. Cumulative number of fish species recorded in the Thames between Key and Gravesend from 1964 to 1980.

to the Thames estuary, the policy of the Thames Water Authority is to (i) avoid nuisance, (ii) be compatible with that of the adjacent estuary, and (iii) present no barrier to the passage of migratory fish where the fresh water discharged to the creek is suitable for such species.

THE WAY AHEAD

In view of the dramatic improvement in the Thames estuary, particularly over the past decade, which culminated in the appearance of some 104 species of fish (Andrews and Rickard 1980), some in considerable numbers (Fig. 4), the Thames Water Authority decided to set up a 17-year program for salmon rehabilitation in three phases. From 1979 to 1986, 50,000 parr were to be introduced into carefully selected sites in the first year and over the remaining six years. This was thought to be sufficient to produce 20,000 smolts, annually. A temporary fish trap will be installed at one of the weirs so that any returning adult fish can be captured. From 1986 to 1991, the rearing program is to continue but with extra work to encourage natural spawning and rearing in the upper waters of the Thames and tributaries. In 1991 and thereafter, if (and only if) the second stage is successful, then the third stage will involve modification of weirs to enable returning salmon to swim directly back to their spawning grounds.

The first phase is well under way now and has met with very considerable success. The first salmon has now been caught by rod and line and the angler was rewarded with a 500 pounds sterling prize. During the last full year (1982), some 130 tagged adult salmon have been caught, mostly by electrofishing at one of the numerous weirs on the freshwater Thames. Since 1979, some 300,000 parr have been introduced into the headwaters of many tributaries of the Thames and in 1983 an additional 40,000 smolts have also been released in the freshwater catchment.

The ending, thus far, demonstrates a successful rejuvenation of one of the world's important rivers. As its filtering action resulted in the concentration of increasing amounts of pollutants, the Thames had been allowed to deteriorate to the point of putrescence before the introduction of pollution control measures. It is a salutary lesson to all of us.

REFERENCES CITED

Andrews, M.J. and D.G. Rickard. 1980. Rehabilitation of the inner Thames estuary. *Marine Pollution Bulletin 11*: 327-332.

Barrett, M.J., P. Casapieri and B.M. Mollowney. 1978. The Thames model, An assessment. *Prog. Wat. Tech. 10*: 409-416.

Casapieri, P. 1980. Guidelines on quality standards and objectives for the Thames estuary. *Chem. and Ind.* 19th April. pp. 316-320.

Casapieri, P. and P.J. Owers. 1979. Modelling the Thames - a Management Use of a Mathematical Model, pp. 381-391. *In: River Pollution Control*, Ellis Horwood Ltd., Chichester, U.K.

Cockburn, A.G. and C.J. Waters. 1978. Pollution in the tidal Thames. *Surveyor 151* :9-10.

Mollowney, B.M. 1973. One-Dimensional Models of Estuarine Pollution, pp. 71-84. *In:* Water Pollution Research Technical Paper No. 13, H.M.S.O., London.

National Water Council. 1976. Report of a Working Party on Consent Conditions for Effluent Discharges to Freshwater Streams. National Water Council, London.

National Water Council. 1978. Report of a Working Party on Consent Conditions for Discharges to Tidal Waters. National Water Council, London.

Pippard, A.J.S. 1961. Pollution of the Tidal Thames. Ministry of Housing and Local Government, H.M.S.O., London.

Royal Commission. 1972. Pollution in Some British Estuaries and Coastal Waters. H.M.S.O., London.

Water Pollution Research Technical Paper No. 11. 1964. Effects of Polluting Discharges on the Thames Estuary. Department of Scientific and Industrial Research, H.M.S.O., London.

INDEX

M

N

O

P

R

S

T